STRUCTURE
AND FUNCTION
RELATIONSHIPS IN
BIOCHEMICAL SYSTEMS

ADVANCES IN EXPERIMENTAL MEDICINE AND BIOLOGY

STRUCTURE AND FUNCTION RELATIONSHIPS IN BIOCHEMICAL SYSTEMS

Edited by

Francesco Bossa
University of Rome
Rome, Italy

Emilia Chiancone
CNR Center of Molecular Biology
Rome, Italy

and

Alessandro Finazzi-Agrò
and Roberto Strom
University of Rome
Rome, Italy

PLENUM PRESS • NEW YORK AND LONDON

Library of Congress Cataloging in Publication Data

Symposium on Structure–Function Relationships in Biochemical Systems (1981: Accademia nazionale dei Lincei)
    Structure and function relationships in biochemical systems.

    (Advances in experimental medicine and biology; v. 148)
    Includes bibliographical references and index.
    1. Enzymes—Congresses. 2. Structure–activity relationship (Pharmacology)—Congresses. 3. Biochemorphology—Congresses. 4. Hemoglobin—Congresses. 5. Sulphur—Metabolism—Congresses. I. Bossa, Francesco. II. Chiancone, Emilia. III. Title. IV. Series. [DNLM: 1. Structure–activity relationship—Congresses. W1 AD559 v. 148/ QU 34 S927 1981]
    QP601.S9493 1981   574.19′25   82-9139
                                                                     AACR2
    ISBN-13:978-1-4615-9283-9   e-ISBN-13:978-1-4615-9281-5
    DOI:10.1007/978-1-4615-9281-5

Proceedings of a symposium on Structure–Function Relationships in Biochemical Systems, celebrating the 75th birthday of Alessandro Rossi Fanelli, held September 28–30, 1981, at the Accademia Nazionale dei Lincei, in Rome, Italy

© 1982 Plenum Press, New York
Softcover reprint of the hardcover 1st edition 1982
A Division of Plenum Publishing Corporation
233 Spring Street, New York, N.Y. 10013

PREFACE

Studies on the elucidation of structure-function relationships
in biochemistry and molecular biology started a few decades ago and
represent an area of continuing interest. The celebration of the
75th birthday of Alessandro Rossi Fanelli, who has made significant contributions in this field, prompted the organization of a
Symposium on 'Structure-Function Relationships in Biochemical Systems'
with the aim of bringing together Rossi Fanelli's students and many
of the scientists who have been connected with the Institute of
Biological Chemistry in Rome. The Symposium was held in Rome on
September 28-30, 1981 at the Accademia Nazionale dei Lincei.

This volume contains the lectures presented at the Symposium
as well as articles covering the main themes of research conducted
in Rossi-Fanelli's Institute. The material is divided into five
sections: Hemoglobin, Myoglobin and Other Respiratory Proteins;
Mechanism of Action of Metal-Containing Enzymes; Bioenergetics, Membrane Structure and Multienzyme Complexes; Cofactor-Dependent Enzymes; Sulfur Metabolism.

We gratefully acknowledge the generous financial support of
the Italian National Research Council and of the University of Rome,
which rendered the Symposium possible. We wish to thank all the con-
tributors and Maurizio Gattoni, Paolo Gerosa and Mario Sanchioni
for their help in the preparation of the illustrative material. Finally we wish to express our gratitude to Professor Rossi Fanelli,
the unfailing animator of our Institute; in his honour we happily
undertook the task of editing this volume.

Francesco Bossa
Emilia Chiancone
Alessandro Finazzi Agrò
Roberto Strom

CONTENTS

CONTENTS

ALESSANDRO ROSSI FANELLI

PROFESSOR ROSSI FANELLI: A PROFILE

Noris Siliprandi

Institute of Biological Chemistry, University of Padua

35100 Padua, Italy

This symposium has been organized to celebrate the 75th birthday of Professor Rossi Fanelli and the formal conclusion of his long
and incomparable academic life.

It is my great privilege, bestowed on me solely because of se-
niority, to depict the personality which brought about the birth and
affirmation of his School.

Milestones of Rossi Fanelli's life were Naples, his birthplace,
Pavia, the seat of his first Chair, and Rome, his permanent abode.
The university career of Professor Rossi Fanelli at the Institute
of Biological Chemistry in the University of Naples began when he
took his degree in Medicine in 1929. He worked in the laboratory
directed by Professor Gaetano Quagliariello until 1942, when he was
appointed to the Chair of Biochemistry in the Faculty of Science at
Pavia University. The scientific activity of the Neapolitan period,
which was developing while Italian biochemistry was taking its first
steps, already outlines the experimental talent and the scientific
originality of Rossi Fanelli. This first appeared in the perfectio-
ning of precise microanalytical procedures for the determination of
various ions and of some amino acids. We today, who are used to such
sophisticated machinery for coping with scientific problems without
worrying too much about analytical methodology, cannot easily un-
derstand what it was to have to transfer to biological materials,
and on a micro-scale, the classic methods of analytical chemistry:
the difficulties, doubts, failures, and cost in energy and time.
The natural leaning of Rossi Fanelli to accurate analytical experiments led him to achieve important results in that period. I re-
member with admiration his great ability - manual too - in setting
up various rather complicated apparatus with "bits and pieces"

1

gleaned here and there, and his undisguised satisfaction when the
work was completed. His innate respect for, and almost, one might
say, loving care, of all apparatus is a direct outcome of those
arduous years.

Amongst results of this Neapolitan period I mention the demonstration that non-hepatic tissues are able to oxidize acetoacetate,
anticipating the nowadays accepted idea that muscle, kidney and
brain use the ketone bodies provided by the liver for energy pur-
poses. Also important has been the demonstration that part of α-ke-
toglutarate that is formed from citrate (in the process that was
controversial at the time and that only after some years Hans Krebs
was to describe as tricarboxylic acid cycle) can be transformed into
glutamate. Thus a hint was given of the transamination process and
of the interconversion of sugars into protides. This was in the
pre-war years and the difficulty, or rather impossibility, of communication especially with the Anglo-American scientific world
negated to these results a certain resonance that, retrospectively,
they merited.

Professor Rossi Fanelli then worked from 1933-34 in Freiburg
with Professor Thannauser and Professor Bohnenkamp on the composition of nucleoproteins and in Prague with Professor Waldschmidt-
Leitz on the activation of enzymes. In these years spent abroad
the young Rossi Fanelli derived great benefit from the very fertile
experience matured in more efficient laboratories and the stimulation of working with established personalities of the international
scientific world. During these laborious years, Professor Rossi
Fanelli found the time to take his degree in Pure Chemistry. In
fact he realized that the degree in Medicine that had revealed to
him the nature of problems which were to become more interesting
within the dominion of biochemistry, did not give him the adequate
methodological formation necessary for an efficient approach.

From his mentor, Professor G. Quagliariello, for whom he always
had great respect and affection, he took his passionate dedication
to the School, and learned his respect for academic values and the
simple aged-old philosophy that comes from "charitas cristiana"
and is expressed with "humilitas". I believe that this devotion to
his Master, coupled with his innate gifts, contributed greatly to
making him an authoritative founder of a School. In fact, the history
of science proves that the scientific elite and the Schools which
have imposed their leadership, have been formed in the communities
that knew how to honour their Masters and at the same time valorize
the talents of their students.

Rossi Fanelli brought from his birthplace his very lively and
witty spirit, a certain taste for disenchanted escapism and an unmi-
stakable aristocratic dignity tempered by a spontaneous simplicity.

In 1942 he was transferred to Pavia, to be joined by his family when the war ended in 1945.

The meeting between myself and Giulio Perri with the man who was to become our Master took place in an euforic peace-time climate.

Having heard of the possibility of reproducing in laboratory animals insulin-deficient diabetes by means of alloxan administration, an easy to prepare substance, Professor Rossi Fanelli applied himself first to the metabolic study of alloxan produced diabetes and the mechanism with which this substance damages specifically the pancreatic cells. This research, even though undertaken with very limited means, produced very good fruit all the same.

However, the great wish to carry on his favourite research on myoglobin, which began during the latter part of his Neapolitan period, matured his decision to study human myoglobin. The willing cooperation of a hospital in Milan treating people with amputated limbs, traumatic conditions or suffering from incurable illnesses, offered a very good starting point. The different solubility in solutions of ammonium sulfate of myoglobin and hemoglobin, already revealed and ably achieved by Rossi Fanelli with a precise choice of experimental conditions, rendered unnecessary the preliminary perfusion of the tissue with the aim of getting rid of the hemoglobin. This simple experimental design led to its success and the red crystals, so long awaited, were deposited on the bottom of the plate. From these poor arts, those stupendous forms of molecular life: mors gaudet succurrere vitam!

The myoglobin crystals systematically re-obtained were submitted to all the investigations permitted by time and place, and led Rossi Fanelli to the first description of the composition and of the salient functional properties of human myoglobin. These results which were published in 1948 in Science, and were the object of a communication at the Barcroft Memorial Conference, oriented his productive future.

Happy years in Pavia, now long ago, but so often recalled!

Pavia, perhaps the city most antithetic to Naples, received the distinguished Professor who had come from the south with vigilant curiosity, but very quickly assimilated him as one of its own. From the University of Lazzaro Spallanzani, Alessandro Volta, and Camillo Golgi, Rossi Fanelli benefited intensely from the influence and stimulation that come out of such institutions so steeped in history. Rossi Fanelli fell under the spell of Pavia's enchanted streets, its antique churches, its mists and its stillness, and when he was called to the Faculty of Medicine in Rome, his farewell was no less sentimentally intense than when he left Naples.

He began to work again in Rome more or less from zero, and the empty laboratories were slowly animated and gradually transformed into the hectic and milling place that is today the Institute of Biochemistry of Rome. The essential ingredient was of course the talent of the young graduates attracted by Rossi Fanelli's reputation and by the call of Biochemistry. First came the very youthful Cerletti and Fasella, then Doriano Cavallini (already a biochemist and coming from General Pathology) and with him De Marco, Mondovì, and then Antonini, and of course many others now gathered around him today.

Now the means and the tools for work were available, the rhythm of work increased in intensity and productivity. Much gratitude is due to the Rockefeller Foundation of U.S.A. for its timely generosity at a moment when our national research funds were insufficient or non-existing. Rossi Fanelli would have very much liked to publicly express his gratitude once again to Dr. Pomerat of the Rockefeller Foundation on this very significant occasion for all the help he has given.

Rossi Fanelli then took up in Rome his research on respiratory chromoproteins in collaboration with Antonini, Cavallini and others. Particularly valuable was the collaboration in these studies of Professor Wyman. Amongst the most important results were: the reconstitution of myoglobin and hemoglobin molecules from globin and heme that permitted amongst other things, the preparation and study of hemoglobins and myoglobins reconstituted with unnatural hemes; studies on association-dissociation equilibria of hemoglobin subunits; the isolation of the polypeptide chains of hemoglobins and the characterization of their functional properties; the important demonstration that subunit reassociation is a spontaneous process that brings about the perfect reconstitution of the native quaternary structure; the exhaustive kinetic and termodynamic studies of the reaction of hemoglobins and myoglobins from different species with oxygen. The high value of this research that will emerge with greater detail in the course of this Symposium has contributed in a determining manner to make the Institute of Biochemistry of Rome a pole of attraction and one of the most well-known international centres of research.

Contemporarily, and with equal interest, Rossi Fanelli dedicated himself, together with myself and others, to the study of diphosphothiamine and flavin coenzymes in order to find new methods of preparation and determination, and, as regards diphosphothiamine, to study also the process of biogenesis and its physiological significance. In the course of his research, triphosphothiamine was identified in animal tissues, and its presence and function in the nervous system has recently been confirmed, thus sanctioning perhaps its role as a second coenzyme of thiamine. From the attempts of analytical separation of thiamine phosphoric esters - in the end happily resolved also thanks to the collaboration of Professor Paul Karrer of Zürich and Prof. Arne Tiselius of Uppsala – emerged the rationale of molecular filtration.

Even in research work in which he was not directly participating, Rossi Fanelli was an untiring animator, and his advice and constructive criticism were never lacking to research workers in his Institute. Everyone found a niche in research work, and all were helped by a first-rate scientific organization. It was he who tirelessly built up the humus from which germinated the numerous and always remarkable results that carry the seal of the Institute of Biochemistry of Rome.

A significant recognition of this scientific growth, which reached its plateau in the 1960s, and that had unquestionable reflections on the progress of Italian biochemistry is to be found in the text of the inaugural statement at FEBS (London 1964). The President, Arne Tiselius, in reviewing the status of biochemistry in member countries defined Italy as a "scientific miracle". In fact, the rapid arrival at a scientific standard comparable with that of the countries in which Biochemistry had been born many years before does merit this definition, and Rossi Fanelli was certainly one of the principal creators of this 'miracle'.

Lastly, his research on neoplasma. In 1963 he organized together with Bruno Mondovì a working group made up of biochemists, cancerologists and medical doctors to study the selective sensibility of neoplastic cells to heat. The application of a new methodology almost free from risks for patients, consented the cure of some types of malignant tumours of the limbs. The heating to 42° of the blood 'in loco' for 2 or 3 hours brings on, in fact, an almost total necrosis of the neoplastic tissue, followed by its disappearance. At 18 years' distance from the first treatment, the success of this new therapeutic application has been proven, apart from the cures directly registered, from its always more extensive clinical applications. Even if the intimate mechanism for which the malignity entails the particular thermal sensitivity still remains obscure, the working group was able to establish that the process of protein synthesis must be involved.

Characteristic of his didactic and scientific activity is Rossi Fanelli's simplicity, his basic, clear and essential simplicity. An impression of clarity and rigour is given out even from reading his papers. Rossi Fanelli is one of the few who has always known how to keep a firm distance from the great "producers" of science, and his antiexhibitionism, his abhorrence of the limelight and of fleeting success, are the logical consequence of this attitude. It is just this natural reservedness which together with his indisputable merits makes him such an outstanding figure.

Professor Rossi Fanelli is a Member of the Accademia Nazionale
dei Lincei, and has been Academic Secretary since 1963. He is a
member of the Accademia dei XL, and an Honorary Member of academic
institutions and foreign scientific societies such as the American
and French Biochemical Societies, the Institute of History of Sciences of the Academy of Science of Buenos Aires, and Honorary Citizen
of New Orleans. Rossi Fanelli was awarded the Feltrinelli Prize for
Medicine, holds the Gold Medal of Merit from the Medical Association,
and is a Commander of the Italian Republic. He has been for many
years a member of the Committee for Biological and Medical Sciences
of the C.N.R., a member of the International Union of Pure and
Applied Biochemistry, and for many years was President of the Italian
Society of Biochemistry. He is at present a Member of the Senior
Council of Health.

Amongst Rossi Fanelli's best known characteristics are his
cordiality and affability, and delightfully roguish sense of humour,
which never leaves him. What I have always admired personally so
much in Rossi Fanelli is his incredible ability to deal so graciously
– and with never a hint of boredom or impatience – with the never-
ending succession of collaborators and guests continually knocking
at his door. Even with the most tedious, his affability, punctuated
frequently by his light-hearted repartee, has never been lacking.

Capable of coming to grips with the essentiality of situations
with unusual rapidity, his counsel has always been determining and
retrospectively exact. His most human quality is, in my opinion, his
unquenchable vigour, not only that which has allowed him such a
decisive grip on so many aspects of academic and non-academic life,
but above all that which he has shown in human endurance, which
normal inpromptu contact does not usually let one explore or get to
know. In a time in which not only faith in Providence, but also that
in man and his institutions seem lost, Rossi Fanelli has always been
firm in his convictions, in those values that, to paraphrase Eliot,
shine like torches in nights bygone and which are kept alive only
by the nostalgia of a few as a poignant never-ending memory. This
strength has also come out in his private life. He loved his family
in a way that we might call old-fashioned. He watched his children
growing up with loving concern, together with his beloved wife, the
splendid and wise Anna, and when tragedy so bitterly struck his
family we all felt the intensity of his desperate but dignified
strength of mind. He will perhaps teach us from which hidden depths
he finds this strength in the years ahead that we all hope to spend
together. We, his pupils, can say of him, paraphrasing the great
Plato: "This our Master, the man who was given to us amongst many
to know, is the best, the wisest, the most just."

If this profile of Professor Rossi Fanelli does not reflect a
true image of him, or that which others hold of him, it is still the
mirror of my sentiment, just the small part that my heart remembers.

A STATISTICAL ANALYSIS OF THE PUBLICATIONS

AND COLLABORATIONS OF ALESSANDRO ROSSI FANELLI

W. E. Blumberg

Bell Laboratories
Murray Hill, N.J. 07974
U.S.A.

In the three and a half decades since World War II, the
scientific career of Alessandro Rossi Fanelli has included study of a
wide variety of subject matter and has involved a great number of
other scientists.   The scale of the effort has progressed from very
modest beginnings, with minimal personnel, funding and equipment, to
spacious, well-equipped laboratories in Rome and a world-wide network
of former students and collaborators.   It is the purpose of this
article to examine the nature of this evolving structure through
statistical and graphical analysis of Professor Rossi Fanelli's
publications.

As a source for this study, a search was made of the
references cited in Chemical Abstracts from 1946 to the present. One
hundred thirty two publications were listed there. A later search of
Index Medicus revealed four not previously included. Undoubtedly
there are others in journals not adequately abstracted by these two
reference serials.

Each reference was entered into a data base as follows: (a)
key words from the title were listed (and sometimes abbreviated), (b)
initials of the authors were discarded except when ambiguity would
result,   (c) the name "Rossi Fanelli" was discarded as it is common to
all entries, (d) the year was shortened to the last two digits, and
(e) the journal, volume, and page were discarded as they are not
relevant to the present investigation. Then the primary data base was
processed by a permuted index generator, which successively cyclically
permutes all the words in a given data entry to the beginning of the
line.   An example of this process is shown in Table 1, where a single
reference generates six lines in the permuted index.   Finally the
entire index is sorted alphabetically.   In this manner all
publications indexed in 1958 will appear together, all publications
co-authored with Antonini will appear together, all publications
mentioning the key word "myoglobin" will appear together, etc.   The
entire index is too lengthy to include here, comprising some 1500
lines of output.

Using the permuted index as a secondary data base, it is easy
to construct histograms and tables of the occurrences of various
authors and key words. The time profile of publications is shown in
Fig. 1, where the number of publications indexed in each year is
plotted as a bar above the year. (Note that, in some cases, the year

Table 1

A. Rossi Fanelli and E. Antonini, "The
Oxygen Equilibrium of Human Myoglobin."
[Journal and page], (1958).

58 Antonini oxygen equilibrium human myoglobin
Antonini oxygen equilibrium human myoglobin 58
equilibrium human myoglobin 58 Antonini oxygen
human myoglobin 58 Antonini oxygen equilibrium
myoglobin 58 Antonini oxygen equilibrium human
oxygen equilibrium human myoglobin 58 Antonini

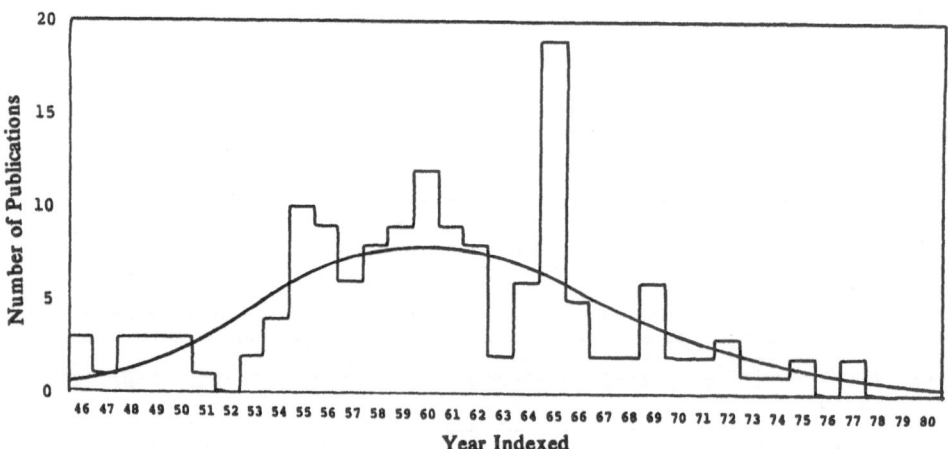

Figure 1

indexed will be the year following the year of publication - no
attempt has been made to correct for this effect.) The continuous
curve in Fig. 1 may be taken as a typical time course of the
publications of a major scientist. The first few years (here 1946-54)
show a self-accelerating or autocatalyitc effect as the laboratory is
growing in personnel and equipment. Next there is a period of
approximately constant output (here 1955-66). Finally there is the
gradual decrease (1967-present) as the scientist (by now an institute
director) becomes more interested in other passtimes - administration,
politics, making money, sailing.

There is a noticeable departure of the histogram from the
smooth curve in Fig. 1 in three places where there was increased
temporary publication activity. These may be referred to as the
"pulses of 1955, 1960, and 1965" or P55, P60, and P65, although each
pulse is several years wide. The causes of these pulses will become
apparent from the analysis below. First let us look at the list of
co-authors, shown in Table 2 arranged chronologically according to
their first year of appearance. An X indicates one or more
publications appearing during that year. As Professor Noris
Siliprandi has indicated in the Introduction to this Symposium, he and
Giulio Perri were Professor Rossi Fanelli's first two students, and

Table 2

```
Co-author             Year
                      4    5        6            7            8
                      6789012345678901234567890123456 7890
(alone)               X.XXXX...X.XX...X..X..............
Siliprandi,N          X.....XXXX..X....................
Perri                 .X...........................,...
Giulotto              ..XX.............................
Fasella               .......XX.X...X....X.X...X........
Cavallini             .......XX...XX....X...X.....X.....
DeMarco               .......XX...XXX....X.............
Merucci               .......XX........................
Mondovì               .......XXXXX....X...X.XXX...X...
Boffi                 ........X........................
Siliprandi,D          .......XXX..X....................
Salvetti              ........X.X......................
Azzone                ........XX.......................
Silvestroni           ........X........................
Trasatti              ........X.X......................
Segre                 ........X........................
Ciccarone             .......XX..X.....................
Antonini              .........XXXXXXXXXXXXX.XX.........
Navazio               .........X.......................
Caputo                .........XXXXX.XX................
Povoledo              ..........X......................
Benerecetti           .........XX......................
DeStefano             .........X.......................
Guacci                .........X.......................
Gibson                .........X.......................
Giuffre               .........X.......................
Ipata                 .........X.......................
Olivo                 .........X.......................
Riva                  .........X...X...................
Wyman                 ..........XX.XXXX.X..............
Bucci                 ..........XX.XX..................
Fronticelli           ..........XX.XX.X................
Zito                  ..........X.XX...................
Moretti               ...........X.....................
Bellelli              ...........XXX...................
Bonacci               ...........XX....................
Bruzzesi              ...........XXX...................
Chiancone             ...........XX....................
Brunori               ...........XX.XX.X...............
Satriani              ...........X.....................
Taylor                ...........X..X..................
Reichlin              ...........XX....................
Wolf                  ...........X.....................
Scioscia-Santoro      ...........X.....XX..............
Bombardiere           ...........X.....................
Guerritore            ...........X.....................
Turini                ...........X.....................
Murawski              ...........X.....................
Carta                 ...........X.....................
Sorcini               ...........X.....................
Tentari               ...........X.....................
Vivaldi               ...........X.....................
Bossa                 ............X....................
Turano                ............X..XX................
Ioppolo               ............X....................
Berger                .............X...X..............
Amiconi               .............X..X...............
Finazzi Agrò          .............X.X................
Rotilio               .............X.X................
Strom                 ..............XXXX.XX.X...
Moricca               ..............X......X...
Cavaliere             ..............XX.....X...
Caiafa                ..............X.X........
Ferraro               ..............X..........
DeSole                ..............X..........
Bozzi                 ..............XXX.X...
Crifò                 ..............XX.X.X...
Renzini               ..............X......
Ravagnan              ..............X......
```

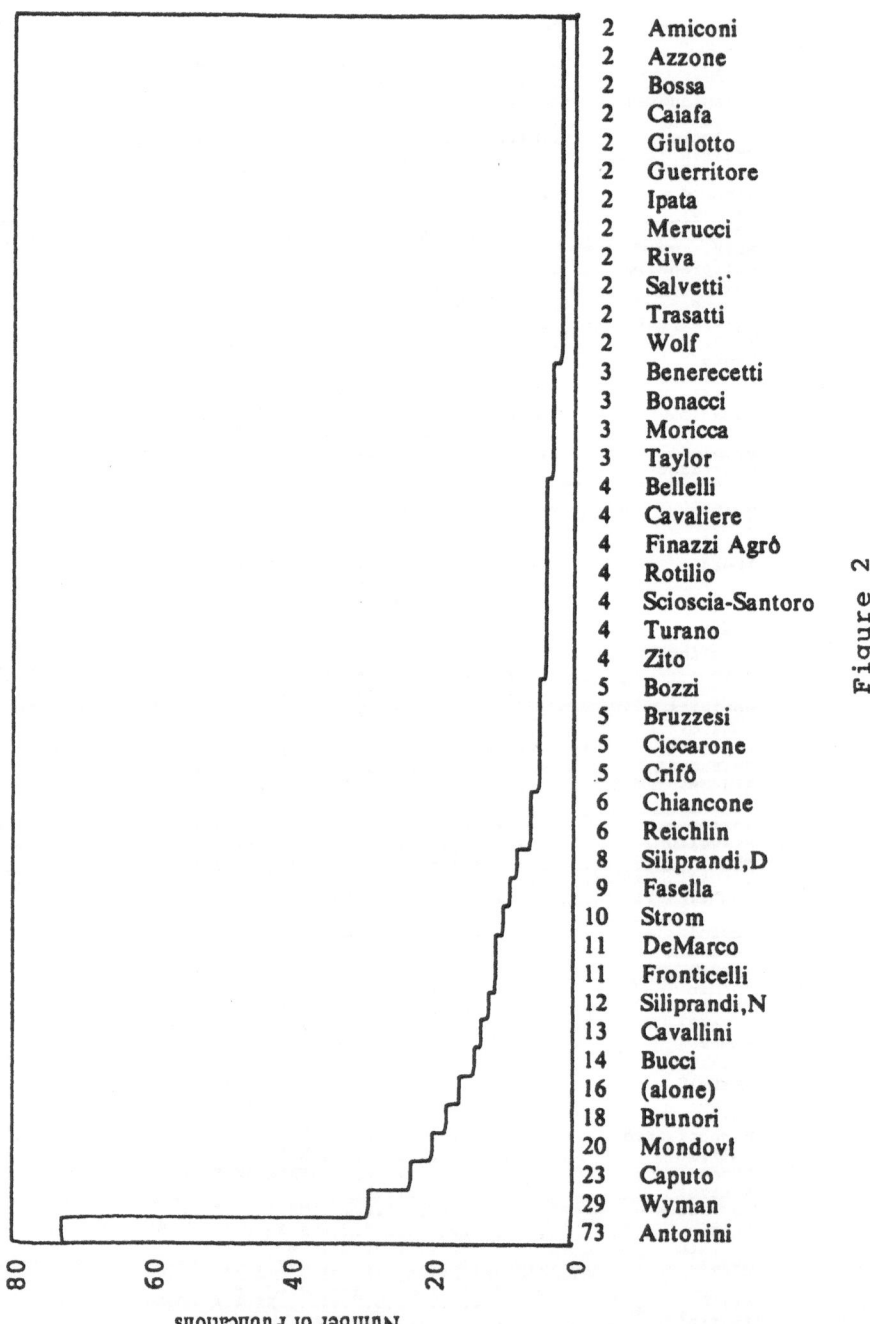

Figure 2

they appear as the first two co-authors. These two cases also
illustrate two different patterns of collaboration which may be seen
in several places in this Table. Professor Siliprandi chose to remain
nearby in Italy and to continue the collaboration over many years,
while Professor Perri chose to seek his fortune a long distance away.
He has, however, now returned to attend this Symposium. In general,
one sees that those people with the longest rows of X´s are the ones
who have remained in Rome. It must be emphasized that this and
succeeding Tables do not represent all the publications of the listed
authors - only those written in collaboration with Professor Rossi
Fanelli.

All the collaborators sharing two or more publications with
Professor Rossi Fanelli are shown in Fig. 2, arranged according to the
number of publications. It is, however, of more interest to examine
when these publications occurred. Publication profiles of ten co-
authors are shown in Fig. 3. Now it can be seen that N. Siliprandi,
Fasella, Cavallini, and Mondovì all contributed to P55. The pulse
centered at 1960 is attributed to Antonini and Caputo, while P65 has
major contributions from Antonini, Caputo, Wyman, Bucci, and Brunori,
among others not shown. The profile of Strom is included to
illustrate recent collaborations between Professor Rossi Fanelli and
the younger scientists in Rome.

Table 3 presents a chronological listing of the appearance of
all the key words describing a biological molecule. It is noteworthy
that Professor Rossi Fanelli was studying two of these in the days
before their molecular identification had been made - Euler´s Z factor
and cocarboxylase had, at the time of first mention, not yet been
attributed to threonine and phosphothiamine, respectively. There is a
gradual progression from small molecules (amino acids and vitamins),
to the globins and to multi-subunit proteins, and finally to very
complex polymers.

As Professor Rossi Fanelli had earned his medical degree
before he started the work for his doctorate in chemistry, he has been
interested in many diseases and abnormalities. These, selected from
the key words of the index, are listed in Table 4 and range from his
early interest in alloxan-induced diabetes to his more recent research
on the heat sensitivity of malignant cells.

Professor Rossi Fanelli has always stressed the examination of
physical and physico-chemical properties of the molecules under study.
To this end, many such techniques were employed, and, indeed, several
of these techniques were developed in Rome specifically for that
purpose. Table 5 shows the techniques and physico-chemical properties
occurring as key words in the index. Heme dissociation and
recombination from heme proteins and the separation and recombination
of the constituent chains of hemoglobin represent unique contributions
of Professor Rossi Fanelli´s laboratory. It should be remarked that
novel techniques are likely to appear as key words in titles of
publications for a few years, but, after the techniques have received
widespread acceptance, they are no longer mentioned.

Biological molecules from many different organisms have been
studied. Table 6 lists the organisms appearing as key words in the
index. (One assumes that the "living animal" in which phosphothiamine
synthesis was studied in vivo would have been further identified in
the text of the article.)

A histogram of the 20 most commonly occurring key words is
shown in Fig. 4. Function and mechanism have been combined, as have
heat sensitivity and stability, since these pairs refer to the same
properties. Alpha and beta always occur together.

Profiles of the occurrence of some selected key words compared
with the publication profile of Professor Rossi Fanelli are given in
Fig. 5. One can see that P55 represents the confluence of activities
on thiamine, myoglobin, and hemoglobin. P60 is correlated with both
hemoglobin and equilibrium (oxygenation curves), while P65 is
concerned almost entirely with hemoglobin (mostly chain properties).

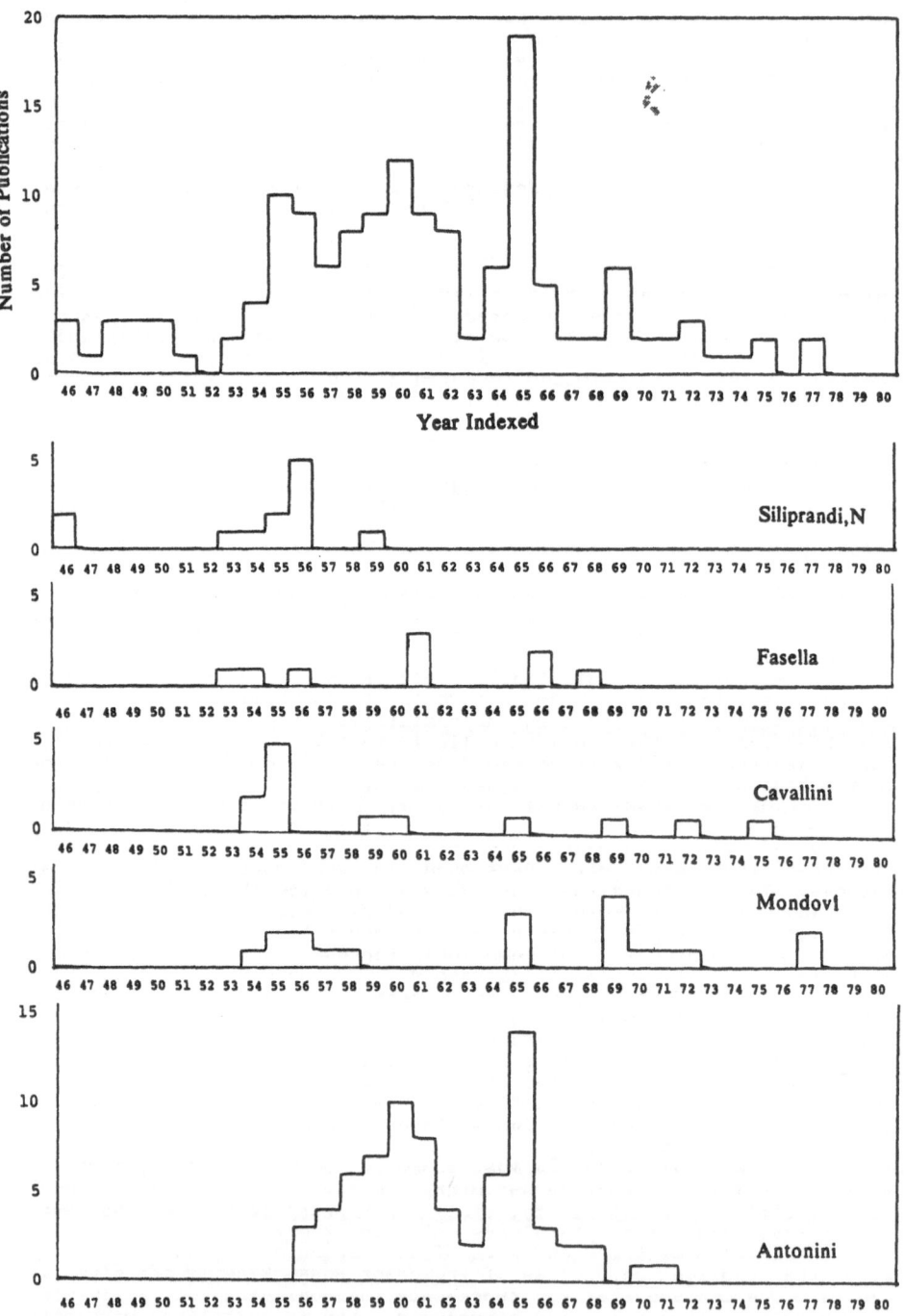

Figure 3

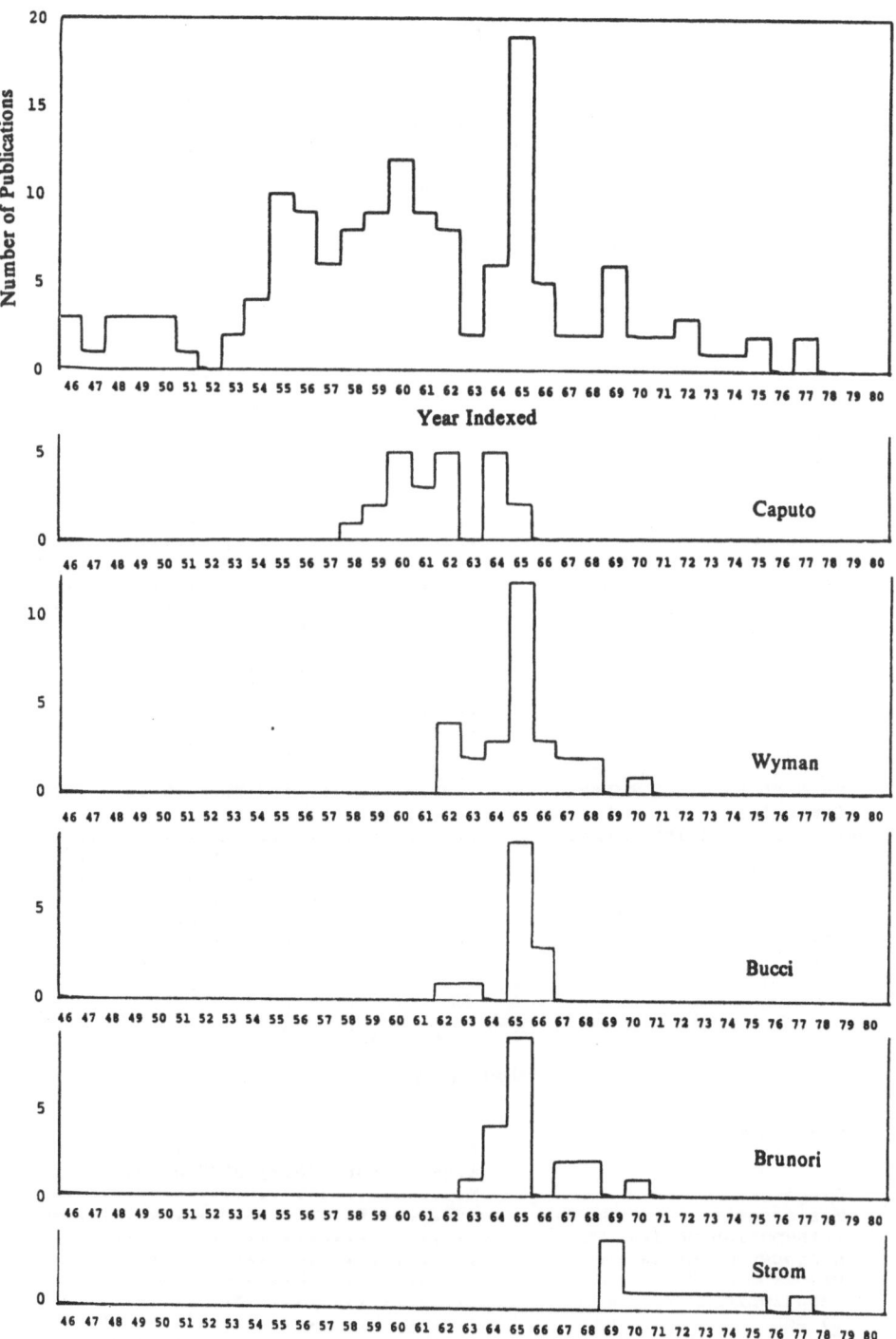

Figure 3 (continued)

Table 3

```
Biological                 Year
Molecule                   4    5         6         7         8
                           67890123456789012345678901234567890
Euler's Z factor           X..................................
Myoglobin                  .XXXX...XXXXXXXX...X..X............
Hemoglobin                 ...XX...XXXXXXXXXXXXXX..XX.........
Threonine                  .....X............................
Thiamine                   .......XXXX..X.X...................
Vitamins                   .......X.X..X.....................
Enzymes                    .......X.X......X.................
Cocarboxylase              ..........X.......................
Ferrimyoglobin             ..........XXX.....................
Ferrihemoglobin            ..........X.......................
Oxythiamine                ..........X.......................
Respiratory pigments       .........X..X....X................
Globin                     .........X.X......................
Protomyoglobin             ..........X.......................
Deuteromyoglobin           ..........X.......................
Coenzymes                  ..........X.......................
Riboflavin                 ..........X.......................
Deuterohemoglobin          ..........X.......................
Protohemoglobin            ..........XX......................
Mesohemoglobin             ..........X.......................
Protoporphyrin globin      ..........X.......................
Hematin                    ..........X.......................
Hemoproteins               ..........X....XX.XX..............
Carboxyheme                ..........X.......................
Chlorocruorin              ............X..X..................
Alpha, beta subunits       ............X..XX.................
Dextran                    ............XX....................
Cottonseed protein         ............X.....................
Carboxymethyl dextran      ............X.....................
Diethylaminoethyl dextran  ............X.....................
Glutamic dehydrogenase     ............X.....................
Mercaptosuccinyl dextran   ............X.....................
Cystamine                  ............X.....................
Aspart. ketoglut. transam. ...............X..................
Malate dehydrogenase       ...............X..................
Antibodies                 ...............X..................
Nucleic acids              ...............X..................
Aspart. aminotransf.       ................X..X..............
Polyene antibiotic         .....................XX...........
RNA                        .....................X.X..........
```

Table 4

```
Disease or                 Year
Condition                  4    5         6         7         8
                           67890123456789012345678901234567890
Diabetes                   X..................................
Thalassemia                .........X.........................
Methemoglobinemia          ...........X.......................
Microcytic anemia          .............X.....................
Hemoglobin A2              .............X.....................
Hemoglobin M               ...............X...................
Cancer                     .....................XX..X...X.....
Ehrlich ascites            ......................XX..X........
Novikoff hepatoma          ..........................X........
```

Table 5

```
Techniques and                  Year
Properties                      4    5         6         7         8
                                67890123456789012345678901234567890
Synthesis                       X....X.........X.......X..XX.......
Crystallization                 .XX.....XXXX...X....................
Solubility                      .X.................................
Optical absorption              ..XX...............................
Denaturation                    ..XXX....XX...XX..X.................
Amino acid composition          ....X...XX.X..X...X.................
Electrophoresis                 ..........X........................
Kinetics                        ..........X........X.X.............
Oxidation-reduction             ...........XXX......X.X.............
Heterogeneity                   ...........X.X.....X................
Oxygen equilibrium              ............XXXXX.X.....X...........
Heme dissoc-recomb              ............XXXX..X.................
Salt effect                     ............X.XX....................
CO equilibrium                  ............X.....X................
Light scattering                ............X......................
Sedimentation                   ............X...X..................
Molecular structure-function    ............XXX..X....X.............
Chain separation                ............X..XX...................
Chain recombination             ............X..X....................
Bohr effect                     ............X..X....................
Proteolytic digestion           ............X.XX....................
Redox potential                 .............XX.X...................
Oxygen-linked                   .............X.....................
Electron microscope             .............X.....................
Heat stability                  ..................X..XXX..X.X.X...
Immunochemistry                 .................XX................
Acid-alkali transformation      ....................X..............
```

Table 6

```
Organism                        Year
                                4    5         6         7         8
                                67890123456789012345678901234567890
Human                           .XX.....XX.XXXXXXXXXXX...X.........
Fetus                           ........XX...XX....X...............
Cow                             ........X..........................
Living animal                   ........X..........................
Sea-fish                        ..........X........................
Mollusk                         ...........X.XX.....................
Invertebrate                    ...........X.......................
Thunnus                         ...........X...X...................
Aplysia                         ............XX.X.......X............
Fish                            ............X......................
Mammal                          ............X............X.........
Mycobacterium                   ...............X...................
Spirographis                    ...............X..X................
Horse                           ................X..X...............
Dog                             ................X..X...............
Cottonseed                      .................XXX...............
Bacillus cereus                 ...........................X......
```

Referring back to Fig. 3, one can see that Caputo is correlated only
with the hemoglobin oxygenation curves (P60) and that Wyman, Bucci,
and Brunori are correlated mainly with the activity on hemoglobin
chain properties (P65). Antonini is represented strongly both in P60
and P65 and has contributed to research on both these subjects.

The Rome laboratory has always attracted visiting scientists,
as Professor Rossi Fanelli has been a most gracious host. Often a
visitor would bring his specific skill or technique to bear on the
problems under study in Rome. An example of this is given in Fig. 6,
where the collaboration with Professor Morris Reichlin (an
immunologist) resulted in a correlation of publications having both
his name and appropriate key words.

The data in the permuted index is also subject to a more
mathematical treatment. One can compute a numerical index to be
called "Specialization" as defined in Table 7. Here one takes the
number of occurrences of the author's name (A) simultaneously with a
subject name (S), the occurrence together denoted by (A+S), and
divides it by all occurrences of A. One can see that Wyman and

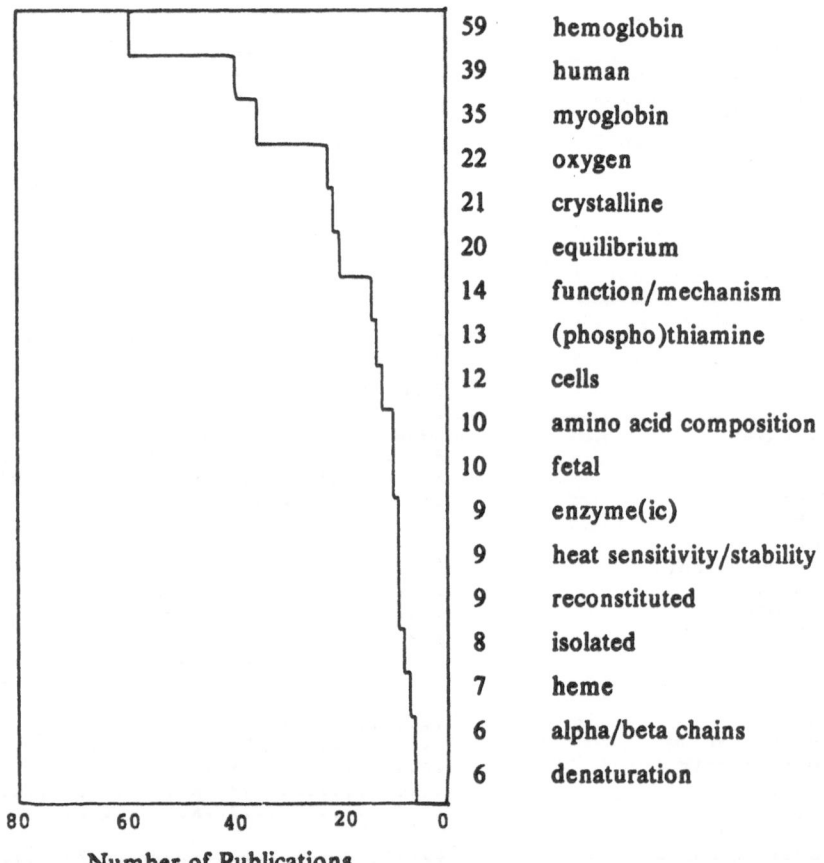

Figure 4

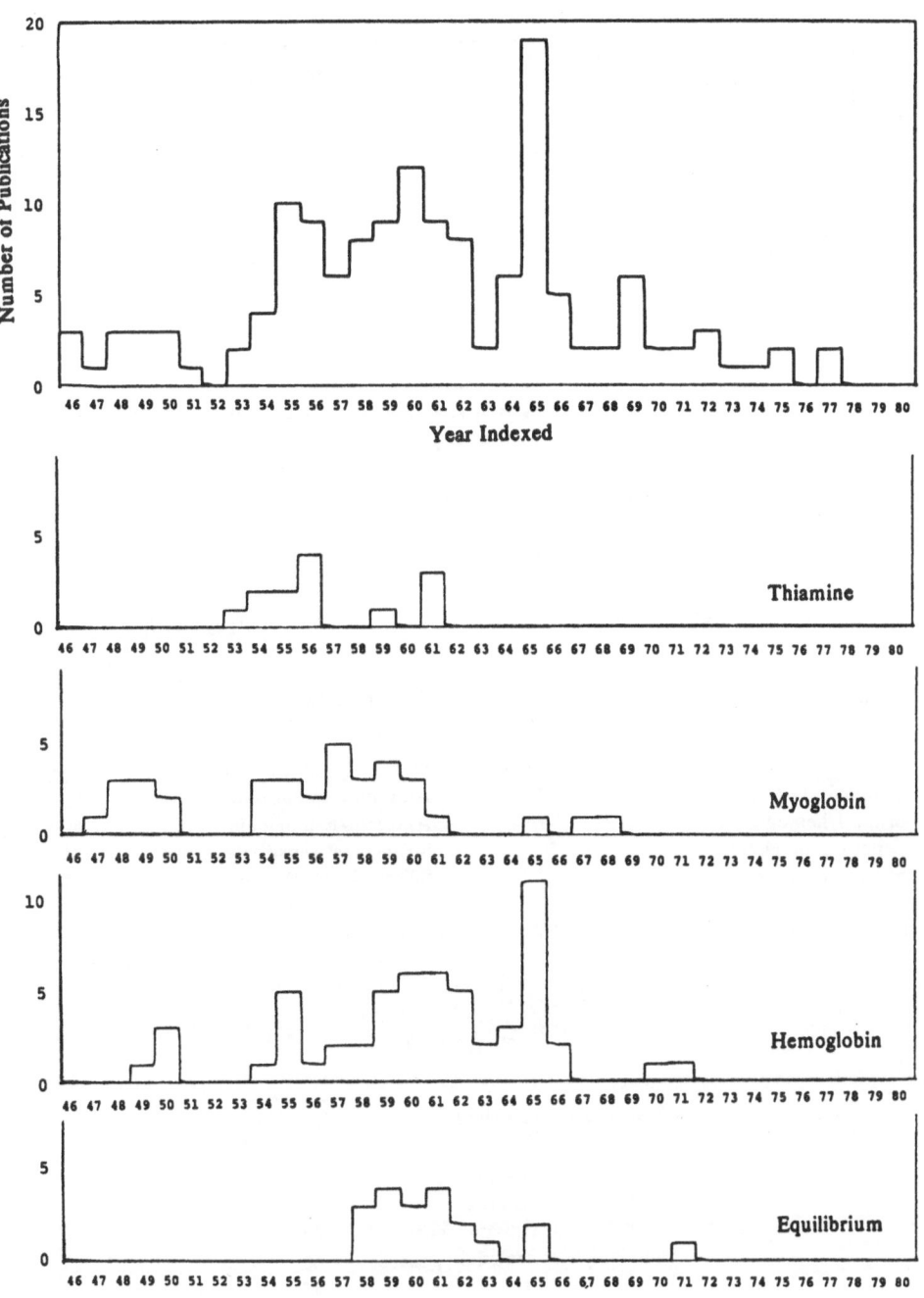

Figure 5

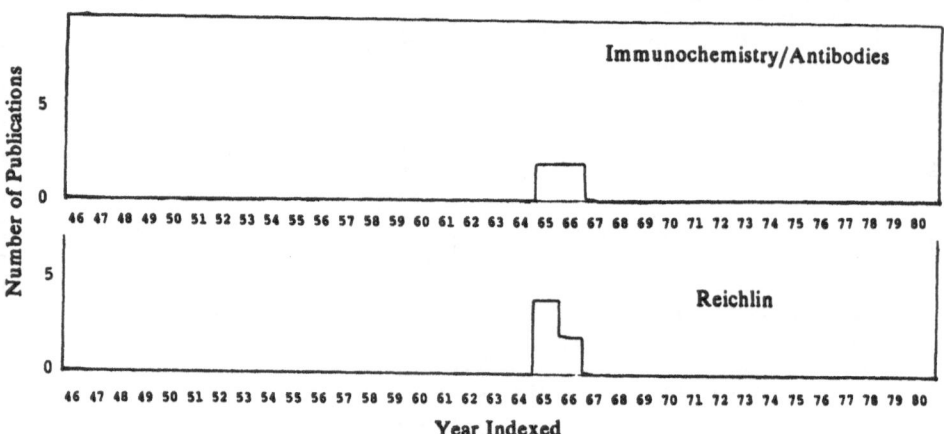

Figure 6

Table 7   Table 8

Specialization	$\dfrac{\Sigma(A+S)}{\Sigma A}$	Accountability	$\dfrac{\Sigma(A+S)}{\Sigma S}$
Wyman/hemoglobin	86%	Mondovi/heat sensitivity, cancer	100%
Siliprandi, N./thiamine	75	Antonini/equilibrium	95
Brunori/hemoglobin	72	Antonini/hemoglobin	80
Caputo/hemoglobin	70	Strom/heat sensitivity, cancer	75
Fasella/thiamine	67	Siliprandi, N./thiamine	67
Antonini/hemoglobin	64	Antonini/myoglobin	48
Strom/heat sensitivity, cancer	60	Fasella/thiamine	46

Table 9

Association Coefficient	$\dfrac{\Sigma(A+S)}{\sqrt{\Sigma A}\cdot\sqrt{\Sigma S}}$
Antonini/hemoglobin	70%
Siliprandi,N./thiamine	67
Strom/heat sensitivity, cancer	67
Mondovi/heat sensitivity, cancer	60
Fasella/thiamine	55
Antonini/equilibrium	49

hemoglobin are correlated 86% according to this index. Than means that, of all the collaborations with Wyman, 86% were on hemoglobin, 14% on other topics.

In like manner a numerical index "Accountability" can be computed as defined in Table 8. The table shows, for example, that no publication of Professor Rossi Fanelli on heat sensitivity and cancer was made that did not also include Mondovì as a collaborator. On the other hand, 95% of the publications on equilibrium included the collaboration of Antonini, etc.

Finally, one may compute many more complicated mathematical expressions. Only one will be illustrated - the "Association Coefficient," which may be defined loosely as the answer to the question "What comes to mind when I mention ´Rossi Fanelli and hemoglobin´?" The answer is "Antonini" 70% of the time. The coefficient is symmetrical. The answer to the question "What comes to mind when I mention ´Rossi Fanelli and Antonini´?" is "hemoglobin" 70% of the time.

In conclusion, I would say that even if I had never met Professor Rossi Fanelli nor had visited his laboratory in Rome, I could have reconstructed the major features of his scientific career from only an examination of his publications and collaborations.

HEMOGLOBIN, MYOGLOBIN, AND OTHER RESPIRATORY PROTEINS

LINKAGE GRAPHS

Jeffries Wyman

CNR Center of Molecular Biology, Institutes of Biochemistry and Chemistry, Faculty of Medicine, University
of Rome, 00185 Rome, Italy

INTRODUCTION

In earlier papers[1,2] I have developed the concept of functional linkage in biological macromolecules in terms of a set of potentials derivable from one another by a group of Legendre transformations. In this note I show how these potentials may be represented
by a set of contour graphs which embody the results of observation
and are transformable into one another by a group of graphical permutations which is equivalent to the Legendre group.

LINKAGE POTENTIALS

The concept of a set of linkage potentials has its origin in
the basic equation

$$dE = T\,d\,S - p\,d\,V + \sum_i \mu_i\,d\,n_i \qquad (1)$$

which incorporates the two laws of classical thermodynamics. Here E
is the total energy, S the total entropy, V the total volume of the
system and the n's give the total amounts of the various components
(including the macromolecule) present in it. This equation shows
that the temperature T, pressure p, and chemical potentials μ_i may
be identified with the first partial derivatives of E with respect
to the extensive variables S, V, and the n_i respectively. By norma-
lizing E with respect to some one component (the macromolecule)

23

chosen as the reference component -- this is justified by the first
order homogeneous property of E -- we obtain an equation of one less
variable, the macromolecule having dropped out:

$$d\bar{E} = T\, d\bar{S} - p\, d\bar{V} + \sum_{1}^{r} \mu_i\, d\bar{n}_i \qquad (2)$$

Here bar denotes the amount of a given extensive quantity per unit
of reference component and r the number of ligands.°

From $\bar{E}$ defined in this way we can, by application of a group
of Legendre transformations, obtain a variety of other potentials
in each of which one or more of the normalized extensive variables
is replaced by the corresponding intensive one. These potentials,
with reversed sign, together with E, constitute the linkage potentials, so called because they are the source of a variety of linkage
relations. They form a closed set, or group,°° of order 2^{r+2}, which
is isomorphic with the group of Legendre transformations by which
they are derived from one another. The group is abelian and has the
symmetry of a 2^{r+2} dimensional rectangle in hyperspace. It may be
noted that one of its members, that in which all the variables are
the intensive ones, is what has been called the binding potential
and denoted by Π (Russian L.).[3]

It will be seen that the number 2 in the exponent of 2^{r+2} comes
from the two physical variables S and V. Since these enter the
equations in exactly the same way as the n's we may treat entropy
and volume as ligands of chemical potential T and -p respectively
to give a total of t = r+2. These 2^t linkage potentials provide
alternative ways of describing the system and each of them is the
source of a set of linkage relations.

° If r = 0, which means that we are dealing with a pure substance,
then of course we have simply $dE = T\, d\bar{S} - p\, d\bar{V}$ and the con-
vention of the bar may be neglected.

°° In earlier discussions we have used the term "group" broadly to
include the potentials themselves as well as the operations by
which they are derived, somewhat as does the crystallographer
in speaking of a crystal as belonging to a given space group.

THE GRAPHICAL REPRESENTATION OF THE LINKAGE POTENTIALS

Clearly, if we had complete a priori knowledge of any one potential as an analytical function of its variables -- if for instance
we had an expression for the normalized energy $\bar{E}$ as a function of $\bar{V}$,
$\bar{S}$, and the $\bar{n}_i$ --, we could, in principle at least, deduce all the
other potentials by Legendre transformation and thereby obtain all
the various linkage relations of which they are the source. Actually
however, we have no such knowledge except in the case of an imagi-
nary and structureless system of perfect gases. Moreover, even if
we had such knowledge the calculations required in applying it would
generally entail solution of intractable equations. It is necessary
therefore to adopt an empirical approach based on direct independent
measurements of corresponding values of intensive and extensive
properties of the system.

Consider the simplest case, where only two ligands come into
play. An example is provided by hemoglobin as it combines with
oxygen (O_2) and proton (H) at constant T and p and fixed values of
all other variables. The system will be equally well defined by any
one of the $2^2 = 4$ potentials

$$P_1(\mu_{O_2} \quad \mu_H) \equiv \pi, \ P_2 \ (\bar{O}_2, \mu_H), \ P_3(\ \mu_{O_2}, \bar{H}) \ P_4(\bar{O}_2, \bar{H})$$

from any one of which the others may be obtained by Legendre transformation. We have experimental methods for measuring simultaneous
values of all four variables and the information obtained can be
equally well represented by any one of four pairs of three dimensional graphs, one pair corresponding to each potential. In each graph
the two independent variables are plotted as the X and Y coordinates.
When projected onto the XZ or YZ plane these graphs give rise to
four pairs of two dimensional countour graphs shown in Fig. 1, where
the contours are for the independent variable identified with the
Y coordinate.

It will be seen that from the information contained in the two
contour graphs corresponding to any one of the four potentials it
is possible to construct all the other graphs and, in particular,
the pair corresponding to each of the three other potentials. Thus
we have here a graphical substitute for an analytical implementation
of the Legendre transformations. In fact, the four pairs of contour

$P_1(\mu_{O_2},\mu_H)$ $P_2(\bar{O}_2,\mu_H)$ $P_3(\mu_{O_2},\bar{H})$ $P_4(O_2,\bar{H})$

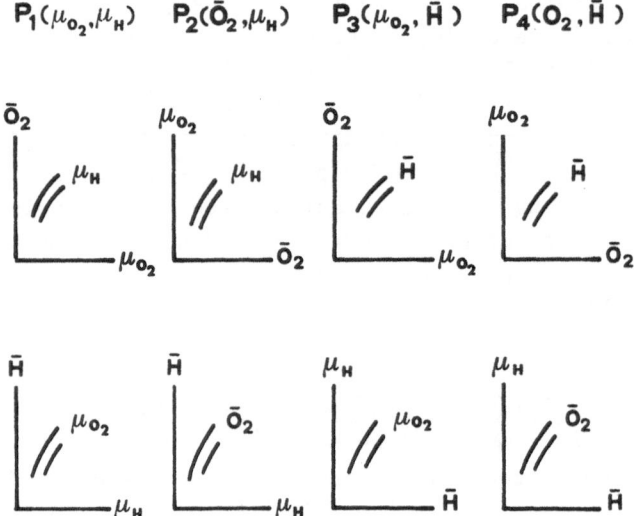

Fig. 1. Linkage graphs for the case of two ligands.

graphs, which we may call linkage graphs,[°] may be seen as a four
order set of which the members may be derived from one another by
a four order group of permutations in the same way as the four
potentials may be derived from one another by a four order group of
Legendre transformations. Thus, for instance, the transition from
the two graphs for P_1 to those for P_2 results from an interchange
of $\bar{O}_2$ and μ_{O_2}; that from P_2 to P_3 from an interchange of $\bar{O}_2$ and
μ_{O_2} together with one of μ_H and $\bar{H}$; and so on. We need go no
further. The multiplication table for this four order permutation

---

[°]Not to be confused with the topological linkage graphs described
previously.[3]

group is the same as that for the four order group of Legendre transformations and both are an instance of the well known four group which described the symmetries of a two dimensional rectangle.

When additional ligands come into play, leading to a larger number of independent variables, the situation becomes more complicated, involving graphs of higher dimensionality. From a practical point of view it is best handled by treating the ligands in pairs as in the example just given, remembering that each pair of potentials represents a subset of the full set of 2^t potentials, and that the four order group of transformations is a subgroup of a 2^t order permutation group.[°]

OPERATIONAL BACKGROUND

Before concluding it may be worthwhile to consider just what is involved in the measurements that underlie such graphical and analytical descriptions of a system as those just considered. At the most general level reflection shows that all binding studies and mass law investigations depend finally on simultaneous measurements of the amounts and chemical potentials of the various components. The reason for this comes from the fact that functional chemistry only comes into being with the recognition and definition of energy; otherwise there is nothing left for us but pure description. In all, or nearly all, cases the measurement of a chemical potential requires the establishment of equilibrium between two or more phases. Thus the simplest way of determining the activity of a component in solution is to measure its vapor pressure. If as an alternative we resort to an electrochemical method (this involves enlarging our expression for the energy by including an electrical work term), we must introduce an electrode, say a hydrogen electrode comprising one or more new phases. Even the measurement of temperature requires the introduction of a second phase in the form of a thermometer, of whatever kind.

[°] We have illustrated the two ligand case by hemoglobin as it combines with oxygen and proton. We might equally well have invoked the case of a single substance subject to change of T and p. The normalized extensive variables, corresponding to $\bar{O}_2$ and $\bar{H}$ then become $\bar{S}$ and $\bar{V}$, namely entropy and volume per unit of substance. S is measured in terms of heat absorbed, starting from some arbitrary level.

At a more experimental level let us consider in detail the
actual procedure employed in determining the upper left hand graph
shown in Fig. 1, which describes the binding of oxygen by hemoglobin as a function of μ_H (or pH). Although this would appear to be
a very special case, it involves considerations of a rather general
nature. First we make up a solution of known composition containing
a strong buffer of desired pH. An aliquot of this is introduced into
a tonometer, or something equivalent to it, where it occupies a
small fraction of the fixed total volume, the rest being occupied
by a gaseous mixture of oxygen and some inert gas, usually nitrogen,
in predetermined proportions, partially saturated with water and
at atmospheric pressure. The tonometer is then placed in a thermo-
stated bath. Temperature, volume, and the total amounts of all chemical components being fixed, the conditions for the attainment of
2 phase equilibrium are clearly satisfied. After allowing sufficient
time for the oxygen to diffuse into the liquid phase and come to
equilibrium with the hemoglobin we remove a sample of the solution
and measure the total amount of oxygen, both dissolved and chemically bound, which it contains -- at least that is what was done in
the earlier experiments. It is assumed, as a close approximation,
that the volume of the gas phase is sufficiently large so that there
has been no appreciable change in the total pressure or composition
of the gas phase and thus no change in the chemical potential of
the oxygen during equilibrium; likewise that the buffer is sufficiently strong so that there has been no appreciable change of pH
(this of course can be checked experimentally); and, finally, that
there has been no significant loss of water by the liquid phase.
On the basis of these assumptions we construct binding curves for
the liquid phase as a whole such as the one shown in Fig. 1. These
curves contain all the information needed by the physiologist in
his studies of respiration. However what we, as molecular biologists,
are concerned about is something different, namely the amount of
oxygen (and proton) truly, or chemically, bound by the macromolecule alone, as distinct from the total, which includes that in
solution.

In order to discover this, in other words to resolve the total
amounts of the ligands into what is truly bound and what is free,
we require, as pointed out earlier, some special pair of "spectacles"
such as were not available to the pioneers. An increasing assortment
of such "spectacles" is now at hand in the form of spectroscopic
techniques, using that word in its broadest sense, and with their

aid it has become possible to achieve the required resolution. What
we really do in effect in making this separation is, conceptually,
to break up the liquid phase into two subphases in equilibrium with
one another. Looking at the liquid phase as a whole we note that
within the limits of our assumptions it is a system containing a
fixed amount of all "components" except oxygen and volume. The
"chemical potentials" of these however are known, namely the total
pressure p and μ_{O_2} . Therefore, as in the larger case of the
whole system, the conditions for a two phase equilibrium are satisfied. On this basis, with the aid of our extra pair of spectacles,
we construct binding curves for the subphase consisting of the
protein alone, from which to draw conclusions as to the mechanism
of the macromolecule, allosteric or other.

CONCLUSION

The foregoing analysis shows how the behavior -- the functional
chemistry -- of a macromolecular system may be represented by a set
of contour graphs which fall into subsets, one corresponding to
each linkage potential. The behavior of the system is described
equally well by any one of these subsets, just as it is equally
well described analytically by any one of the linkage potentials.
The graphs or potentials may be thought of as different ways of
looking at the system, which remains the same from whichever side
it is viewed. The graphs may be derived from one another by a group
of permutations which is the exact equivalent of the group of Legendre transformations by which the potentials may be derived from one
another. The macromolecular system might be likened to a book --
some great book such as "War and Peace," with all its characters
and episodes -- which is translated back and forth between many
languages. Each version corresponds to a potential, or to an equivalent subgroup of graphs. The rules of translation from one lan-
guage to another are the group operations (Legendre transformations,
graphical permutations). No matter how many times and into how many
different languages the book is translated, it remains the same.

All this brings home the place of the group concept in dealing
with a macromolecular system -- or in fact any system of interacting
components. And it is provocative to learn that, at the opposite
extreme of magnitudes, the bizarre world of subatomic particles gives
hints that it, also, may be susceptible to better understanding in
terms of group concepts.

REFERENCES

1. J. Wyman, A group of thermodynamic potentials applicable to
   ligand binding by a polyfunctional macromolecule, Proc. Nat.
   Acad. Sci. U.S.A. 72: 1464 (1975).
2. J. Wyman, The cybernetics of biological macromolecules. Biophys.
   Chem. (1981) in press.
3. J. Wyman, The binding potential, a neglected concept. J. Mol.
   Biol. 11: 631 (1965).

NATURE OF THE IRON-OXYGEN BOND AND CONTROL OF OXYGEN AFFINITY OF

THE HAEM BY THE STRUCTURE OF THE GLOBIN IN HAEMOGLOBIN

M.F. Perutz

Medical Research Council
Laboratory of Molecular Biology
Hills Road, Cambridge CB2 2QH, England

Spectroscopic and chemical evidence speak in favour of the iron-oxygen bond being polar. X-ray analysis shows that the oxygen molecule is inclined at an angle of about 115° to the haem plane. Cooperative binding of oxygen by haemoglobin is attributable to an equilibrium between two alternative structures that differ in oxygen affinity by the equivalent of 3-3.5 kcal/mol. The author has proposed that in the low-affinity structure the globin opposes the movement of the iron atom from its pentacoordinated pyramidal geometry in the haem of deoxyhaemoglobin to its hexacoordinated planar geometry in the haem of oxyhaemoglobin, while in the high-affinity structure this restraint is absent. Recent evidence supporting this mechanism is described.

This paper discusses the Fe-O bond in myoglobin and haemoglobin and the origin of the cooperativity of the reaction of haemoglobin with oxygen. The nature of the Fe-O bond has been the subject of speculation and experiment. The oxygen molecule has a spin of $S = 1$ and the ferrous iron in deoxyhaemoglobin has a spin of $S = 2$, yet when the two combine to form oxyhaemoglobin Pauling and Coryell found the compound to be diamagnetic; they argued that the Fe-O bond should have the resonating structure.

$$\text{Fe}^{-} - \overset{..}{\underset{..}{\text{O}}} \overset{\overset{..}{\text{O}}:}{{}^{\diagup}} \ , \ \text{Fe} = \text{O} \overset{\overset{..}{\text{O}}:^{-}}{{}^{\diagup}{}^{..}}$$

that makes all electrons paired (1). In modern terms this means that the two $1\pi g^*$ orbitals no longer have the same energy, so that their electrons pair in the single πy^*, which has a lower energy

than πx^*, because it lies at right angles to the Fe-O-O plane. On
Pauling's model, the bond between the iron and oxygen would be made
by hybridization between the π orbitals of the oxygen and the dxz
and dyz orbitals of the iron, with some net transfer of charge from
the oxygen to the iron.   This model was challenged by J.J. Weiss,
who suggested that the bond might be ionic between a ferric ion and
a superoxide ion, net charge being transferred from the iron to the
oxygen ($Fe^{3+}O_2^-$) (2).   Experimental support for Weiss' model was
first advanced by Misra and Fridovich (3).   They showed the autoxidation of oxyhaemoglobin to be a first-order reaction depending
only on [HbO_2]; when epinephrin was added to the solution it was
oxidized to adrenochrome, but this oxidation was inhibited in the
simultaneous presence of superoxide dismutase and catalase, which
suggests that superoxide ion is liberated on autoxidation of oxyhaemoglobin.   Recently, Demma and Salhany (4) have shown that lib-
eration of oxygen by flash photolysis of oxyhaemoglobin reduces cytochrome c, and this, too, is inhibited by superoxide dismutase and
catalase.   It could be argued that the superoxide ion is the result of an excited state induced by the flash but solvent effects
also speak in favour of a polar ($Fe^{3+}O_2^-$) bond.   Brinigar et al.
(5) synthesized a haem with a covalently attached pyridyl base that
allows it to combine reversibly with molecular oxygen at -45°.   In
 dimethylformamide (ϵ = 36) half saturation of this complex requires
an oxygen pressure of only 5 torr, in 10% N-methylpyrrolidine-90%
toluene 28 torr, and in pure toluene (ϵ = 2.4)   about 400 torr.
The ($Fe^{3+}O_2^-$) structure is also supported by spectroscopic evidence.
The IR O-O stretching frequency in oxyhaemoglobin is 1107 cm^{-1} (6),
which is in the superoxide ion range (1150-1100cm^{-1}), much lower
than that of the oxygen molecule (1556 cm^{-1}), and higher than that
of a single O-O bond ($\sim$800 cm^{-1}).   X-ray fluorescence also points
to the presence of unpaired electron density on the iron atom (7).
Cobalt porphyrins and haemoglobins combine reversibly with molecular
oxygen.   Again solvent effects speak in favour of a polar Co-O
bond.   Stynes and Ibers (8) found that the oxygen affinity of
cobalt-porphyrin complexes rises with polarity of the solvent.   At
-23°C Co(II)protoporphyrin IX dimethylester methylimidazole in dimethylformamide requires an oxygen pressure of 12.6 torr for half
saturation; substitution of toluene as a solvent raises that pressure to 417 torr.   The unpaired electron of the cobaltous d^7 ion
provides a useful probe for exploring the nature of the Co-O bond.
This electron gives an ESR signal with nuclear hyperfine splitting
from which the unpaired electron density can be located.   In the deoxy
derivatives hyperfine splitting from the cobalt is combined with
that of a single nitrogen atom of the proximal histidine, which
shows that the unpaired electron occupies the d_z^2 orbital pointing
towards the histidine.   In the oxy derivatives, the nitrogen splitting disappears and the separation of the hyperfine lines due to
cobalt is reduced to about a third.   Since that separation is
directly proportional to the unpaired electron density on the

cobalt, it is inferred that a substantial fraction of the density
has gone to the oxygen (9,10,11).   This has been confirmed by a
similar experiment in reverse, using $CoHb^{17}O_2$. ^{17}O has a nuclear
spin of $I = ^5/_2$, so should give rise to hyperfine splitting if the
unpaired electron density is transferred from the Co to O_2.   This
is indeed observed, and its magnitude suggests that about 60% of the
unpaired electron density is transferred (12).   It could be argued
that the metal—oxygen bonds might be different in the Fe and Co derivatives, but the similarity of the O—O stretching frequencies,
1107 cm^{-1} for Fe and 1106 cm^{-1} for Co, suggests that they are similar (13).   Unfortunately, Weiss died before his prediction was
confirmed experimentally.   The $Fe^{3+}O_2^-$ model can be reconciled
with diamagnetism or weak paramagnetism of the complex by postulat-
ing that the transferred d electron of the iron pairs with one of
the two π^* electrons of the oxygen, and that the spin of the other
π^* electron is paired with the odd d electron left behind on the
iron by antiferromagnetic coupling.   The diamagnetism of oxyhaemoglobin has been challenged recently by Cerdonio et al., who produced
evidence of a low-lying triplet state that makes it weakly paramagnetic at room temperature (14,15).   Their observations, though
apparently flawless, do raise problems.   For example, if oxyhaemoglobin at room temperatures had a molar susceptibility per haem of
+2460 x 10^{-6} cgs/mol, as they report, one would expect its NMR spectrum to exhibit hyperfine-shifted haem proton resonances, but these
have not been observed.

The oxygen adducts of the picket-fence complex definitely are
diamagnetic (16).   The oxygen molecules are bent to the haem axis
and lie in four alternative orientations.   Because of that disorder
it has not been possible to determine the coordinates of the terminal
oxygen as accurately as those of the other atoms.   In the 1-methyl-
imidazole (1-MeIm) complex O–O = 1.16 Å and Fe–O–O = 131^+_-, and in
the 2-MeIm complex O–O = 1.22 $\pm$ 0.02 Å and Fe–O–O = 129 $\pm$ 1°.   The
authors state that they may have underestimated the O–O distance
in the unhindered complex by as much as 0.15 Å.   The geometry agrees
with Pauling's prediction of a bent FeO_2 bond, and the O–O distance
is close to that of 1.27 Å, predicted by him in a recent paper.   It
is slightly shorter than that of 1.34 Å in the superoxide anion,
in agreement with the ESR results, which show that no more than
two-thirds of the density of one electron is transferred from the
metal to the antibonding π^* orbitals of the oxygen.

The structures of sperm whale deoxy- and oxymyoglobin have been
refined to a resolution of 1.6 Å (17,18).   In deoxymyoglobin the
iron atom lies at the apex of a pyramid with the four nitrogen atoms
at its base.   The displacement of the iron from the mean plane of
the four porphyrin nitrogens is 0.4 Å and from the mean porphyrin
plane, including the α carbons, 0.55 Å.   These displacements are
the same, within error, as in the model compound (2-MeIm) meso-

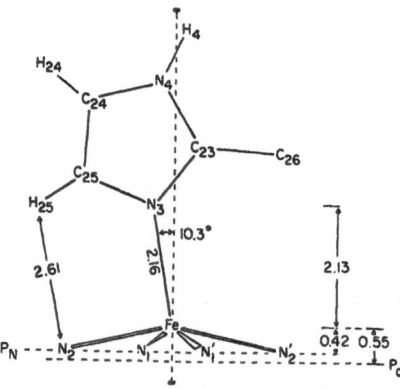

Figure 1. Coordination of the iron in (2-methylimidazole)mesotetra-
        phenylporphinate Fe(II). As far as can be judged at the
        resolution of the deoxymyoglobin and haemoglobin Fouriers,
        the stereochemistry of the iron is the same in these pro-
        teins as in this model compound. P_N is the mean plane of
        the porphyrin nitrogens; P_C the mean plane of the porphy-
        rin nitrogens and carbons (53).

tetraphenylporphinato Fe(II) shown in Figure 1 and also the same as
in deoxyhaemoglobin (19).   On the other hand, unlike the picket-
fence haems, the haem in oxymyoglobin is not planar, but the iron
is displaced from the plane of the four porphyrin nitrogens by 0.18 Å
towards the proximal histidine.   The difference may be caused by
the dihedral angle between the plane of the histidine imidazole and
the Fe-pyrrole bonds.   In the picket-fence complex the imidazole
lies at about 45° to the N-Fe-N bonds (16), which minimizes van der
Waals repulsion and maximizes overlap of the Fe d orbitals with the
π orbitals of the imidazole nitrogen.   In oxymyoglobin, on the
other hand, the plane of the imidazole makes an angle of only 7°
with the N-Fe-N bonds.   In this eclipsed orientation repulsion is
maximized and overlap minimized so that the proximal histidine tends
to pull the iron away from the porphyrin plane.   The oxygen molecule
occupies a single ordered position with Fe-O-O = 115° and Fe-O = 1.8
Å.   The imidazole of the proximal histidine and Fe-O-O are approximately coplanar, the oxygen being constrained to that orientation
by steric hindrance of the distal histidine, valine, and phenylalanine.   N(ε) of the distal histidine is in contact with the first,
iron-bound oxygen atom, but it is not clear from the x-ray data
whether this is a van der Waals contact or a hydrogen bond (Figure 2).
Chemical evidence speaks in favour of the former.

      I now come to cooperative oxygen binding, also known as haem-
haem interaction, which is exhibited by all vertebrate haemoglobins.

These haemoglobins are tetrameric and have sigmoid oxygen equilibrium
curves, which means that their oxygen affinity rises with increasing oxygen saturation.   This cooperative behaviour is attributable
to a transition between two alternative structures in equilibrium,
one with a low and the other with a high oxygen affinity (20,21,22,
23).   They are distinguished by the internal structure of the four
subunits, by the mutual arrangement of the subunits, and by the
number and energy of the bonds between them.   In the oxy, or R
structure, the iron atoms are hexacoordinated and the structure
appears to put no significant constraints on the haem that are not
also present in free subunits; the oxygen affinity of this structure
is only slightly higher than that of free subunits.   In the deoxy,
or T structure, the iron atoms are pentacoordinated and the haems
are constrained by additional bonds within and between the subunits;
the oxygen affinity of that structure is lower than that of free
subunits or that of the R structure by the equivalent of 1.5-3.5
kcal/mol Fe, depending on the strength of the constraining bonds.

   For understanding haem-haem interaction, the two basic questions
are: how does combination of ligands with the haem irons change the
quaternary structure of the globin from T to R?; and conversely,
how does the change from R to T lower the ligand affinity of the
haem iron?   I proposed that the equilibrium between the two structures is governed by the displacement of the iron atoms and the proximal histidines from the plane of the porphyrins and by the steric
effect of the ligand on the distal valines in the β subunits (24).
By the laws of action and reaction, if movement of the iron and the
proximal histidine towards the porphyrin on ligand binding changes
the structure from T to R, then a transition from R to T must pull
the iron and histidine away from the porphyrin.   In that case, the
T structure should exercise a tension on the haem that restrains
the iron from moving into the porphyrin plane (25).   The existence
of such a restraint should be detectable by physical methods.

   To study the influence of the quaternary structure of the globin
on the state of the haem, we needed a method of changing that structure without changing the sixth ligand at the haem.   There are two
ways of doing this: one is to use a valency hybrid in which the
haems in either the α or the β subunits are ferric; combination of
the ferrous haems with ligand is used to change the quaternary
structure and the effect on the ferric haems is studied spectroscopically.   Alternatively, addition of the allosteric effector
inositolhexaphosphate (IHP) may switch the quaternary structure
from R to T.   In certain fish haemoglobins the transition may be
accomplished by merely lowering the pH.

   Gibson (26) first observed a change in the Soret (γ) band on
transition from what are now known to have been deoxyhaemoglobin

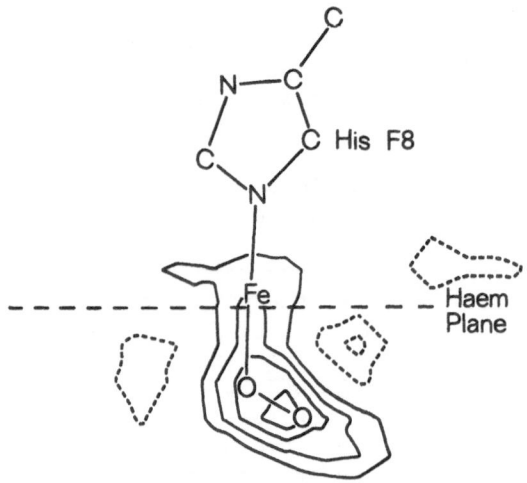

Figure 2a. Difference Fourier synthesis of oxymyoglobin, showing
          the electron density for the bound oxygen. The Fourier
          synthesis was computed with |F(observed)| − |F(calculated)|
          as coefficients, the calculated structure amplitudes
          were derived from the positions of all the atoms except
          the two oxygens.

dimers, which are equivalent to the R structure, to tetramers in
the T structure; fuller descriptions of the spectral changes have
been given by Brunori et al. (27) and by Perutz et al. (28). They
consist of blue shifts of the Soret, visible, and near IR bands,
together with the appearance of a shoulder at 590nm flanking the
peak at 556nm. The spectral changes in mammalian and fish haemoglobins are similar. Sugita (29) showed that they are caused almost entirely by the haems in the α subunits. Perutz et al. (30)
suggested that these shifts may be attributable to an increase in
Fe-N bond distances on transition from the R to the T structure,
but it was not clear if this interpretation was correct.

The evidence from resonance Raman spectra is conflicting. At
first such spectra of deoxyhaemoglobin in the R and T structures
seemed indistinguishable, but with improved techniques differences
have been found. The most striking difference occurs in the low
frequency region which has been explored by Nagai et al. (31) and
by Desbois et al. (32). It consists of a shift of a band at 215
cm^{-1} in the T structure to 220 cm^{-1} in the R structure. Hori
and Kitagawa have assigned this band to the Fe-N(ε) stretching frequency (33). Using difference Raman spectroscopy Shelnutt et al.
have found differences also in the high frequency region (34). Bands
that lie at 1357, 1471, 1567, and 1605 cm^{-1} in the T structure shift

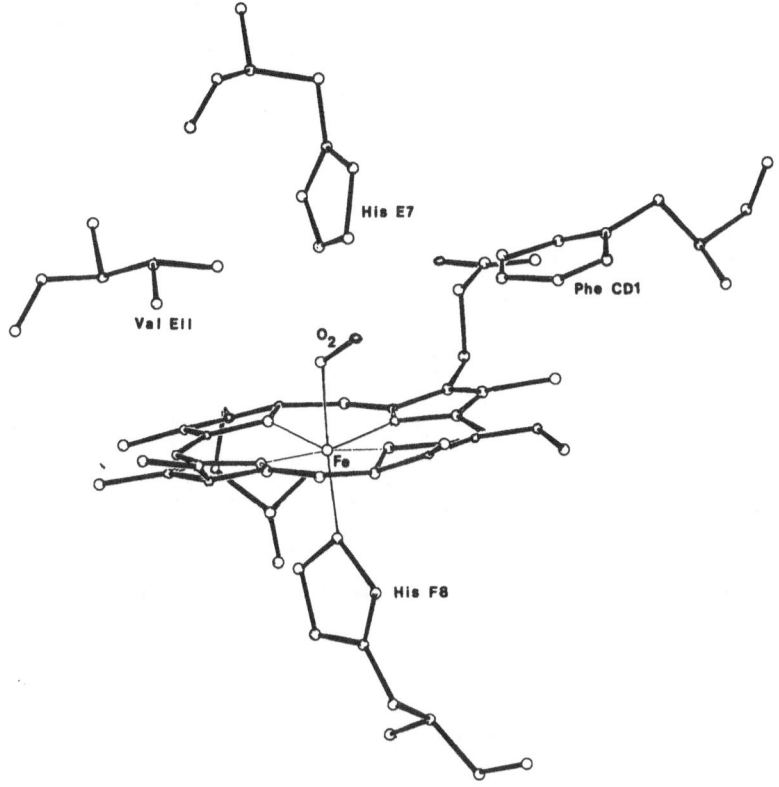

Figure 2b. Stereochemistry of the haem pocket with bound oxygen (17).

to lower frequencies by between 1.3 and 2.2 cm^{-1} in the R structure (Figure 3), but no shifts occur in any of the bands that are believed to be markers for the Fe-N(porph) bond lengths. The shifts occur in bands known to be sensitive to the occupation of the π* antibonding orbitals of the porphyrin, and correspond to a charge transfer to these orbitals on going from the T to the R structure. Since the binding of oxygen requires charge transfer from the haem to the oxygen molecule, the presence of extra charge in the π* orbitals of the haem would favour higher oxygen affinity. At this stage it is not clear where the charge comes from, nor how much this effect contributes to the free energy of haem-haem interaction.

Nitrosylhaemoglobin (HbNO) proved the most revealing of the ferrous derivatives. The R→T transition gives rise to blue shifts and reduction in intensity of the α,β, and γ bands, and the appearance of "high spin" bands at 495, 518 and 603nm, though the complex remains low spin ($S = \frac{1}{2}$) (35,36). Studies of hybrid haemoglobins show that these spectral changes are mainly caused by the α subunits,

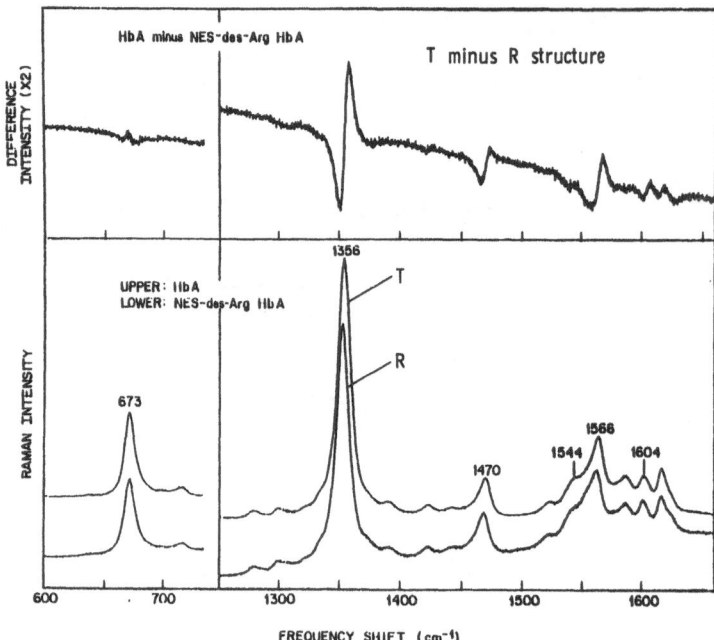

Figure 3. Resonance Raman spectra of human deoxyhaemoglobin in the
          T and R structures with difference spectrum, showing the
          shifts to lower frequency on transition from T to R (34).

just as in deoxyhaemoglobin.   Nitrosylhaemoglobin in the R structure
shows a single IR stretching frequency characteristic for hexacoordinated nitrosyl haems; transition to the T structure causes the
appearance of a second IR band, of intensity equal to the first,
characteristic of pentacoordinated haems (37).   Similarly, nitrosylhaemoglobin in the R structure shows an ESR spectrum similar to
that of hexacoordinated nitrosyl haems, while nitrosylhaemoglobin
in the T structure shows a composite of penta- and hexacoordinated
nitrosyl haems (38).   Similar results have been obtained from comparisons of the resonance Raman spectra of penta- and hexacoordinated
nitrosyl haems with those of nitrosylhaemoglobin in the R and T
structures (39,40)(Figure 4).   Taken together, these results imply
that in the R structure all four haems are hexacoordinated but in
the T structure only the haems in the β subunits remain hexacoordinated; those in the α subunits become pentacoordinated, the iron-
histidine bond having been broken by the restraints that impede the
movement of the proximal histidine towards the porphyrin.   This experiment corroborates the restraint at the haems in the T structure,
but it does not determine its energy equivalent, though kinetic experiments showed that an activation energy of 17 kcal/mol was needed
to break the Fe-N(ϵ) bond (35,41,42).   The bond breaks because
occupation of the antibonding Fe dz^2 orbital by the unpaired NO
electron has weakened it.

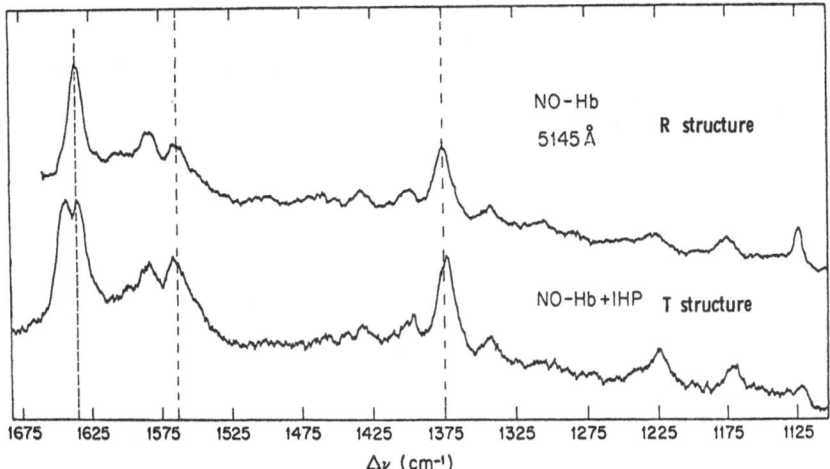

Figure 4. Resonance Raman spectra of human nitrosylhaemoglobin in the T and R structures. The R structure shows a single band at 1634 cm^{-1} characteristic of hexacoordinated nitrosyl haem. The T structure shows two bands of equal intensity, one at 1634 cm^{-1} and the other at 1644 cm^{-1}, characteristic of hexa- and pentacoordinated haems, respectively (40).

We now come to the class of ferric derivatives that have provided the most useful information concerning the effect of changes of quaternary structure on the state of the haem.   These are the mixed-spin derivatives in which there exists a thermal equilibrium between the spin states of $S = {}^5/2$ and $S = 1/2$. The derivatives investigated include hydroxyl, azide, thiocyanate and nitrite methaemoglobin (34, 44).   The spectral changes induced by the R→T transition in these derivatives include blue shifts of the Soret band, increases in intensity of the visible high-spin bands and of the charge-transfer bands in the near IR, and decreases in intensity of the low-spin bands.   The most striking spectral changes were seen in nitrite methaemoglobin of carp, which has the red colour characteristic of low-spin ferric haemoglobins in the R structure and the brown colour characteristic of high-spin ferric haemoglobins in the T structure.

Magnetic measurements of human haemoglobins by NMR indicated that the R→T transition caused the paramagnetic susceptibility of hydroxymethaemoglobin to rise by 45% and that of thiocyanatemethaemoglobin by 11%.   Human azide methaemoglobin cannot be converted to the T structure, but two abnormal haemoglobins form valency hybrids that allowed the effect of the R→T transition on the paramagnetic susceptibility of the α and β subunits to be measured separately by IR absorption spectroscopy.   This showed that in the R

structure the α haems are less than 10% high spin, while in the T
structure with IHP the high-spin fraction rises to 27%.   The high-
spin fraction of the β haems in the R structure is probably about
10%; in the T structure with IHP it rises to 35%.   In trout IV and
carp azide methaemoglobin measurements of the magnetic susceptibility
and of the azide stretching frequencies showed a rise of the high-

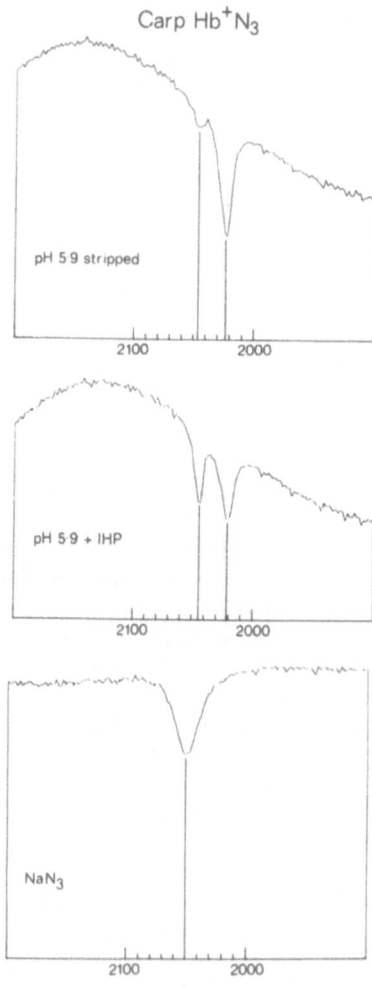

Figure 5. IR absorption spectra of carp azide methaemoglobin at
        20°C. In the R structure the azide stretching frequencies
        at 2023 cm^{-1}, characteristic of low-spin azide haems, do-
        minates. In the T structure two peaks of equal intensity
        appear at 2023 and 2046 cm^{-1}, characteristic of low- and
        high-spin haems, respectively (43).

spin fraction in the tetramer from about 10% in the R structure to 50% in the T structure, which is equivalent to a free energy change of approximately 1 kcal/mol haem (Figure 5).   All these measurements were done at a single temperature near 20°C.   Qualitatively, these results showed that in mixed-spin derivatives the R→T transition causes a change to higher spin, equivalent to a stretching of the iron-nitrogen bonds (43).

Thermal spin equilibria in methaemoglobin derivatives have been studied extensively (for review see Ref. 45), but measurements have been confined to derivatives in the R structure.   Messana et al. (44) have determined the effect of the R→T transition on spin equilibria by measuring magnetic susceptibilities between 300 and 90 K with a high-resolution, superconducting magnetometer.   The derivatives used were carp azide, nitrite, and thiocyanate methaemoglobin. The authors expected to find a dependence upon 1/T like the theoretical curves for mixed-spin derivatives, that is, a low-spin ground state followed by a gradual transition to higher spin above some critical temperature, but the actual results were rather different.

At the lowest temperatures all the plots of χ vs. 1/T were linear (Figure 6).   From the linear parts of the curves the effective magnetic moments, $\mu(e)$, could be derived:

$$\mu(e) \quad = \quad \sqrt{\frac{3\chi\kappa T}{N\beta^2}} \quad = 2.828 \ \sqrt{\chi T}$$

where N is Avogadro's number, κ, the Boltzmann constant, and β, the Bohr magneton.   For all but one of the haemoglobins in the R structuse, the moments have values characteristic of   low-spin haem complexes, but for all haemoglobins in the T structure and for the thiocyanate derivative in the R structure, their values are intermediate between low and high spin.   In all instances the effective magnetic moment for the frozen haemoglobins in the T structure is larger than in the R structure.   In solution, several derivatives show reverse Curie behaviour characteristic of compounds in which there exists a thermal equilibrium between two spin states.   This behaviour persists after freezing down to temperatures of between 250 and 200 K.   The magnetic moments in solution in the T structure are also larger than in the R structure; the rise just above the freezing point varies from 1.26-fold in thiocyanate to 1.84-fold in the nitrite derivative.   The 1.52-fold rise in carp azide methaemoglobin at room temperature corresponds to a 2.4-fold rise in magnetic susceptibility, which agrees with the 2.5-fold increase found in the closely related trout IV azide methaemoglobin by NMR and also with the increase calculated from the change in the relative intensities of the high- and low-spin IR azide stretching frequencies of carp azide methaemoglobin mentioned above (43).

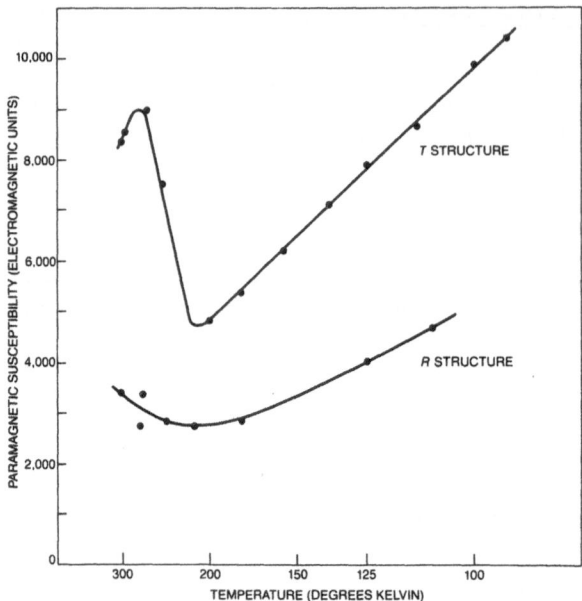

Figure 6. Direct measurement of the magnetic changes observed on
switching carp azide methaemoglobin from the R to the T
structure. The paramagnetism of the iron atoms is higher
in the T than in the R structure at all temperatures. In
the R structure at low temperature all the iron atoms are
low spin and the Curie Law is obeyed. At about 200 K the
thermal energy approaches ΔE, the difference in energy
between the $^6A_1$ and $^2T_2$ states, the iron atoms begin to
oscillate between them, and the susceptibility rises with
rising temperature. In the T structure at low temperatures
a random mixture of high- and low-spin iron atoms is frozen
in. At about 250 K the fraction of high-spin iron atoms
rises sharply, only to drop again at higher temperatures
for reasons that are not clear (44).

Figure 7 shows a plot of spin equilibria calculated from the
data in Figure 6.   The two sloping lines correspond to the temperature range where a thermal spin equilibrium exists.   From the
difference in height between them, the free energy difference between the equilibria in the R and T structures is calculated as
1.0-1.2 kcal/mol Fe, in agreement with the value derived from the
IR measurements mentioned above, and with a recently published
determination of the spin equilibrium by resonance Raman
spectroscopy (46).

So much for the physical measurements to determine the energy
equivalent of the restraint that opposes transition to the low-
spin state of the T structure.   The first chemical evidence that
restraint of the kind I had envisaged could actually diminish the
ligand affinity of haem came from an experiment by Rougee and

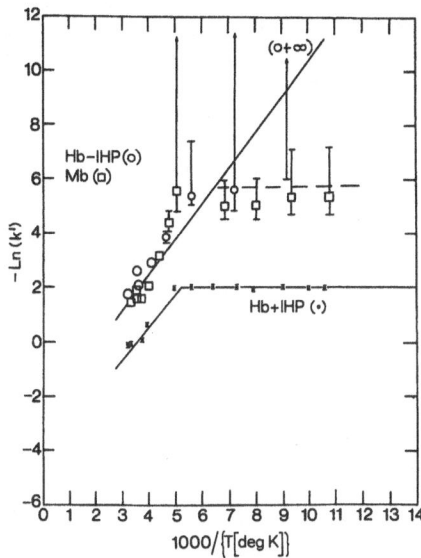

Figure 7. Experimental points and theoretical curves replotted as
the negative logarithm of the apparent spin equilibrium
constant $K'=8\chi T - \mu_L^2)/(\mu_H^2 - 8\chi T)$, where $-\frac{1}{2}\ln K' = \Delta G/RT$.
χ is the paramagnetic susceptibility, $\mu_L^2 = 4.0 \mu_B^2$ and
$\mu_H^2 = 35 \mu_B^2$; (---) indicates the low temperature asymptote
of $-\ln K'$ for azide metmyoglobin. Error bars are drawn for
an arbitrarily assumed error of $\pm 3\%$ in χ for all data
and tend to become very large when μ_e^2 approaches μ_L^2.
Hence, the low temperature data are not useful for determining ΔG (44).

  Brault (47).   They measured the carbon monoxide affinity of
deuterohaem in benzene at $25^\circ C$.   That affinity dropped 200-fold on
substitution of the sterically hindered 2-methylimidazole for imidazole as the distal base.   Collman, Traylor and their co-workers
(48–51) later asked whether restraint of the kind I had envisaged
could actually lead to a reduced oxygen affinity.   They measured
the thermodynamic constants of oxygen binding to some of the cobalt
and iron picket-fence complexes with either N-methylimidazole or
1,2-methylimidazole as the fifth ligand.   The former combines with
the cobalt or iron atom without steric hindrance; the latter is re-
strained by close contact of the 2-methyl group with the porphyrin,
so that it opposes the movement of the metal atom into the plane of
the porphyrin on ligation with oxygen.   This is the same effect as
I imagined the globin has on the T structure of haemoglobin.   The
results show that the mean oxygen affinity (p_{50}) for the unhindered
N-methyl cobalt complex is 150 torr, compared with 960 torr in the
hindered 1,2-methylimidazole cobalt complex.   In the iron complexes
the corresponding values are 0.59 and 38 torr, corresponding to a
difference in free energy of oxygen binding of 2.5 kcal/mol,

comparable with the free energy of haem-haem interaction in haemoglobin.   The answer to the question asked at the beginning of this
paragraph is therefore in the affirmative, restraint does lead
to reduced oxygen affinity.

On the basis of this, or indeed of any theory of haem-haem interaction, one would have expected the Fe-O bond to be weaker in
the T than in the R structure, weaker bonds being associated normally
with lower affinity.   Nagai et al. (31) have tested this point by
resonance Raman scattering, but found the Fe-O stretching frequency
of oxyhaemoglobin to be within 5 cm^{-1} the same in the two structures.
They have now repeated this experiment with the oxygenated picket-
fence complex, comparing the Fe-O stretching frequency in the
unhindered 1-methylimidazole complex, which has an oxygen affinity
comparable with that of the R structure, with that in the hindered
1,2-methylimidazole complex, which has an oxygen affinity comparable
with that of the T structure.   The result was the same: the Fe-O
stretching frequencies were identical.   They were not able to detect the Fe-N(Im) stretching frequency.   Since the Fe-O stretching
frequency is 570 cm^{-1}, compared with 215 cm^{-1} for the Fe-N(Im) bond
(in deoxyhaemoglobin), the latter is clearly the weaker of the two
and therefore may be the one that is stretched by the constraints
of the T structure or the steric hindrance of the picket-fence
complex.   Therefore, this bond is the one where a major part of
the free energy of haem-haem interaction is likely to be stored,
but so far it has not been possible to measure its stretching frequency in any liganded haemoglobin derivative.

CONCLUSION

Ferrous iron in haemoglobin binds oxygen end on, with an Fe-O-O
angle of 115°.   The ligand pocket is constructed so it can accommodate an oxygen molecule in this orientation and constrain it within
narrow limits to one azimuth.   The Fe-O bond is polar with transfer
of charge from the iron to the oxygen.   The distances of the iron
atom and the proximal histidine from the porphyrin plane play a
vital part in determining the equilibrium between the T and R structures of the haemoglobin molecule.   The low oxygen affinity of the
T structure is associated with a restraint or tension that opposes
the movement of the iron atom towards the porphyrin plane.   The
resulting strain is distributed between the haem and the protein in
varying proportions in different derivatives.   In oxyhaemoglobin
no significant effect of the T structure on the haem has been detected so far, which suggests that most of the strain is stored in
the protein.   In deoxyhaemoglobin it seems to be associated with
a stretching of the Fe-N(ϵ) bond.   In nitrosylhaemoglobin it causes
the haems in the α subunits to be torn from their bonds with the
proximal histidines of the globin.   In derivatives where there
exists a thermal spin equilibrium the restraints of the T structure

shift that equilibrium towards higher spin, that is, towards the form with the longer iron–nitrogen bonds.

Model compounds designed to mimic the steric restraint of the proximal histidine in the T structure have lower oxygen affinities than do similar model compounds lacking such restraint, and the degree by which the oxygen affinity is lowered is of the same magnitude as in the T compared with the R structure, which shows that the concept of steric restraint is viable chemically.

REFERENCES

1. L. Pauling and C. Coryell, Magnetic properties and structure of hemoglobin, oxyhemoglobin and carbonmonoxy hemoglobin, Proc. Natl. Acad. Sci. USA 22:210 (1936).

2. J. J.Weiss, Nature of the iron-oxygen bond in oxyhemoglobin, Nature 202:83 (1964).

3. H. P. Misra and I. Fridovich, The generation of superoxide radical during the autoxidation of hemoglobin, J. Biol. Chem. 247:6960 (1972).

4. L. S. Demma and J. K. Salhany, Direct generation of superoxide anions by flash photolysis of human oxyhemoglobin, J. Biol. Chem. 252:1226 (1977).

5. W. S. Brinigar, C. K. Chang, J. Geibel and T. G. Traylor, Solvent effects on reversible formation and oxidative stability of heme-oxygen complexes, J. Am. Chem. Soc. 96:5597 (1974).

6. C. H. Barlow, J. C. Maxwell, W. J. Wallace and W. S. Caughey, Elucidation of the mode of binding of oxygen to iron in oxyhemoglobin by infrared spectroscopy, Biochem. Biophys. Res. Commun. 55: 91 (1973).

7. A. S. Koster, Electronic state of iron in the oxygen and carbon monoxide adducts of heme proteins, J. Chem. Phys. 63:3284 (1975).

8. H. C. Stynes and J. A. Ibers, A pronounced solvent effect on the reversible oxygenation of a cobalt (II) porphyrin system, J. Am. Chem. Soc. 94:5125 (1972).

9. B. M. Hoffmann and D. H. Petering, Coboglobins: oxygen-carrying cobalt-reconstituted hemoglobin and myoglobin, Proc. Natl. Acad. Sci. USA 67:637 (1970).

10. J. C. W. Chien and L. C. Dickinson, Electron paramagnetic resonance of single crystal oxycobaltmyoglobin and deoxycobalt-myoglobin, Proc. Natl. Acad. Sci. USA 69:2783 (1972).

11. T. Yonetani, H. Yamamoto and T. Iizuka, Studies on cobalt myoglobins and hemoglobins III. Electron paramagnetic resonance studies of reversible oxygenation of cobalt myoglobins and hemoglobins, J. Biol. Chem. 249:2168 (1974).

12. R. K. Gupta, A. S. Mildvan, T. Yonetani and T.S. Strivastava, EPR study of ^{17}O nuclear hyperfine interaction in cobalt-oxyhemoglobin: conformation of bound oxygen, Biochem. Biophys. Res. Commun. 67:1005 (1975).

13. J. C. Maxwell and W. S. Caughey, Infrared evidence for similar
     metal-dioxygen bonding in iron and cobalt oxyhemoglobins,
     Biochem. Biophys. Res. Commun. 60:1309 (1974).
14. M. Cerdonio, A. Congiu-Castellano, F. Mogno, B. Pispisa, G. L.
     Romani and S. Vitale, Magnetic properties of oxyhemoglobin,
     Proc. Natl. Acad. Sci. USA 74:398 (1977).
15. M. Cerdonio, A. Congiu-Castellano, L. Calabrese, S. Morante,
     B. Pispisa, and S. Vitale, Room-temperature magnetic pro-
     perties of oxy- and carbonmonoxyhemoglobin, Proc. Natl.
     Acad. Sci. USA 75:4916 (1978).
16. J. P. Collman, Synthetic models for the oxygen-binding hemopro-
     teins, Acc. Chem. Res. 10:265 (1977).
17. S. E. V. Phillips, Structure of oxyhemoglobin, Nature 273:247
     (1978).
18. S. E. V. Phillips, private communication.
19. G. Fermi, Three-dimensional Fourier synthesis of human deoxy-
     haemoglobin at 2.5 A resolution: refinement of the atomic
     model, J. Mol. Biol. 97:237 (1975).
20. M. F. Perutz, Structure and mechanism of haemoglobin, Br. Med.
     Bull.32:237 (1976).
21. M. F. Perutz, Hemoglobin structure and respiratory transport,
     Sci. Am. 239:92 (1978).
22. M. F. Perutz, Regulation of oxygen affinity of hemoglobin:
     influence of structure of the globin on the heme iron,
     Ann. Rev. Biochem. 48:327 (1979).
23. J. M. Baldwin, Structure and function of hemoglobin, Prog.
     Biophys. Mol. Biol. 29:225 (1975).
24. M. F. Perutz, Stereochemistry of cooperative effects in haemo-
     globin, Nature 228:726 (1970).
25. M. F. Perutz, Nature of haem-haem interaction, Nature, 237:495
     (1972).
26. Q. H. Gibson, The photochemical formation of a quickly reacting
     form of haemoglobin, Biochem. J. 71: 293 (1959).
27. M. Brunori, E. Antonini, J. Wyman, and S. R. Anderson, Spectral
     differences between haemoglobin and isolated haemoglobin
     chains in the deoxygenated state, J. Mol. Biol. 34: 357
     (1968).
28. M. F. Perutz, J. E. Ladner, S. E. Simon and C. Ho, Influence
     of globin structure on the state of the heme. I. Human
     deoxyhemoglobin, Biochemistry 13:2163 (1974).
29. Y. Sugita, Differences in spectra of α and β chains of hemo-
     globin between isolated state and in tetramer, J. Biol.Chem.
     250:1251 (1975).
30. M. F. Perutz, E. J. Heidner, J. E. Ladner, J. G. Beetlestone,
     C. Ho and E. F. Slade, Influence of globin structure on the
     state of the heme. III. Changes in heme spectra accompanying
     allosteric transitions in methemoglobin and their implicat-
     ions for the heme-heme interaction, Biochemistry 13:2187
     (1974).
31. K. Nagai and T. Kitagawa, Differences in Fe(II)-Nε (His-F8)

stretching frequencies between deoxyhemoglobins in the two alternative quaternary structures, Proc. Natl. Acad. Sci. USA 77:2033 (1980).

32. A. Desbois, M. Lutz and R. Banerjee, Low-frequency vibrations in resonance Raman spectra of horse heart myoglobin. Iron-ligand and iron-nitrogen vibrational modes, Biochemistry 18:1510 (1979).

33. H. Hori and T. Kitagawa, Iron ligand stretching band in the resonance Raman spectra of ferrous iron porphyrin derivatives. Importance as a probe band for quaternary structure of hemoglobin, J. Am. Chem. Soc. 102:3608 (1980).

34. J.A. Shelnutt, D. L. Rousseau, J. L. Friedman and S. R. Simon, Protein-heme interaction in hemoglobin: evidence from Raman difference spectroscopy, Proc. Natl. Acad. Sci. USA 76: 4409 (1979).

35. M. F. Perutz, J. V. Kilmartin, K. Nagai and S. R. Simon, Influence of globin structures on the state of heme. Ferrous low spin derivatives, Biochemistry 15: 378 (1976).

36. R. C. Cassoly, Relation between optical absorption spectrum and structure of nitrosylhemoglobin, R. Acad. Sci. Ser. D. 278: 1417 (1974).

37. J.C. Maxwell and W. S. Caughey, An infrared study of NO bonding to heme B and hemoglobin A. Evidence for inositol hexaphosphate induced cleavage of proximal histidine to iron bonds, Biochemistry 15: 388 (1976).

38. A. Szabo and M. F. Perutz, Equilibrium between six- and five-coordinated hemes in nitrosylhemoglobin: interpretation of electron spin resonance spectra, Biochemistry 15: 4427 (1976).

39. A. Szabo and L. B. Barron, Resonance Raman studies of nitric oxide hemoglobin, J. Am. Chem. Soc. 97: 660 (1975).

40. J. D. Stong, J. M. Burke, P. Daly, P. Wright and T. G. Spiro, Resonance Raman spectra of nitrosyl heme proteins and of porphyrin analogues, J. Am. Chem. Soc. 102: 5815 (1980).

41. J. M. Salhany, S. Ogawa, and R. G. Shulman, Spectral-kinetic heterogeneity in reactions of nitrosyl hemoglobin, Proc. Natl. Acad. Sci. USA 71: 3359 (1974).

42. J. M. Salhany, S. Ogawa, and R.G. Shulman, Correlation between quaternary structure and ligand dissociation kinetics for fully liganded hemoglobin, Biochemistry 14: 2180 (1975).

43. M. F. Perutz, J. K. M. Sanders, D. H. Chenery, R. W. Noble, R. R. Pennelly, L. W. Fung, C. Ho, I. Giannini, D. Pörschke and H. Winkler, Interactions between the quaternary structure of the globin and the spin state of the heme in ferric mixed spin derivatives of hemoglobin, Biochemistry 17:3640 (1978).

44. C. Messana, M. Cerdonio, P. Shenkin, R. W. Noble, G. Fermi, R. N. Perutz and M. F. Perutz, Influence of quaternary structure of the globin on thermal spin equilibria in different methemoglobin derivatives, Biochemistry 17:3652 (1978).

45. T. Iizuka and T. Yonetani, Spin changes in hemoproteins Adv.
    Biophys. 1: 157 (1970).
46. D. M. Sholler and B. M. Hoffmann, Resonance Raman and electron
    paramagnetic resonance studies of the quaternary structure
    change in carp hemoglobin. Sensitivity of these spectroscopic
    probes to heme strain, J. Am. Chem. Soc. 101: 1655 (1979).
47. M. Rougee and D. Brault, Influence of trans weak or strongfield
    ligands upon the affinity of deuteroheme for carbon monoxide.
    Monoimidazole heme as a reference for unconstrained five-
    coordinate hemoproteins, Biochemistry 14: 4100 (1975).
48. J. P. Collman, J. J. Brauman, K. M. Dowsee, T. R. Halbert, S.
    E. Hayes and K. S. Suslick, Oxygen binding to cobalt porphy-
    rins, J. Am. Chem. Soc. 100: 2761 (1978).
49. J. P. Collman, J. J. Brauman, K. M. Dowsee, T. R. Halbert and
    K. S. Suslick, Model compounds for the T state of hemoglobin,
    Proc. Natl. Acad. Sci. USA 75: 564 (1978).
50. J. Geibel, J. Cannon, D. Campbell and T. G. Traylor, Model
    compounds for R-state and T-state hemoglobins, J. Am. Chem.
    Soc. 100: 3575 (1978).
51. D. K. White, J. B. Cannon, and T. G. Traylor, A kinetic model
    for R-state and T-state hemoglobin. Flash photolysis of
    heme-imidazole-carbon monoxide mixtures, J. Am. Chem. Soc.
    101: 2443 (1979).
52. A. Warshel, Energy-structure correlation in metalloporphyrins
    and the control of oxygen binding by hemoglobin, Proc. Natl.
    Acad. Sci. USA 74: 1789 (1977).
53. J. L. Hoard, private communication.

THE FUNCTION OF HIGH HEMOGLOBIN IN LARGE FISH

Quentin H. Gibson and Francis G. Carey

Department of Biochemistry, Molecular and Cell Biology
Wing Hall, Cornell University, Ithaca, NY 14853
Woods Hole Oceanographic Institution
Woods Hole, MA 02543

Over the years, a considerable body of evidence has accumu-
lated which shows that fish hemoglobins not only differ from
mammalian hemoglobin, but also differ greatly from one another
(1,2,3). A large amount of work is involved in characterizing
the function of a hemoglobin in detail so relatively few have
been studied fully. It seems, however, that there may be three
general classes of fish hemoglobins. These are first, the Root
effect hemoglobins, in which one chain has a very low affinity
for ligand in the T-state, and which show a marked change in
affinity with pH (4). Second, and coexistent in the same fish
with the Root effect hemoglobins, are the pH invariant and
effector independent hemoglobins, studied first by Hashimoto and
Matsuura (5) and since examined in great detail here in Rome as
trout 1. The third group includes the shark hemoglobins which
are cooperative, have chains which are alike in the T-state, and
respond only moderately to pH change and to effectors (6,7,8).
In looking for an explanation of this functional diversity it is
natural to think first of the transport function of hemoglobin
in fishes. There is, however, a good deal of evidence to suggest that many fishes can manage quite well without it. Rather
more than 50 years ago, Nicloux (9) converted the hemoglobin of
several species of fishes to COHb, and came up with the surprising result that their behavior was not visibly affected. These
observations were, of course, made under aquarium conditions, and
the fish might be handicapped in nature. There are, however,
several species of fishes with blood free from hemoglobin which
occur in nature (10,11). This situation contrasts so strongly
with experience in mammals as to justify more detailed
investigation.

Mammals often have high hemoglobin values with hematocrit readings of 40 to 45% which seem to represent a practical upper limit apparently imposed by increasing viscosity of the blood. With a few, but interesting exceptions, fishes have significantly less hemoglobin in their blood (12). It follows that, although they can transport the oxygen required by their tissues with a smaller cardiac output than would be needed if they had no hemoglobin, the saving in cardiac output associated with the extra oxygen combining power of the carrier is correspondingly smaller than in mammals. It is not possible to say just how much less, because the calculation would require much fuller understanding of respiration in fishes than is available at present, but, if mammals have enough hemoglobin to allow circulation to be reduced 20 to 30-fold, fish may have enough to permit a five-fold reduction. This is still a large advantage, but it must be weighed against the needs of the fish expressed in terms of cardiac output, or since many fishes have a high proportion of muscle in their bodies, as the blood flow in muscle. The data for cardiac output are, necessarily, for aquarium conditions. They agree in giving very small figures for output, at least by mammalian standards, and this is true even of vigorous fast-swimming fishes such as the salmon with values on the order of 1 ml. per gram muscle per hour (13). For a 70 kg. man at 50% of maximum cardiac output the value would be 20 times greater. On the basis of these numbers, a fish with no hemoglobin at all in its blood could supply all the oxygen it needed to its tissues with a cardiac output of 5 ml./g.hr., or one fourth the cardiac output of a man working at 50% of maximum capacity.

We have recently been able to supplement the data for circulation in fishes in aquaria with observations on large free swimming fish such as the blue shark and the swordfish. The principle is simple - the difficulty lies in putting it into effect. When a fish becomes large, the rate at which its body temperature would change when the temperature of the water it is swimming in changes would be quite slow if all heat exchange took place by conduction through the tissues. To a first approximation the time needed for the temperature at the center to make half of its full response to an external temperature change would increase as the square of the radius of the fish. For a really large fish such as a giant bluefin tuna (500 kg) this time is of the order of a day, and will be several hours even for a fish of a tenth that weight. This theoretical result agrees well with the results of measurements on temperature changes in dead fish where the deep tissues take a long time to reach the temperature of their post-mortem environment (14). In the living free-swimming fish, when the temperature is followed by telemetry, things are quite different, and a change in water temperature is followed quite quickly by a change in tissue temperature. This is because the circulation in the living fish acts not only as a

transporter of oxygen but as a heat transfer medium. All that
is needed to calculate blood flow is the rate of change of tissue
temperature for any known difference between tissue and water
temperature. The only assumption is that the blood reaches
temperature equilibrium with the tissues and in the gills, an
assumption that is almost surely justified since the diffusion
of heat is 50 times faster than the diffusion of oxygen. As
equilibrium is closely approached for oxygen, it must be effec-
tively established for heat transfer. A small correction is
required to allow for the production of metabolic heat in the
tissues, which is given by the caloric equivalent of the oxygen
content of the arterial blood multiplied by the coefficient of
oxygen utilization. The numerical value in degrees is from 0.2
to 0.5.

The apparatus used to obtain the data consists of a thermistor probe at the end of a harpoon which carries a battery pack,
electronics, and transducer required to convert the temperature
reading to a train of acoustic pulses whose frequency is proportional to temperature. The pulses are picked up with a hydro-
phone on the research vessel and timed with a stop-watch. Water
temperature may either be recorded on the same principle, or the
depth measured by a pressure-sensitive transducer and the temperature derived by reading it from a bathythermograph record
renewed at regular intervals. The research vessel is maneuvered
to stay near the fish, and in favorable weather conditions a
single fish may be followed for as long as four or five days.
At the conclusion of the experiment the fish is lured back
to the research vessel, again harpooned and killed to establish
its dimensions and the anatomical detail of the thermistor
placement. It will be recognized that the problems are of the
character summarized in Mrs. Beeton's cookbook recipe beginning:
"First catch your hare....". In most places the sea temperature
at any given depth does not vary rapidly from point to point, and
as the method only works when the tissue and water temperatures
are different from one another it is essential that the fish
should make dives from time to time. Some of the large fish
which have been followed have dived at regular intervals, but
others have failed to do so, remaining at or near the surface for
days on end.

A record obtained from a blue shark followed in November some
50 miles south of Martha's Vineyard is shown in Figure 1. This
animal made almost uniformly spaced excursions through the thermocline separated by 80 minutes, and on each passage muscle
temperature lagged behind water temperature in both upward and
downward directions. It should be noted that the muscle temperature began to move as soon as the water temperature changed,
an indication that circulation and not conduction through the
muscle is involved, since at the thermistor depth of 11 cm.

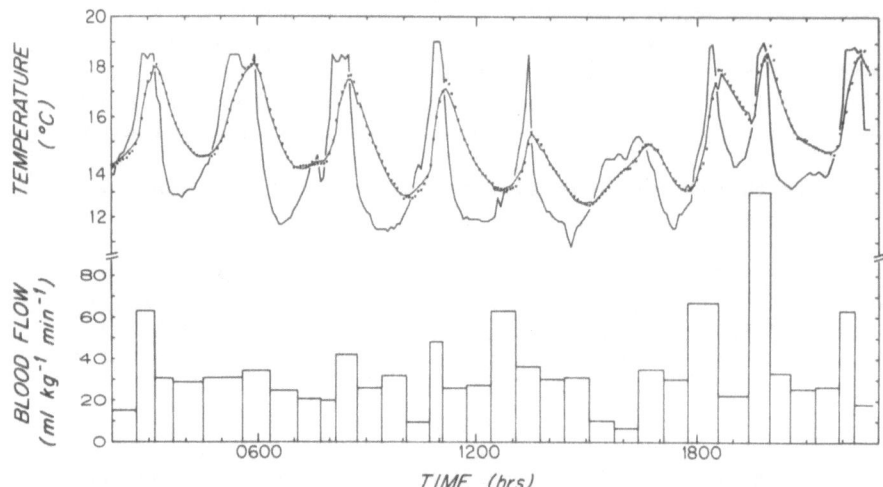

Figure 1. Muscle and water temperature records for an 80 kg
blue shark, <u>Prionace glauca</u>. The data were collected by
telemetry as described in the text. The line with the largest
excursion is water temperature. The muscle data are shown as
points near a solid line calculated from the water temperature
measurements over the period of time and with the value of blood
flow indicated in the blocks in the lower part of the record.

a change at the surface would need an hour or so to have a measurable effect on tissue temperature. In the case of the specific
shark illustrated in Figure 1, if conduction were the only means
of thermal exchange between the tissue and the exterior, the
temperature in the tissue would scarcely change during short
dives. This is illustrated in Figure 2 where the tissue temperature changes are calculated for a cylinder of suitable size
exposed to the environmental temperatures measured.

There are several methods for calculating blood flow from
records such as those of Figure 1. The simplest is to average
the temperature of the water and the tissue for a pair of points
where the tissue temperature is widely different from water temperature and to calculate blood flow as the change in tissue
temperature during the interval divided by the mean difference
between water and tissue temperatures. The results tend to be
erratic because of the effect of small errors in measuring tissue
temperature which seldom changes by more than a degree during a
five minute period. A better method which has a smoothing effect

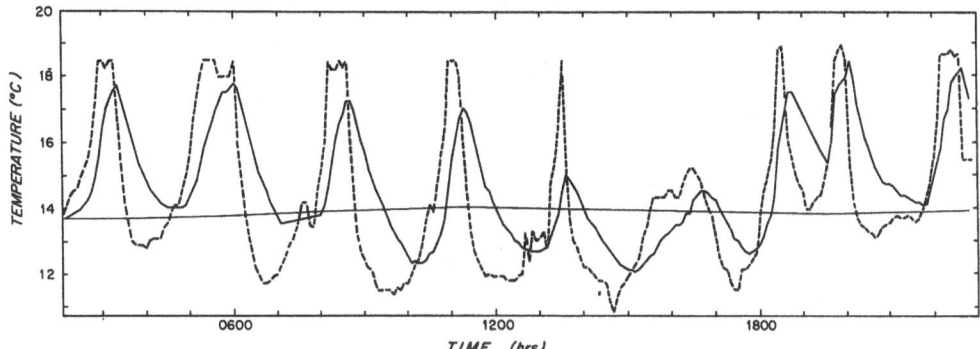

Figure 2.   Calculated and observed muscle temperature at a depth
of 10 cm. for the blue shark assuming that heat exchange between
the fish and the water takes place by conduction only.   The fish
was represented by a uniform cylinder divided into 20 concentric
rings and the water temperature was taken from the experimental
record of Figure 1.   The coefficient of thermal diffusion was
set at 0.001 cal./sec. deg. cm.^{2} and numerical integration
performed over the whole period of the record.   Water temperature, dashed line; observed muscle temperature, solid varying
line; calculated muscle temperature, solid horizontal line.

is to use the water temperature and an assumed value for blood
flow to calculate the time course of tissue temperature for
several periods, adjusting the blood flow value to give an optimal fit.   The inescapable drawback of this procedure is that
blood flow must be assumed to be constant over the period
examined.   It turns out that for the record considered the blood
flow changes in a regular way being greater while the shark is
rising and less while it is diving.   Attempts to fit both ascending and descending limbs of the curves with a single value of
blood flow give systematic misfits to the experimental results.
A slightly more elaborate model represents the shark as a series
of concentric circular laminae and includes the effects of conduction through the tissue and of metabolism as well as of circulation.   In the present case there is little difference between
the results of the two methods which show consistently that the
blood flow is about twice as great while the shark is rising and
give an overall average value of blood flow of 2 ml./g. hr.

This result is quite consistent with the low values obtained
by direct measurement of cardiac output in laboratory experiments
on small fish and with the everyday observation that fresh fish
muscle is predominantly white and simply cannot have large numbers of patent capillaries.   The low blood flow figures are also

consistent with the low rate of travel recorded for blue sharks.
The net translational speed of the tracking ship has regularly
been on the order of one knot, and although it is, of course,
possible that the shark was actually swimming rapidly in close
circles, there is no evidence of this. Direct observation as the
shark approached to accept a mouthful of mackerel at the surface
showed leisurely swimming motions and a low speed quite consistent with the net travel at one knot. The shark may be pictured
as swimming just enough to balance its net negative buoyancy and
waiting for prey to come to it rather than seeking it out. If
this behavior does not contravene the economic principle that
there is no free lunch, it is at least a low-cost meal in terms
of circulation.

A part of a similar set of data for a swordfish followed for
several days during which it made a number of excursions through
the thermocline is presented in Figure 3. In this case the fish
spent much longer periods in warm and cold water so we could be
sure that a steady state had been reached with tissue temperature
a fraction of a degree above water temperature.

This small difference in temperature is central to the conversion of muscle temperature records to blood flow and must be
discussed further. It is an anatomical fact that arteries and
veins supplying a muscle mass commonly lie close together, and
that the venous blood in a fish will be 0.3 to 0.5 degree warmer
than the arterial blood because of the metabolic heat generated
in the tissues. Suppose that the contact between the arteries
and veins is close enough to allow one half of this metabolic
heat to be transferred from the veins to the arteries. Then,
when a steady state is reached, the venous blood must be, in
this example, 0.6 to 1 degree warmer than the arterial blood.
The increased temperature difference will allow the same amount
of heat to be transferred in spite of the heat exchange between
veins and arteries. The only difference is that the muscle is
slightly warmer. Suppose that the fish now enters water at a
lower temperature. The arterial blood will become colder, but
by heat exchange with the venous blood, will be warmed to a
point half-way between tissue and water temperatures. The
tissue will therefore be cooled by the blood at only one half
the rate which would be seen if there were no heat exchange
between arterial and venous blood. As estimated blood flow is
proportional to the rate of cooling of the tissue, it follows
that any heat exchange will lead to underestimation of tissue
blood flow.

In the specific case of the swordfish, however, the small
steady state difference between water and tissue temperature
shows that there was no significant exchange of heat between the
arterial and venous streams of blood entering and leaving the

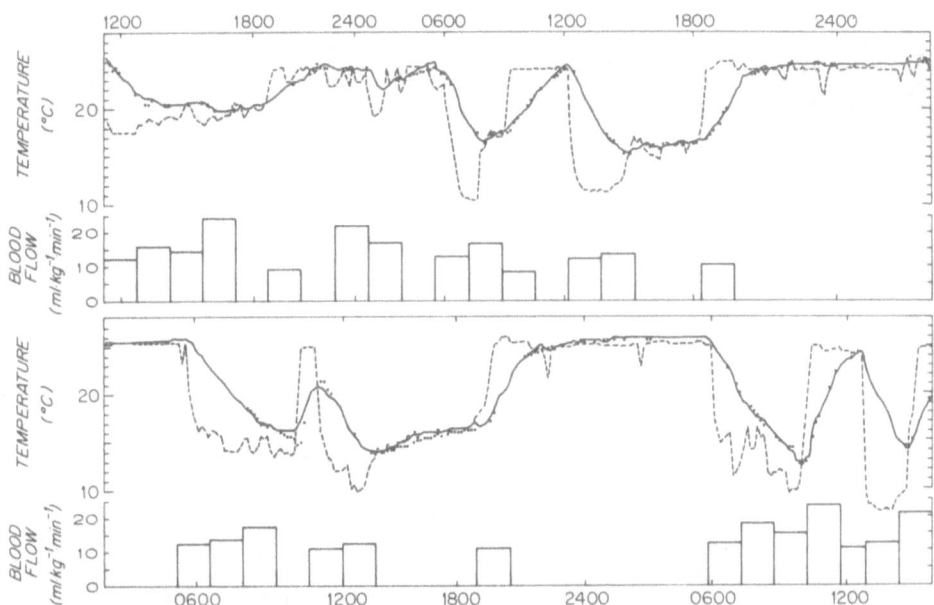

Figure 3. Muscle and water temperature records for the swordfish <u>Xiphias gladius</u>. Data were collected as described in the
text and represented by the model of Figure 2 but with the in-.
clusion of blood flow and the production of metabolic heat. A
uniform interval defined by 20 experimental points was used for
the calculations. The points for muscle temperature are experimental, the line is calculated with the values for blood
flow represented by the blocks in the lower part of the figure.
The dashed line is water temperature.

muscle. The blood flow in the swordfish does not depend on
whether it is rising or sinking in the water, and is rather less
than half that in the blue shark. Again, the net motion of the
fish through the water was very slow. At this point the experiments show that two species of large fishes swimming freely have
a low rate of blood flow through their muscles, and anatomical
arrangements which do not lead to significant heat exchange
between arteries and veins. There are, however, a number of
examples where extensive heat exchange does occur and is put to
physiological use by the animal.

The brain of the swordfish is unusual in its anatomical
arrangement. The brain of a large (say 200 kg) fish is the size
of a cherry, but is contained in a bony brain cavity about the
size of a small orange. This cavity is largely filled with
avascular fat, but contains, in addition to the brain, a central

highly vascular core and an organ apparently made up of a mass
of mitochondria. The brain temperature in freshly caught swordfish is 7 to 10° above water temperature, so it is reasonable to
suggest that the vascular center is a countercurrent heat
exchanger which conserves metabolic heat generated by the mitochondria in the brain cavity.

It is obviously impractical to place a harpoon probe in the
brain of a swordfish, but a probe was located in the fat about
1.5 cm. from the brain and vascular area itself, and after
returning the fish to the water the temperature was followed for
more than five days. The temperature records for the first 33
hours of the experiment are given in Figure 4. The placement of
the probe is shown diagrammatically in Figure 5. In contrast to
Figure 3 (data for swordfish muscle) the temperature in the
brain cavity does not tend towards water temperature but towards
a point about 7° above. To model the system the head of the
fish was taken to be cylindrically symmetrical and 16 cm. in
radius. The inner 4 cm. were taken to be vascular with a heat
diffusion coefficient of 0.001 cal./sec.cm.^{2} degree, and the
outer 12 cm. were represented as avascular with a diffusion
coefficient of 0.0003 cal./sec.cm.^{2} degree. The thermistor
was placed 5.5 cm. from the center. All the dimensions given
were chosen to reproduce the anatomical data given in Figure 5.

This simple model has only two adjustable parameters, blood
flow, and the efficiency of the heat exchanger. Of the two,
blood flow has little effect on the final temperature reached
under steady state conditions (as at the end of the record)
because although more metabolic heat is available from a higher
blood flow, heat is removed proportionately more rapidly from the
vascular organ in the brain cavity. Blood flow is, however,
highly significant in modeling the rate of response to change in
water temperature, a large blood flow producing rapid temperature changes in the tissues. The introduction of a heat
exchanger in the circulation reduces the rate of change of temperature in the tissue in the same proportion that heat is conserved. For example, if 90% of metabolic heat is retained, the
rate of change of tissue temperature is reduced to 10% of the
value which would be seen in the absence of a heat exchanger.
As a result, the model requires a high value for blood flow.

In numerical trials it was found that a blood flow of about
100 ml./g.hr. was needed to allow the relatively rapid changes
at the 0500 and 1400 hours to be reproduced. The results are
very sensitive to changes in the efficiency of the heat
exchanger, and it was clear that the brain temperature could not
be modeled over the whole period with a single value of exchanger
efficiency. The observations were broken up into shorter blocks
as shown in Figure 4. An excellent fit was then obtained with a

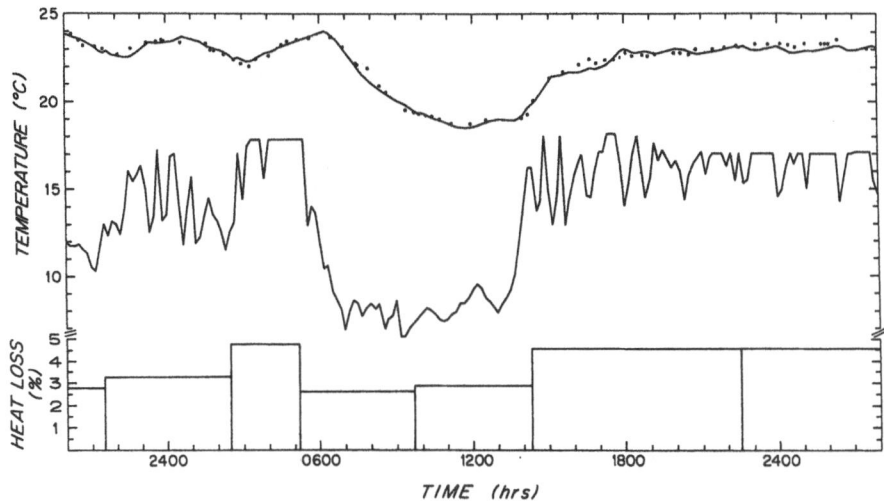

Figure 4. Water temperature and cranial temperature in the swordfish. Data were obtained by telemetry in a fish which had been caught by long-lining and maintained on deck by artificial respiration for stereotactic placement of the thermistor probe in the pericerebral fatty tissue. The lower continuous line is water temperature. The experimental data are shown as points, and the upper continuous line was obtained by calculation from a model in which the head of the fish was represented as a cylinder of 16 cm. diameter. The inner 4 cm. were treated as vascular and metabolically active, receiving their blood supply through a heat exchanger which allowed the escape of the pro-portion of the metabolic heat shown by the blocks in the lower part of the figure. The remaining tissue was treated as meta-bolically inert with thermal diffusion coefficient 0.0003 cal./sec. deg. cm.^{2}. The calculated temperature is for a point 5.5 cm. from the center.

high efficiency of heat exchange when the fish was in colder water, and low efficiency when it was in warmer water. Numeri-cally, the exchanger allows 2.7% to 3.3% of the metabolic heat to escape with the venous blood when the fish is in cold (12°) water, rising to 4.5% to 4.8% when the water temperature is 18°. In other words, the fish not only keeps its brain warmer than the water, but regulates brain temperature. Expressed in another way, a 10° change in water temperature produces only a 6° change in brain temperature.

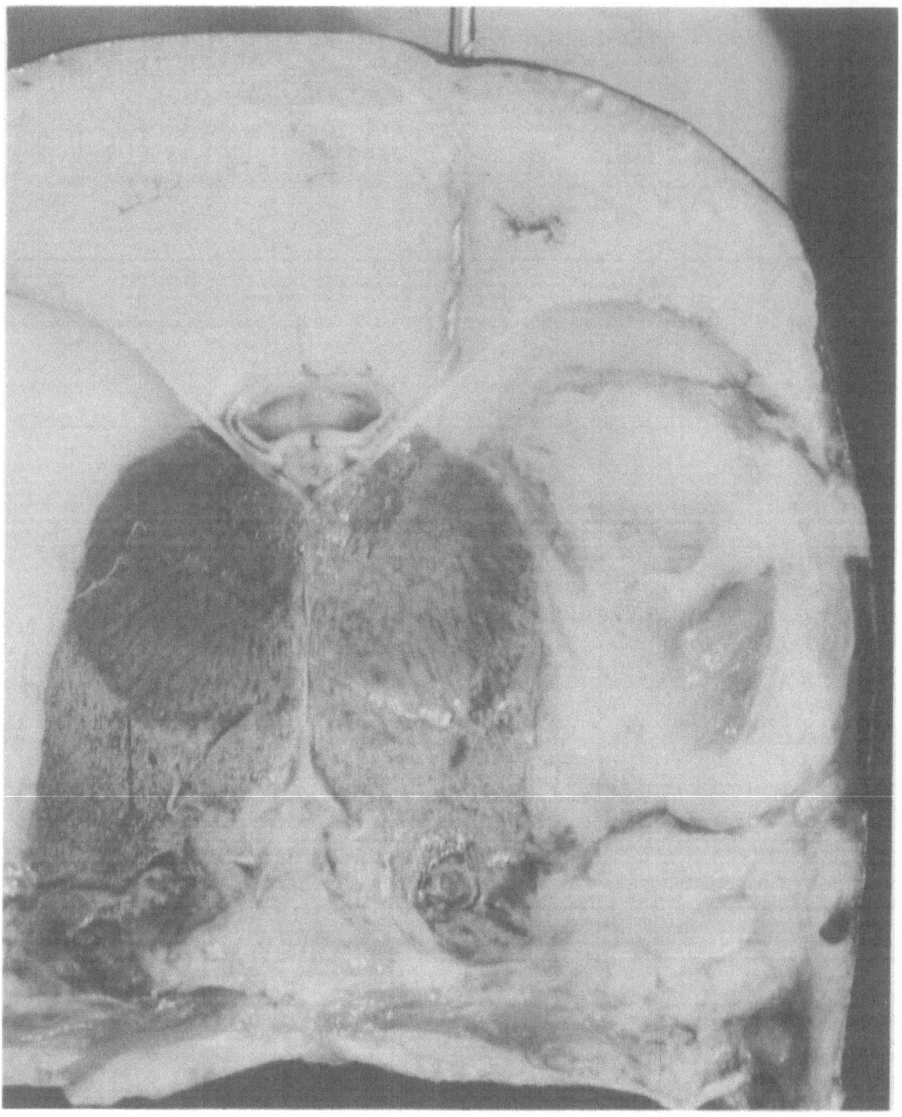

Figure 5. Head of a 20 kg swordfish cut in cross section just behind
the eyes. The mitochondria-rich thermogenic tissue appears as the
central dark mass. The lighter area just beneath it is the vascular
heat exchanger which arises from the carotid artery. The small brain
lies at the bottom of the fat-filled cranial cavity, partly imbedded
in the thermogenic tissue. A drill has been twisted through the roof
of the cranium and pushed down so that its tip lies on the thin floor
of the cranial cavity, the same position as that of the thermistor
used to measure temperature by telemetry from a free-swimming fish.
The relative dimensions indicated in this section were used to model
temperature changes in the cranial cavity.

The example of the swordfish shows that vascular mechanisms can be used to retain metabolic heat and to regulate internal temperature. There is a catch in this, however, as the flow of heat from the venous to the arterial blood must be associated with a corresponding flow of oxygen from the arterial to the venous blood. This represents a metabolic loss to the animal and is a part of the cost of having a warm tissue. For the swordfish, this cost is unimportant as the mass of warm tissue is small and the animal is large. The case of the Atlantic bluefin tuna is different for practically all of the tissues are warm and oxygen losses apply to the whole of the cardiac output instead of to a small part of it. It is not easy to give an accurate estimate of the loss of oxygen in the heat exchangers of the tuna, but it is quite appreciable. Indeed, if it were not that oxygen diffuses at only 1/55th of the rate that heat does, it would scarcely be practical for any fish to warm all of its tissues. Even so, some of the large warm bodied fishes have hemoglobins with unusual properties. The muscle of bluefin tuna is quite warm, and may be as much as 15° warmer than the water. When account is taken of the heat losses by conduction to the surface and in Joule heating of the water, Carey and Gibson (15) concluded that no more than 1% of the metabolic heat could pass through its heat exchanger. This is highly advantageous from the point of view of keeping the temperature constant, and the data of Carey and Lawson (Ref. 16, Fig. 11) show that very little internal relaxation had occurred even after 5 hours in water 15° colder than the initial internal temperature, while faster fluctuations are smoothed out altogether. In a highly efficient heat exchanger the calculated loss of oxygen is of the order of a tenth of the total available to the muscle, and would be even greater if it were not for the special properties of the hemoglobin. Again, Rome is a fitting place to discuss these, for it was here that the unusual temperature dependence of the oxygen affinity of the hemoglobin of the bluefin tuna was first noticed in 1960 - too soon, unfortunately for its significance to be appreciated, or for the mechanism to be understood (17).

It has been known ever since the early work of Barcroft that the affinity of hemoglobin for oxygen decreases as the temperature is raised, and all mammalian hemoglobins conform to this pattern. In the tuna, however, the oxygen affinity actually increases with rising temperature. The affinity of the hemoglobin is unusually low in the cold, and the effect of temperature is easily measured even when the hemoglobin is equilibrated with room air. The effect depends on pH as well as on temperature, and is influenced by saturation also. At low fractional saturation at any pH, and at any saturation at low pH (pH 6), the reverse temperature dependence disappears, and is replaced by a normal temperature effect (i.e., in the same direction as for mammalian hemoglobin). Some of these points are illustrated in Figures 6, 7 and 8.

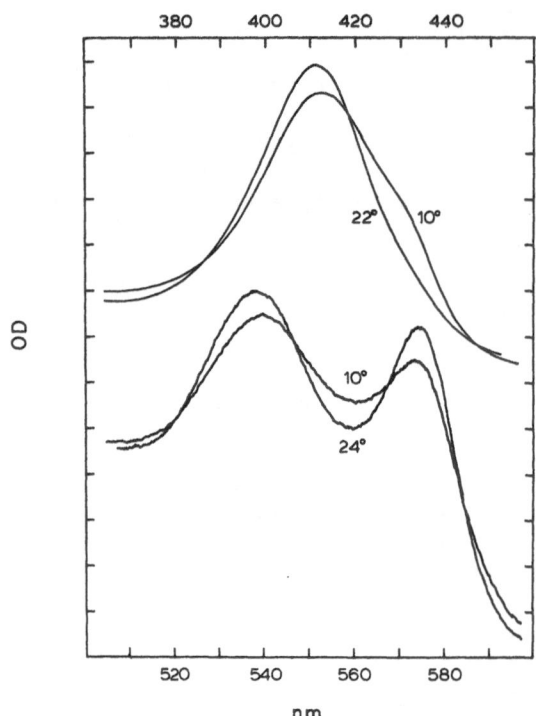

Figure 6. Spectra of tuna hemoglobin at two different temperatures. Phosphate buffer, pH 7.5, oxygen 250 uM, heme concentration 5.5 uM. The upper curves and wavelength scale show the
results obtained in the Soret region, the lower curves and
wavelength scale present data for the visible region of the
spectrum at a 10-fold expansion of the vertical scale. Cooling
by 12°C produces a significant decrease in oxygenation.

    The molecular mechanism which produces these effects has now
been examined in some detail, and appears to depend on two main
factors. The first is a large difference in affinity between the
two types of chain in the T-state. In the bluefin tuna this is
100-fold, and so not much less than the difference between the R
and T-states in mammalian hemoglobin. The second is an extreme
dependence of the R-T equilibrium on temperature with an enthalpy for the change of approximately 100 kcal. These factors work
together so that in the cold the transition from T to R does not
take place on ligand binding. The result is that the properties
of the lower affinity chain in the T-state are expressed and the
hemoglobin is only partly saturated even at high oxygen. A
relatively small increase in temperature, acting on the T to R
equilibrium, may permit switching to R. The effect on ligand

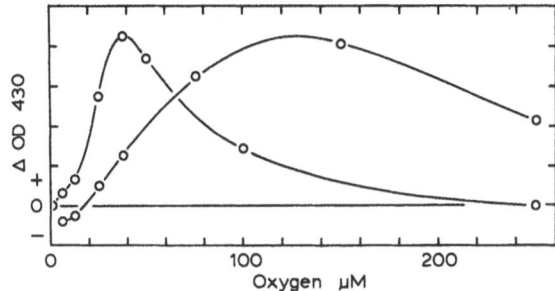

Figure 7. The effect of the initial oxygen concentration on the difference spectrum produced by lowering the temperature from 23° to 14°C. The maximum change in bound oxygen was 13% of capacity at pH 7.8. At low oxygen concentrations, the effect is reversed in the pH 7.5 solution. 0.1 M phosphate buffer, 8.5°C temperature change. The ordinate is in arbitrary optical density units at 430 nm where positive-going indicates increased deoxyhemoglobin with cooling.

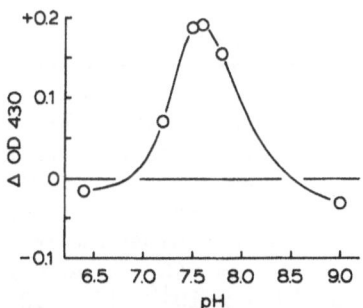

Figure 8. Effect of pH on the magnitude of the temperature difference spectrum. 6 μM hemoglobin in phosphate buffer, pH 6.4 through 7.8, borate buffer pH 9.0. Reference at 23°, sample at 14°. Oxygen concentration was 250 uM except at pH 9 where a lower concentration, 6 μM was required to observe the effect. The ordinate is in optical density units at 430 nm where positive-going indicates increased deoxyhemoglobin with cooling. The temperature effect is greatest in the physiological range of pH.

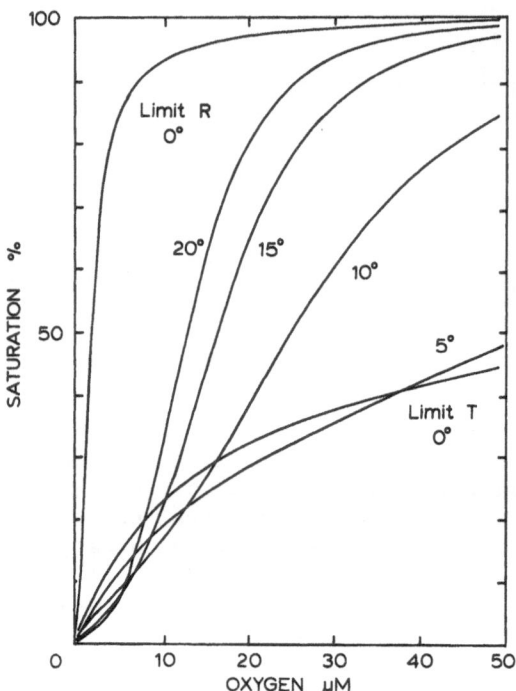

Figure 9. Calculated O_2 equilibrium curves for dilute buf-
fered solutions of tuna hemoglobin, pH 7.5, 0.1 M phosphate.
The curves were generated using the two state model extended to
take account of chain differences. Data for the kinetics of
oxygen binding at 0° measured by flash photolysis (Morris and
Gibson, 1981), were used for the parameters K_T and K_R. Data
for L, which is ligand independent were derived from analysis of
the course of CO-binding as measured by stopped flow and flash
photolysis methods (Saffran and Gibson 1978). The numerical
values used are: α_T 14 μM, β_T 640 μM, α_R 0.2 μM, β_R 0.15 μM at 0°.
The values of L are: 0°, 1.4×10^{10}; 5°, 7.8×10^8; 10°, 1.7×10^6;
15°, 1.1×10^5; 20°, 1.3×10^4.

affinity is of the order of 500-fold, and the displacement of the
curve is correspondingly large, quite outweighing the normal
effect of temperature on the intrinsic values of the R and T
affinities themselves (Figure 9).

This simple mechanism easily accommodates the effect of pH
which acts by altering the value of the allosteric constant so

that a temperature change which will cause switching at one pH
may fail to do so at another. The normal temperature dependence
at low saturation and at low pH is also immediately explained
since in both these cases we are dealing with the T-state only,
and its behavior with temperature is quite normal. It is
scarcely possible to say how much difference the properties of
the hemoglobin make to the respiratory economy of the fish
because the hemoglobin data have been obtained for dilute solutions in phosphate buffers in the absence of CO_2, whereas in
the blood there are important changes in pH, in CO_2, and in the
ionic composition in the interior of the red cell which cannot
yet be modeled soundly. Carey and Gibson (15) suggested that the
oxygen loss was approximately halved.

In contrast to the tuna, the swordfish, although equipped
with an efficient heat exchanger for its brain, has an apparently
normal temperature dependence of oxygen affinity, with a Q_{10} of
2.0 (18). Since the mass of warm tissue is so small the loss of
oxygen in the exchanger does not require a significant increase
in cardiac output. The hemoglobin content, however, is high by
comparison with many other species of fish.

Tuna blood has a hematocrit of about 45% which is substantially higher than that of most other fishes. This should provide a dividend for the fish in terms of reduced cardiac output,
but the point cannot be checked numerically because the high
efficiency of the tuna heat exchanger prevents us from measuring
blood flow by the method used for the blue shark and the swordfish. The high hemoglobin is in line with the life-style of the
tuna which is believed to swim actively during most of its
life. It must be admitted, however, that most tuna tracked by
telemetry have not maintained a high speed over the ground.

It may be that the high temperature maintained in the tuna
is more directly related to its high hemoglobin. As pointed out
earlier, the temperature rise obtained from a heat exchanger does
not depend strongly on the blood flow through it. The rise does
depend, however, and directly so, on the hemoglobin concentration, and hence the oxygen carrying capacity of the blood. If
the hemoglobin is doubled, either twice as great a difference
between tissue and water can be maintained, or the same rise can
be generated with an exchanger of only half the efficiency. It
is therefore reasonable to look for a correlation between hemoglobin and the existence of heat exchange mechanisms keeping all
or part of a fish at or above water temperature. So far, observations of tissue temperature have usually been made on muscle,
and the correlation between raised temperature and high hemoglobin is excellent. If the arguments which have just been presented are sound, it is likely that any large fish with a high
hemoglobin, say one half of that found in man, may have a heat

exchanger somewhere in its body, even if its muscles are not warm. The Atlantic bluefin tuna is an extreme case with a hemoglobin content similar to that of man, and heat exchangers in the blood supply both to the muscles and to the viscera. The porbeagle shark which commonly lives in cold water is another example with an unusually high hemoglobin and warm muscles and viscera.

We propose that the presence of high hemoglobin in many large fishes may be only secondarily related to its transport ability, but may rather be a part of a temperature elevating mechanism. This is parallel to the suggestion that the function of Root effect hemoglobins is to operate in gas glands to release oxygen against high partial pressures either into the swim bladder (19) or into solution in the ocular media (20). The property of extreme response to pH which permits this is not immediately related to oxygen transport to active tissues such as muscle. It is likely that exceptions to our proposal will be found in fishes which swim actively during most of their lives and are therefore dependent on continuous oxygen supplies to their muscle. An example is the menhaden which swims around all the time with its mouth open. Finally, it should be stressed that the temperature elevating function we propose is available only to large fishes of 50 kg or so, unless like the skipjack, they are exceptionally active. With this limitation, temperature regulation may deserve consideration as a reason for unusually high hemoglobin in fishes, and as a possible contributor to the functional diversity of fish hemoglobins.

This work was supported by National Science Foundation Grant BMS 74-08233 and United States Public Health Service Grant GM-14276-14 to Quentin H. Gibson and National Science Foundation OCE-8018674 to Francis G. Carey. Contribution No. 5062 from the Woods Hole Oceanographic Institution.

REFERENCES

1. A. Krogh and I. Leitch, The respiratory function of the blood in fishes, J. Physiol. (London) 52: 288 (1919).
2. C. Prosser and F. A. Brown, "Comparative Animal Physiology," 2nd Ed., Saunders, Phila., PA (1961).
3. A. Riggs, Properties of Fish Hemoglobins, in Vol. 4, p. 209, "Fish Physiology," W. S. Hoar and D. J. Randall, eds., Academic Press, N.Y. (1970).
4. R. Root, The respiratory function of the blood of marine fishes, Biol. Bull. 81: 307 (1931).

5. K. Hashimoto and F. Matsuura, Comparative studies on two
    hemoglobins of salmon. IV. Oxygen dissociation curve,
    Bull. Japan Soc. Sci. Fish. 26: 827 (1960).
6. C. Albers and K. Pleschka, Effect of temperature on CO_2
    transport in elasmobranch blood, Res. Physiol, 2:
    261 (1967).
7. M. Dickinson and Q. H. Gibson, Functional properties of
    the hemoglobin of the porbeagle shark, Biochem. J.
    (London), (1981) In press.
8. R. R. Penelly, R. W. Noble and A. Riggs, Equilibria and
    ligand binding kinetics of hemoglobins from the sharks
    Prionace glauca and Carcharhinus milberti, Comp.
    Biochem. Physiol. 52: 83 (1975).
9. M. Nicloux, Action de l'oxyde de carbone sur les poissons
    et capacite respiratoire du sang de ces animaux, Compt.
    Rend. Soc. Biol. 89: 1328 (1923).
10. J. Schlicher, Vergleichende physiologische Untersuchungen
    der Blutkörperchenzahlen bei Knochenfischen, Zool.
    Jahrb. 43: 121 (1927).
11. J. T. Ruud, Vertebrates without erythrocytes and blood
    pigment, Nature (London) 73: 848 (1954).
12. D. J. Randall, Gas exchange in fish, in Vol. IV, p. 253,
    Fish Physiology, W. S. Hoar and D. J. Randall, eds.,
    Academic Press, NY (1970).
13. D. J. Randall, The circulatory system, in Vol. IV, p. 133,
    Fish Physiology, W. S. Hoar and D. J. Randall, eds.,
    Academic Press, NY (1970).
14. F. G. Carey. Unpublished observations.
15. F. G. Carey and Q. H. Gibson, Reverse temperature depen-
    dence of oxygenation of tuna hemoglobin, Biochem.
    Biophys. Res. Commun. 78: 1376 (1977).
16. F. G. Carey and K. D. Lawson, Temperature regulation in
    free-swimming bluefin tuna, Comp. Biochem. Physiol. 44A:
    375 (1973).
17. A. Rossi-Fanelli and E. Antonini, Oxygen equilibrium of
    hemoglobin from Thunnus thynnus, Nature (London), 186:
    895 (1960).
18. M. F. Andersen, J. S. Olson, Q. H. Gibson and F. G. Carey,
    Studies on ligand binding to hemoglobins from teleosts
    and elasmobranchs, J. Biol. Chem. 248: 331 (1973).
19. R. Fänge, Physiology of the swimbladder, Physiol. Rev. 46:
    299 (1966).
20. J. B. Wittenberg and B. A. Wittenberg, The choroid rete
    mirabile of the fish eye. I. Oxygen secretion and
    structure: Comparison with the swimbladder rete
    mirabile, Biol. Bull. 146: 116 (1974).

INVERTEBRATE HEMOGLOBINS:

THE DIMERIC HEMOGLOBIN FROM THE MOLLUSC SCAPHARCA INAEQUIVALVIS

Eraldo Antonini, Emilia Chiancone and Franca Ascoli[°]

CNR Center of Molecular Biology, Institutes of Chemistry
and Biochemistry, Faculty of Medicine, University of Rome,
Rome, Italy
[°]Laboratory of Molecular Biology, University of Camerino,
Camerino, Italy

The dimeric hemoglobin from the mollusc Scapharca inaequivalvis represents a beautifully simple model system for the study of co-operative phenomena in hemoglobins. The mollusc was brought to our attention due to an ecological problem. S. inaequivalvis is an indopacific species which has settled along the Middle Adriatic coast only in recent years. Its demographic increase has since been explosive and has been associated with the progressive disappearance of the native species Venus gallina. The adaptation and the survival of S. inaequivalvis can be easily explained in terms of the presence of erythrocytes in its coelomic fluid and of the periodical limitations in oxygen content of the Middle Adriatic Sea due to the process of eutrophization.[1]

Studies on molluscan hemoglobins are scarce[2,3,4] although bivalve lamellibrancs have been known for a long time to possess nucleated red cells.[5] These are shaped as biconvex discs of fairly uniform size (about 20 μm in diameter) and contain cytoplasmic granules. The red cells of S. inaequivalvis conform to these characteristics and are shown in Fig. 1. The hemolysate of molluscan red cells contains two hemoglobin components, which differ not only in electrophoretic mobility, but also in sedimentation velocity. Thus component I (HbI) has a $s_{20,w}$ value of $\sim$ 2.8 S, typical of hemoglobin dimers, and component II (HbII) has a $s_{20,w}$ value of $\sim$ 4.5 S, typical of hemoglobin tetramers. Both kinds of hemoglobin show co-operative oxygen binding and lack the alkaline Bohr effect.[2,3,4]

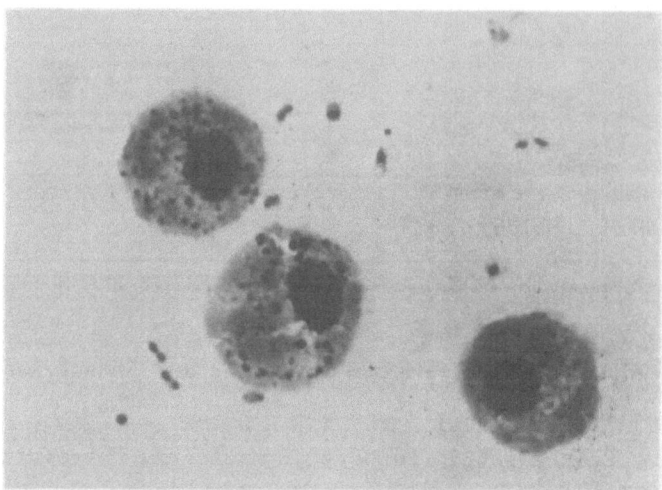

Fig. 1.   The hemoglobin containing cells of S. inaequivalvis.
Magnification 3000.

STRUCTURAL PROPERTIES

The dimeric hemoglobin from S. inaequivalvis is constructed
from two identical polypeptide chains each associated with a heme
group. This is shown by the presence of a single band in the SDS/
polyacrylamide gel electrophoresis pattern of the hemoglobin itself
and in the electrophoresis patterns of the globin moiety in the
presence of 8 M urea. In the SDS gels the mobility of the polypeptide chain is very similar to that of the α and β chains of human
hemoglobin and corresponds to an apparent molecular weight of
$\sim$ 16000 in fair agreement with the estimated minimum molecular
weight on a heme basis, which is around 17000,[1] and with the stoichiometry of the titration of the apoprotein with hemin.[6]

In the heme pocket two histidine residues constitute the
proximal and distal ligands of the iron atom, as indicated by the
EPR spectra of the protein reconstituted with cobalt containing
porphyrins. These spectra also show that the deoxygenated derivative
is characterized by a constrained structure of the heme site which
results from the distorted coordination of the hindered proximal
histidine.[6] A similar structure has been proposed previously for
the α chains in deoxy cobalt HbA.[7] The existence of such a restraint
is reflected also in the absorption and circular dichroism spectra
of the deoxygenated derivative. In particular, a marked shoulder
appears at 590 nm in the optical spectrum and the corresponding
transition has an unusually high optical activity.[5]

It is important to emphasize that the sedimentation velocity

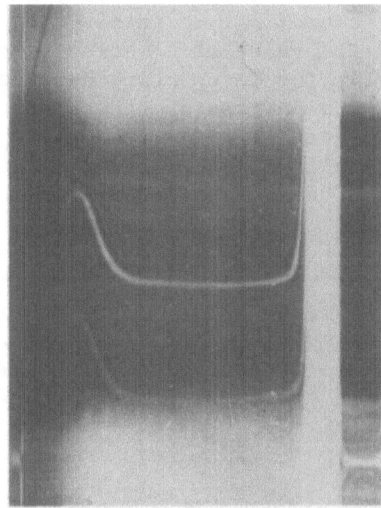

Fig. 2. Sedimentation velocity patterns of oxy-(upper) and deoxy-
(lower) S. inaequivalvis dimeric hemoglobin. The ultra-
centrifuge run was performed at a protein concentration of
7 mg/ml in phosphate buffer of I = 0.1 M, pH 7.0 at 10°C.
Photograph taken 102 min after reaching full speed of 56000
revs/min.

of oxy HbI is constant at 2.7 S over the pH range 5 to 9 and does
not change when the protein is deoxygenated (Fig. 2).

LIGAND BINDING EQUILIBRIA

Oxygen binding

The oxygen binding properties of Scapharca HbI are characterized by strong cooperative interactions (h=1.5, at 20°C) and by the
absence of a Bohr effect in the pH range 5.5 to 9.2. The oxygen
affinity at half saturation, P_{50}, is rather low, its value being
7.8 mm Hg at 20°C. The values of h and P_{50} are insensitive to the
addition of NaCl or organic phosphates. Thus, oxygen binding does
not show any evidence for heterotropic interactions.

The oxygen equilibrium can be described by a two stage Adair
equation characterized by two (intrinsic) Adair constants K_1 and
K_2. These can be easily obtained by precise measurements over the
whole oxygenation range and must be consistent with the value of h.
Moreover the study of the temperature dependence of K_1 and K_2 allows
to determine the intrinsic ΔH and ΔS values at each oxygenation
step, since no corrections need to be made for the contributions

of non heme ligands. In 0.1 M phosphate buffer at pH 7.8 the value
of the median oxygen pressure, P_m, increases from 4.2 to 16.0 mmHg
between 10 and 35°C and the value of h increases significantly from
1.43 to 1.53. Accordingly, the temperature dependence of K_1 is
greater than that of K_2. Comparison of the enthalpy and entropy
contributions to the difference in the free energy of oxygenation
between the second and the first step brings out that the origin of
cooperativity in oxygen binding is primarily entropic in nature.[8]

Carbon monoxide binding

The equilibrium binding data for carbon monoxide have not been
determined directly. However, the partition constant between oxygen
and carbomonoxide is about 100 and appears to correspond to a simple
equilibrium. On this basis the value of P_{50} for carbon monoxide is
about 0.1 mm Hg at 20°C.

KINETICS OF LIGAND BINDING

The dimeric HbI from S. inaequivalvis seems to be ideally suited
to reconcile equilibrium and kinetic measurements in a cooperative
hemoglobin after a rigorous study of the kinetics of the reactions
with ligands. The available kinetic data are far from being complete,
but suffice to show that the main kinetic features of ligand binding
are qualitatively similar to those of mammalian hemoglobins.

Oxygen dissociation

The time course of oxygen dissociation from oxy HbI, as measured
in rapid mixing experiments with sodium dithionite solutions, corresponds to a first order reaction (Fig. 3). The oxygen dissociation rate constant, k, is around 100 sec^{-1} at 20°C and is independent
of pH in the range 7 to 9; the apparent activation energy is $\sim$ 10
Kcal/mol.

Carbon monoxide combination

The combination reaction of deoxy HbI with carbon monoxide measured in stopped flow experiments has an autocatalytic time course
which may be taken as the reflection of the cooperativity in ligand
binding observed at equilibrium. It can be described in terms of
two rate constants $1_1'$ and $1_2'$, the ratio of the final to the initial
rate being about 2 (Fig. 4).

The time course of carbon monoxide combination after complete
photodissociation of carbonyl HbI is the same as that measured in
flow experiments. On partial photodissociation a quickly reacting
form appears. Fig. 5 shows that the initial rate is linearly related
to the degree of photodissociation.

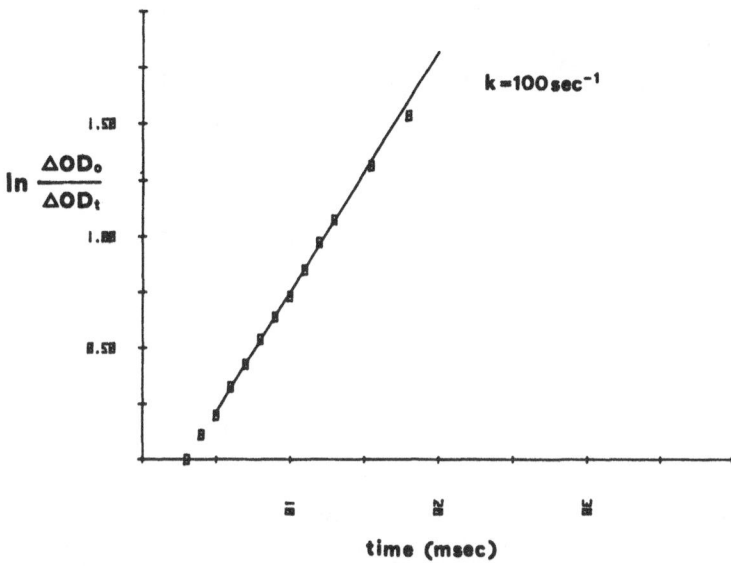

Fig. 3.   Time course of oxygen dissociation by sodium dithionite
           for S. inaequivalvis dimeric hemoglobin. Protein concentra-
           tion 3 μM in heme (after mixing) in phosphate buffer of
           I = 0.1 M, pH 7.0; observation wavelength 432 nm; tempera-
           ture 20°C.

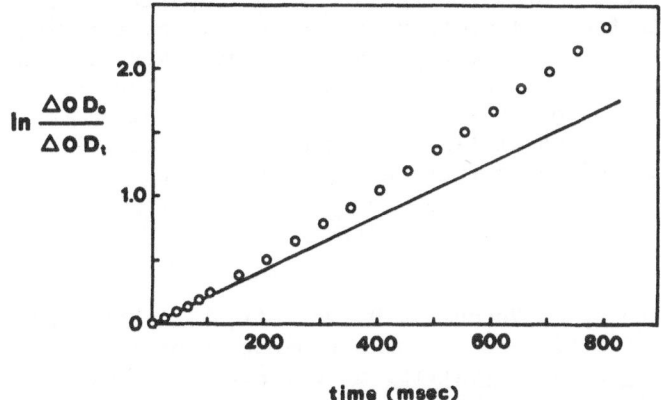

Fig. 4.   Time course of carbon monoxide combination with S. inaequi-
           valvis dimeric hemoglobin as measured in stopped flow expe-
           riments. Protein concentration 2 μM in heme (after mixing)
           in phosphate buffer of I = 0.1 M, pH 7.0; CO concentration
           (after mixing) 25 μM; observation wavelength 432 nm;
           temperature 20°C.

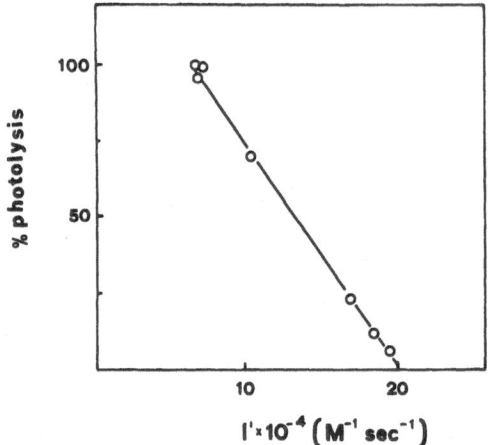

Fig. 5.   Second order rate constant for carbon monoxide recombina-
          tion after flash photolysis of S. inaequivalvis dimeric
          hemoglobin as a function of the percentage of photodisso-
          ciation. Protein concentration 7 μM in phosphate buffer
          of I = 0.1 M, pH 7.0; CO concentration 90 μM; observation
          wavelength 436 nm.

CONCLUSIONS

     The structural and functional properties of the naturally
occurring dimeric hemoglobin from the mollusc S. inaequivalvis make
it an ideal model system for the study of cooperative effects in
hemoglobin. In fact the protein is made up by two identical subunits
and is dimeric in both the oxygenated and deoxygenated state. It
shows cooperativity in oxygen binding, but no heterotropic effects.
Thermodynamically cooperativity arises mainly from entropic contributions. The kinetics of ligand binding shows some of the characteristic features of the corresponding reactions of mammalian hemoglobins.

REFERENCES

  1. E. Chiancone, P. Vecchini, D. Verzili, F. Ascoli, and E. Anto-
      nini, Dimeric and tetrameric hemoglobins from the mollusc
      Scapharca inaequivalvis. Structural and functional properties,
      J. Mol. Biol.152: 577 (1981).
  2. S. Ohnoki, Y. Mitomi, R. Hata, and K. Satake, Heterogeneity of
      hemoglobin from Arca (Anadara satowi). Molecular weights
      and oxygen equilibria of Arca HbI and II, J. Biochem. 73:
      717 (1973).
  3. H. Furuta, M. Ohe, and A. Kajita, Subunit structure from ery-
      throcytes of the blood clam, Anadara broughtonii, J. Biochem.
      82: 1723 (1977).

4. J. S. Djangmah, P. A. Gabbott, and E. J. Wood, Physico-chemical characteristics and oxygen-binding properties of the multiple hemoglobins of the West African blood clam _Anadara senilis_ (L.) _Comp. Biochem. Physiol._ 60B: 245 (1978).

5. A. B. Dawson, Supravital studies on the colored corpuscles of several marine invertebrates. _Biol. Bull. mar. biol. Lab., Woods Hole_ 64: 233 (1933).

6. D. Verzili, R. Santucci, M. Ikeda-Saito, E. Chiancone, F. Ascoli, T. Yonetani and E. Antonini, Studies on _Scapharca_ hemoglobins. Properties of the dimeric protein reconstituted with Fe- and Co-porphyrin, _Biochim. Biophys. Acta_, in press.

7. T. Inubushi, M. Ikeda-Saito and T. Yonetani, Subunit inequivalence in allosteric constraints of prosthetic groups in T-state hemoglobins, _in_: "Interaction between iron and proteins in oxygen and electron transport", C. Ho and W.A. Eaton, eds., Elsevier North Holland Publ. Co. New York, New York (1981).

8. M. Ikeda-Saito, T. Yonetani, E. Chiancone, F. Ascoli, D. Verzili and E. Antonini, Studies on _Scapharca inaequivalvis_ hemoglobins. Thermodynamic properties of the oxygen equilibria, in preparation.

EFFECTS OF HEAVY METALS ON THE RESPIRATORY PROTEINS

OF MARINE ORGANISMS IN RELATION TO ENVIRONMENTAL POLLUTION

Joseph Bonaventura, Celia Bonaventura and Marius Brouwer

Marine Biomedical Center
Duke University Marine Laboratory
Beaufort, NC 28516 U.S.A.

Seawater is a fluid medium containing measurable concentrations of many heavy metals that are potentially toxic to organisms. From the diversity of species and the complexity of life histories of marine animals, it is obvious that biochemical, physiological and behavioral adaptations have evolved to allow these animals to exist in what is a potentially toxic environment. In marine organisms the pathological effects of heavy metals have only been observed at locations where the activities of man have significantly increased the concentrations of heavy metals in the marine environment and in laboratory systems where animals have been isolated and subjected to increased levels of metals in seawater or in their diet. In unstressed or unpolluted areas, marine organisms have evolved the capability to deal with the fluctuations of metal availability in the environment. Adaptation to metal pollution sometimes occurs if the change to higher metal concentrations takes place slowly. A better understanding of the detoxification process utilized in normal and metal-stressed environments is necessary for proper interpretation of the consequences to man's health of using metal-dosed organisms as part of the food supply. It is important to recognize that there is an intricate relationship between the role of metals as essential nutritional elements on one hand and as potentially toxic chemicals on the other. Part of the complexity in the analysis of this relationship lies in the interplay among metals. Synergistic as well as antagonistic interactions between metals and the environment enter into these considerations.

Previous studies indicate that there is a tremendous variability in the sensitivity of marine organisms to specific concentrations of heavy metals. Furthermore, within any one species the sensitivity of each stage in the life history can vary. Fish, cru-

staceans, and molluscs have been intensively investigated in terms
of the stages and processes of development. Those stages which are
least affected by elevated concentrations of heavy metals would
appear to provide model systems for the study of the protection
mechanisms of marine organisms against heavy metal pollution. Transformation of metals into various nontoxic products may occur. Thus,
marine organisms can respond to metal stress by forming a metabolic
product, which can bind or immobilize the compound so that the
complex can be excreted. Alternatively, the organism may incorporate
the trace metal into some biologically inert structures, such as
intracellular calcium-phosphate concretions. Finally, metal binding
proteins similar to mammalian metallothioneins have been demonstrated in a variety of marine organisms[1-3]

In this paper the role of blood in trace metal detoxification
will be addressed. The first question which needs to be answered
concerns the extent to which blood may be considered a primary
target in trace metal toxicity or just a metal carrier. Respiratory
proteins, which reversibly bind oxygen and other ligands, act at
the interface between the respiring animal and its environment and
therefore are exposed to metals coming from the environment itself.
In the marine organisms the potential for interacting with metals
is high especially in the case of extracellular respiratory proteins.
In fact when the proteins are contained in red cells the cell membrane represents a barrier to the penetration of xenobiotics.

The major classes of respiratory proteins are the globins, the
hemerythrins and the hemocyanins. Structural and functional diversity
in both the organism and the environment is parallelled by a remarkable variety of functional behaviors of respiratory proteins as is
exemplified in Fig. 1. In vivo fine tuning of the respiratory function frequently occurs and is achieved by the allosteric nature of
the proteins.[4] Interaction of respiratory proteins with trace metals
may adversely affect the oxygen transport function or, alternatively,
respiratory proteins may act as carriers of trace metals.

It is not clear whether the oxygen carrying proteins are themselves primary targets of xenobiotic action. Ongoing investigations
may help clarify this matter. In fact Kuiper et al.[5] have clearly
shown that mercury binds to hemocyanin from Panulirus interruptus
and that the oxygen binding properties of this extracellular respiratory protein are adversely affected by mercury insofar as cooperativity is decreased and finally lost.

In order to determine whether respiratory proteins in the blood
are involved in the transport of trace metals, they must be isolated
and the trace metal interactions measured. Methods and procedures
for the isolation and purification of hemoglobins and hemocyanins
of marine organisms, as depicted schematically in Figure 2, are in
routine use.

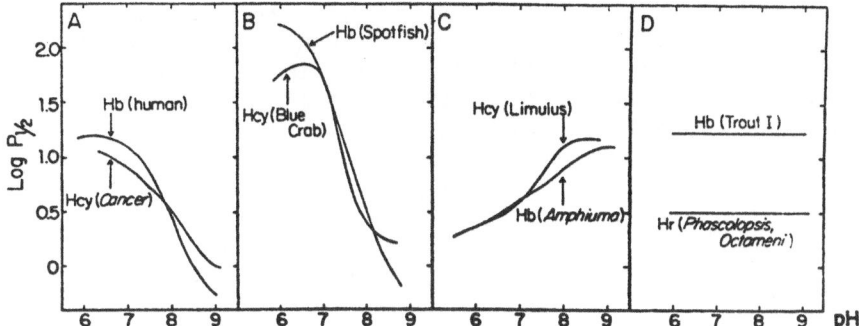

Fig. 1. Oxygen Bohr effect curves for heme and non-heme oxygen binding proteins. These curves illustrate the functional similarity and plasticity of oxygen transporting proteins.

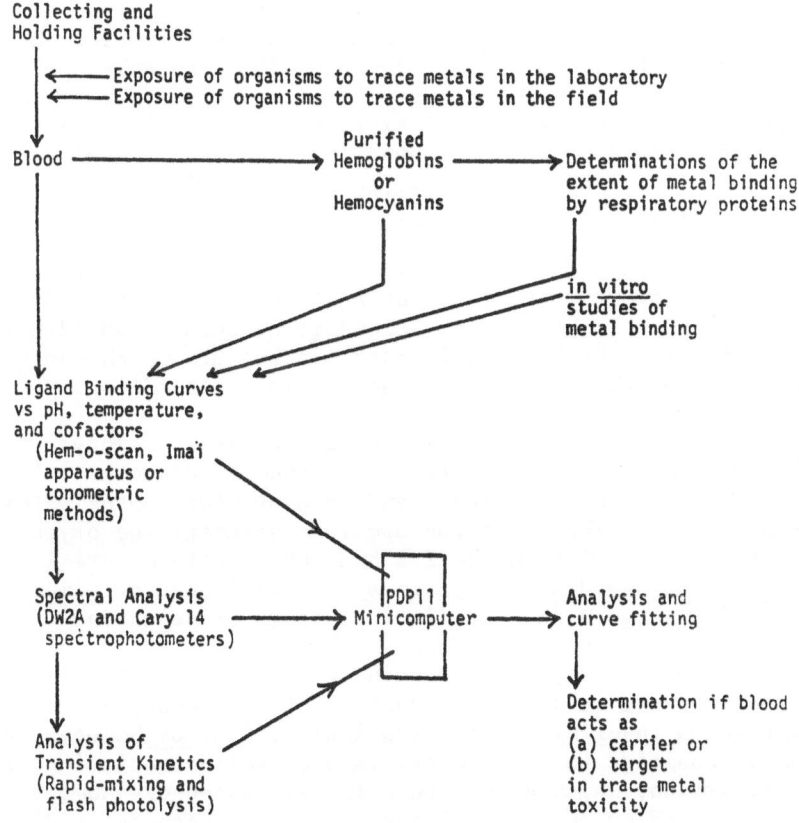

Fig. 2. Schematic diagram for the analysis of the effects of heavy
metals on extracellular respiratory proteins.

Several studies have shown that heavy metals can bind to both hemoglobins and hemocyanins in vitro.[5-9] Techniques which have been utilized to measure the functional perturbations brought about by heavy metal binding include tonometric measurements of oxygen binding, automated oxygen dissociation analysis, rapid-mixing spectrophotometry and flash photolysis. Computer processing of data plays an important part in these methodologies. Determination of the impairment of the oxygen delivery process utilizes a combination of the above methods of analysis of protein function and relies on atomic absorption spectrophotometry for determination of the extent of metal binding to the respiratory protein under investigation.

TYPE CASES FOR THE INTERACTION OF HEAVY METALS WITH HEMOCYANINS: LIMULUS POLYPHEMUS AND CALLINECTES SAPIDUS

In carrying out our investigations of the effects of heavy metals on hemocyanins, we have concentrated on the hemocyanin systems of the horseshoe crab, Limulus polyphemus and the blue crab, Callinectes sapidus. The rationale for this is as follows: the hemocyanin system of Limulus polyphemus has been extensively studies and characterized; the Callinectes hemocyanin, while less characterized, represents an oxygen carrying system from an important line of arthropods whose hemocyanins differ rather remarkably from that of Limulus. Additionally, we have found that the heavy metals we have used in our studies sometimes have contrasting effects on the hemocyanin of Limulus and Callinectes.[10]

Limulus hemocyanin, as it exists in the hemolymph of the organism, is a 60S structure having a molecular weight of 3.3×10^6. The 60S structure is built from somewhere between 8 and 12 distinct polypeptide types which assemble into a structure which can be thought of as an association of 8 hexameric units.[9-11]

Callinectes sapidus, the blue crab, possesses hemocyanin in its hemolymph which is a mixture of 22S and 16S structures.[12,13] The 25S structure is a dodecamer and accounts for approximately 70% of the hemocyanin. The 16S structure is a hexamer. The oxygen binding properties of both the 25S and 16S structures are identical. The Callinectes sapidus hemocyanin system like that of Limulus is composed of a large number of distinct polypeptide chains.

Binding of mercury, copper, zinc and cadmium to Callinectes hemocyanin appears to alter the surface of the hemocyanin molecule and lead to an indefinite self-association. Limulus hemocyanin is even more sensitive to heavy metal induced self-association. This self-association, which occurs both in oxy- and-deoxy hemocyanin, can be monitored by observing light scattering changes at 310 nm, a wavelength corresponding to a minimum in the optical absorption spectrum of hemocyanin. The kinetics of self-association, however, are different in that the deoxygenated hemocyanin associates more

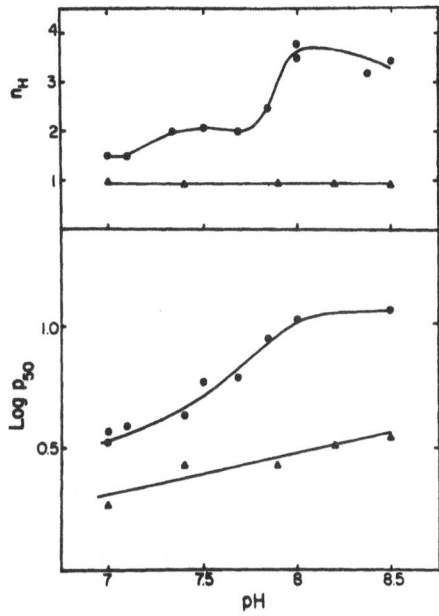

Fig. 3. Oxygen binding by <u>Limulus</u> 60S hemocyanin in 50 mM Tris, 10 mM CaCl$_2$,ionic strength 0.13 M. Solid circle = control; solid triangle = control + 0.33 mM HgCl$_2$. Experiments done at 20°C.

rapidly that the oxygenated derivative. Figure 3 shows the effect of mercury on the 60S hemocyanin of <u>Limulus polyphemus</u>. It can be easily seen that 0.33 mM mercuric chloride greatly increases the oxygen affinity of the hemocyanin. Simultaneously, there is a large reduction in the Bohr effect. Cooperative oxygen binding is also lost. The slope of the Hill plot, n_h , in fact, is less than 1. The less than unity slope of the Hill plot is a reflection of the intrinsic differences in oxygen affinity between the various subunits which build the 60S structure. Figure 4 shows the effects of mercuric chloride and cadmium chloride on <u>Callinectes</u> hemocyanin. The conditions of this experiment are such that the 25S and 16S hemocyanin components exist in the same proportions as in the native hemolymph. Figure 4 shows that both mercury and cadmium reduce the cooperativity of oxygen binding in <u>Callinectes</u> hemocyanin, but have opposite effects on oxygen affinity below pH 8.0. It should be noted that the effect of cadmium decreases with increasing pH because the cadmium ion activity decreases.

Figure 5 shows oxygen binding experiments carried out with <u>Limulus</u> hemocyanin under conditions where the molecule is always the 48-mer 60S structure. This figure shows that mercury irreversi-

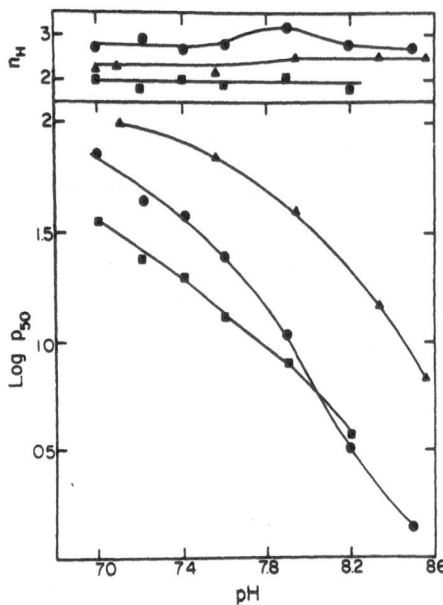

Fig. 4. Oxygen binding by <u>Callinectes</u> hemocyanin in 50 mM Tris, 10
mM $CaCl_2$, ionic strength 0.13 M. Solid circle = control; solid
triangle = control + 0.24 mM $HgCl_2$ (free Hg^{2+} concentration
= 3.6 x 10^{-16}M); solid square=control + 0.5 mM $CdCl_2$ (free
Cd^{2+} ion activity = 3 x 10^{-5}M). Experiments done at 20°C.

bly fixes the 60S structure of Limulus hemocyanin in a non-coopera-
tive high oxygen affinity conformation. This result is reminiscent
of similar data reported by Kuiper et al.[5] on <u>Panulirus</u> hemocyanin.
Even after extensive dialysis of the mercury treated sample, there
is little change in its oxygen binding properties. Changes in absorbance at 250 nm suggest that the mercury is binding to a sulfhydryl group forming a mercaptide bond. Atomic absorption spectrometry
indicates that 2 mercury atoms are bound per subunit in the <u>Limulus</u>
60S molecule.

Figure 6 shows the effects of cadmium and mercury on the oxygen
binding properties of <u>Callinectes</u> hemocyanin. Again, the conditions
of these experiments are such that the 25S and 16S structures exist
in the same concentration as in the native hemolymph. It should be
recalled, however, that the oxygen binding properties of these two
structures are identical. Inspection of Figure 6 shows that mercury
and cadmium not only change the $p_{\frac{1}{2}}$, but also the affinity of the
T-state. Mercury also decreases the oxygen affinity of the R-state.

<u>Callinectes</u> 25S hemocyanin at pH 7 in the presence of calcium

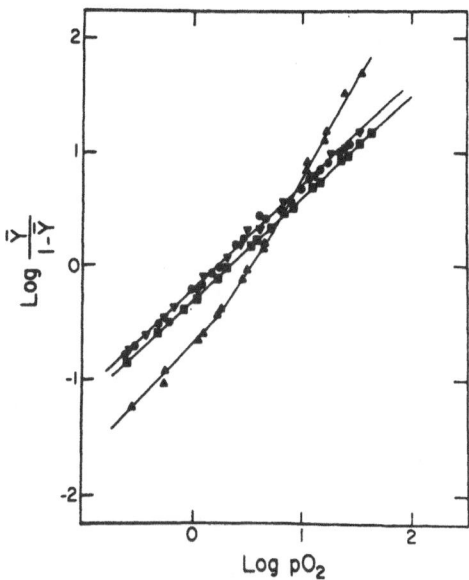

Fig. 5. Oxygen binding by <u>Limulus</u> 60S hemocyanin. (▲) 60S hemocyanin
in 50 mM Tris, 10 mM CaCl$_2$, ionic strength 0.13 M, pH 7.
(●) same buffer + 0.33 mM HgCl$_2$ (free Hg^{2+} ion concentration
= 5 x 10^{-16}). (▼) previous solution dialyzed vs. 50 mM Tris,
10 mM CaCl$_2$, 2 mM EDTA, pH 7, followed by second dialysis
vs. 50 mM Tris, 10 mM CaCl$_2$, ionic strength 0.13 M, pH 7.
Experiments performed at 20°C.

has a $p_{\frac{1}{2}}$ of 72 mmHg and n_h of 2.8. Adding 0.24 mM HgCl$_2$ increases
the $p_{\frac{1}{2}}$ to 102 mmHg and decreases n_h to 2.1. Dialysis of this mercury
treated sample vs. 50 mM Tris, 10 mM CaCl$_2$, 2 mM EDTA pH 7, followed
by a second dialysis of the same buffer without EDTA leaves the 25S
structure intact. Oxygen binding experiments with this material
show that the mercury induced changes in oxygen binding characteristics of <u>Callinectes sapidus</u> hemocyanin are completely reversible.
Atomic absorption spectrometry of the mercury treated <u>Callinectes</u>
hemocyanin as well as absorbance changes at 250 nm convincingly show
that 1 mercury atom is attached to a sulfhydryl group in <u>Callinectes</u>
hemocyanin and hence is not dialyzable at pH 7. This sulfhydryl
bound mercury is not bound at a site that perturbs either the oxygen
binding site or allosteric sites on the molecule. This is to be
contrasted to the situation in <u>Limulus</u> 60S hemocyanin where the
mercury effects are not at all reversible.

In summary, these and other[5] experiments have clearly shown
that hemocyanins are useful molecules in probing the effects of heavy

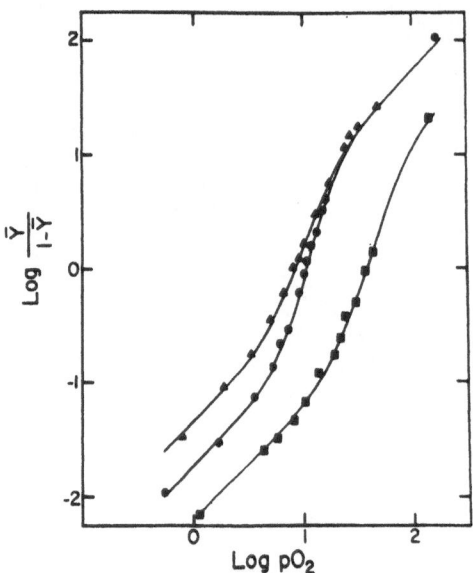

Fig. 6. Oxygen binding by <u>Callinectes sapidus</u> hemocyanin in 50 mM
Tris, 10 mM $CaCl_2$, ionic strength 0.13 M, pH 7.9. Solid
circle = control. Solid triangle = control, plus 0.5 mM
$CdCl_2$, solid square = control + 0.24 mM $HgCl_2$. Experiments
done at 20°C.

metals on respiratory protein function in vitro. However their importance as in vivo probes under conditions of heavy metal pollution
stress, remains to be assessed. In addition, this type of studies,
which began with a strong impetus from environmental health concerns,
has shown that new and exciting information can be obtained on hemocyanin structure, function and assembly at the same time as gaining
insight in the area of marine biomedicine.

REFERENCES

1. P. A. Rainbow and A. G. Scott, Two heavy metal-binding proteins
      in the midgut gland of the crab, <u>Carcinus maenas</u>, <u>Mar.Biol</u>.
      55: 143 (1979).
2. V. Talbot and R. J. Mages, Naturally-occurring heavy metal bin-
      ding proteins in invertebrates. <u>Arch. Environ. Contam. Toxi-
      col</u>., 7: 73 (1978).
3. J. W. Ridlington and B. A. Fowler, Isolation and partial cha-
      racterization of a cadmium-binding protein from the American
      oyster (<u>Crassostrea virginica</u>). <u>Chem. Biol. Interactions</u>,
      25: 127 (1979).

4. J. Wyman, Regulation in macromolecules as illustrated by haemo-
    globin, Quart. Rev. Biophys. 1: 35 (1968).
5. H. A. Kuiper, L. Zolla, L. Calabrese, P. Vecchini, S. Costanti-
    ni and M. Brunori, Effects of mercuric chloride on the struc-
    tural and functional properties of Panulirus interruptus
    hemocyanin, Comp. Biochem. Physiol. 69C: 253 (1981).
6. E. Antonini and M. Brunori "Hemoglobin and myoglobin in their
    reactions with ligands," North Holland Publishing Co.,
    Amsterdam (1971).
7. E. Antonini and M. Brunori, Transport of oxygen: respiratory
    proteins, in:"Molecular Oxygen in Biology" O. Hayashi, ed.
    North-Holland, Amsterdam (1974).
8. C. Bonaventura and J. Bonaventura, Anionic control of function
    in vertebrate hemoglobins, J. Exp. Zool., in press.
9. J. Bonaventura and C. Bonaventura, Hemocyanins: relationships
    in their structure, function and assembly, Amer. Zool. 20:7
    (1980).
10. M. Brouwer, C. Bonaventura and J. Bonaventura, Heavy metal
    interactions with Callinectes sapidus hemocyanin: Structural
    and functional changes induced by a variety of heavy metal
    ions. Biochemistry, submitted for publication.
11. M. Brenowitz, J. Bonaventura, C. Bonaventura and E. Gianazza,
    Subunit composition of a high molecular weight oligomer:
    Limulus polyphemus hemocyanin. Arch.Biochem.Biophys. Sub-
    mitted for publication.
12. L. M. Hamlin, and W. W. Fish, The subunit characterization of
    Callinectes sapidus hemocyanin, Biochim. Biophys. Acta 491:
    46 (1977).
13. T. T. Herskovitz, L. J. Erhunonwunsee and R. C. San George,
    Subunit structure and dissociation of Callinectes sapidus
    hemocyanin, Biochim. Biophys. Acta, 667: 44 (1981).

ACKNOWLEDGEMENTS

     We thank Mr. John Dixon and Ms. Thea Brouwer for skillful
technical assistance. We also thank Mr. Patrick Whaling for the
metal analyses. The grants NIEHS Center Grant ESO 1908, NOAA Grant
NA8ORADO0063, NIH Grant HL 15460, NSF Grant PCM 79-06462 are gratefully acknowledged.

MECHANISMS OF ACTION OF METAL-CONTAINING ENZYMES

INTERMEDIATES IN THE REDUCTION OF DIOXYGEN BY LACCASE AND CYTOCHROME *c* OXIDASE

Bo G. Malmström

Department of Biochemistry & Biophysics
University of Göteborg and Chalmers Institute
of Technology, S-412 96 Göteborg (Sweden)

INTRODUCTION

There are more than 200 enzymes known which catalyze reactions in which molecular oxygen (dioxygen) is one of the substrates.[1] Only four of them can, however, achieve the four-electron reduction of dioxygen to two molecules of water, namely the so-called blue oxidases (ascorbate oxidase, ceruloplasmin and laccase) and cytochrome *c* oxidase. In recent years there has been considerable progress in our understanding of the detailed catalytic mechanism of these enzymes, notably with laccase and cytochrome oxidase.[2] In this paper I will briefly summarize some of the contributions of my own research group to the characterization of reaction intermediates in the reduction of dioxygen by these two enzymes.

It is quite appropriate that my recent work on the catalytic mechanism of oxidases should be described at a symposium in honor of Professor A. Rossi Fanelli, as the studies to be summarized here are a direct outgrowth of investigations with laccase and cytochrome oxidase that I carried out at the Istituto di Chimica Biologica in Rome in the period 1968 - 1971. In August 1968 I came to Rome to carry out a kinetic study of fungal laccase by rapid-reaction techniques in collaboration with Professor E. Antonini, a stopped-flow apparatus not being available in my own laboratory at that time. Much to our surprise we found that, whereas the fully reduced enzyme reacts extremely rapidly with O_2, the reoxidation of the partially reduced enzyme was exceedingly slow.[3] This led us

for the first time to suggest that dioxygen reduction in
laccase and related oxidases involves multi-electron
steps made possible by the cooperation of the electron-
accepting sites. It was our conviction that the same
mechanistic concept would apply also to cytochrome
oxidase, an idea which we could confirm experimentally a
few years later.[4]

LACCASE

The blue oxidases and cytochrome oxidase are more
complex with regard to their prosthetic group composition
than other oxidases and oxygenases.[1,2] These enzymes are
structurally and functionally asymmetric[5] with four or
more metal ions in at least three distinct coordination
environments.

Spectroscopic and Redox Properties

A molecule of laccase in its oxidized state contains
4 Cu^{2+} ions.[6] As Cu^{2+} has a d^9 configuration, the
isolated ion has one unpaired electron and should be
detectable by EPR. Surprisingly there are, however, only
two EPR-detectable copper ions in oxidized laccase. These
two ions are bound in different ways to the protein, as
reflected by differences in their EPR parameters, and are
designated type 1 and type 2, respectively. The type 1
Cu^{2+} ion is responsible for the cerulean colour of the
enzyme, which stems from an unusually strong charge-
transfer absorption band around 600 nm.[7] Type 1 Cu^{2+}
ions are also found in some small proteins with a single
metal ion, such as azurin and plastocyanin, and for these
the structural basis of the anomalous properties is known.[8]
Key features are a thiolate ion from a cysteine residue
as one ligand, functioning as the electron donor in the
charge-transfer transition, and a flattened tetrahedral
ligand geometry. Type 2 Cu^{2+} has more normal spectroscopic properties but has an anomalously high affinity
for certain anions, notably F^-, which act as inhibitors.

The absence of an EPR signal from the two remaining
copper ions, generally named type 3, can be explained on
the basis that these consist of an antiferromagnetically
coupled $Cu^{2+}-Cu^{2+}$ pair.[6] The coupling in laccase is so
strong that the pair does not display any paramagnetism
in the temperature range 40-300 K. Under some circumstances it is, however, possible to produce a type 3 Cu^{2+}
EPR signal by differential reduction of one of the ions.[9]

In addition, the reactions of the enzyme with NO support the presence of a copper pair.[10]

Laccase takes up electrons from reductants one at a time, the primary electron acceptor being the type 1 Cu^{2+}.[6] The single electron in the reduced type 1 site cannot be transferred to the type 3 pair before the type 2 Cu^{2+} has also been reduced, because the type 3 site functions as a cooperative two-electron acceptor. The type 2 Cu^{2+} does not react with the reducing substrate until the type 1 copper has been reduced, in agreement with much evidence that this site is not available in the oxidized enzyme.[2,6] The reduction of the type 3 copper pair occurs by an intramolecular transfer of two electrons, once the type 1 and 2 sites have both been reduced.

The Catalytic Reaction

The type 3 copper pair constitutes the O_2-reducing site in laccase.[6] It undoubtedly reacts with O_2 as soon as it has been reduced by two electrons, transferred from the type 1 and 2 sites, as described in the previous section. The intermediate so formed probably has a structure similar to that found in oxyhemocyanin[11]: $[Cu^{2+}-O^--O^--Cu^{2+}]$. An optical intermediate found[12] in 1973 was first suggested to be this peroxide intermediate. It was, however, later shown[13] that the optical intermediate is associated with an EPR signal from an oxygen radical, demonstrating that O_2 in this state must be $1e^-$- or $3e^-$-reduced. Later work[14] with laccase depleted in type 2 copper has shown that the intermediate represents a $3e^-$-reduced state. Thus, it probably has the structure $[Cu^{2+}-O^--Cu^{2+}]$. This would be consistent with the unusual relaxation properties associated with the EPR signal of the oxygen-radical intermediate. The paramagnetic intermediate is definitely part of the catalytic cycle, as a considerable portion of the enzyme is in this state during catalytic turnover.[15] It is further reduced to H_2O by the transfer of $1e^-$ from the reduced type 1 site.

CYTOCHROME c OXIDASE

Cytochrome c oxidase is the most complicated of all the enzymes having O_2 as a substrate. Not only does it catalyze the four-electron reduction of O_2 to $2H_2O$, like the blue oxidases, but it also couples the electron-transfer reactions to the translocation of protons across an energy transducing membrane.[16]

The Heme and Copper Prosthetic Metal Ions

The functional unit of cytochrome oxidase, like that of laccase, contains four metal-ion prosthetic groups.[17] Two of these are heme a bound in different ways to the protein, cytochrome a and a_3, whereas two are copper ions in separate chemical environments, Cu_A and Cu_B. In oxidized cytochrome oxidase two components only, cytochrome a and Cu_A are detectable by EPR. The EPR signal from cytochrome a is typical for a low spin heme Fe^{3+}.[18] The absorption at $g2$ is generally ascribed to Cu_A^{2+}. The absence of any EPR signals from oxidized cytochrome a_3 and Cu_B is attributed to an antiferromagnetic coupling between these components, in a similar way as in the type 3 copper ions in laccase. This model is established by the demonstration[19] that the high spin Fe^{3+} signal at $g6$, which can be observed on partial reduction, can be assigned to cytochrome a_3^{3+}. In addition, it has recently been possible to observe an EPR signal from Cu_B^{2+} under some conditions.[20,21]

The Dioxygen Reaction

Cytochrome a is the primary acceptor of electrons from the reducing substrate, ferrocytochrome c. Much evidence[22] suggests the following sequence of electron transfer:

$$c \rightarrow a \rightarrow Cu_A \rightarrow (Cu_B, a_3) \rightarrow O_2$$

Thus, the bimetallic unit $[Cu_B a_3]$ constitutes the O_2-reducing site, similar to the situation in laccase. Its reduction involves the intramolecular electron transfer of two electrons from cytochrome a and Cu_A. At room temperature the reaction of the reduced $Cu_B a_3$ site is too rapid[23] for any intermediates to be detected. A number of intermediates have, however, been found by the application of a low-temperature trapping technique introduced by Chance and his associates.[24] With this method they[25] as well as others[26,27] have observed a number of intermediates when the solubilized enzyme reacts with O_2, starting either from the fully reduced or the mixed valence state:

$$\text{Fully reduced} + O_2 \longrightarrow I \longrightarrow II \longrightarrow III \dashrightarrow$$
$$\qquad\qquad\qquad\quad (A_1) \qquad\qquad\quad (B)$$

$$\text{Mixed valence} + O_2 \longrightarrow I_M \longrightarrow II_M \longrightarrow III_M \ -- \rightarrow$$

$$(A_2) \hspace{4cm} (C)$$

(The designations are those used by my own group, but the nomenclature of the Chance group is shown in parenthesis.)

Chance first suggested[25] that the successive intermediates involve electron redistributions only within the $[Cu_B O_2 a_3]$ unit, but no additional electron transfer to this unit. Quantitative EPR measurements have established that this is true only in the case of the mixed-valence oxidase.[26,27] The three intermediates formed in the reaction of the fully reduced enzyme with O_2 involves sequential electron transfers from reduced cytochrome a and Cu_A, on the other hand. Thus, intermediate I has O_2 reduced to the peroxide level and presumably has the structure $[Cu_B^{2+}-O^--O^--a_3^{3+}]$.

Intermediate I is reduced by $1e^-$ to yield intermediate II. Recently this intermediate has been found[28] to give a Cu_B^{2+} EPR signal with unusual properties. An analysis of the EPR spectrum indicates that Cu_B^{2+} in the intermediate interacts with another paramagnetic species. The paramagnetic group should involve cytochrome a_3 and one atom of O, as the O-O bond is undoubtedly broken on the addition of the third electron. A comparison with a structure given for intermediate I then shows that the formal valence of cytochrome a_3 in intermediate II should be $[FeO]^{2+}$. There are several possible electron distributions, but only one of them, namely the ferryl ion, $Fe^{4+}=O$, would be expected to be paramagnetic. Consequently the structure of intermediate II can be written $[Cu_B^{2+} O=a_3^{4+}]$. This species is further reduced by the intramolecular transfer of $1e^-$ to yield intermediate III.

CONCLUDING REMARKS

The investigations with laccase and cytochrome oxidase summarized here have shown that a common feature of the few enzymes reducing O_2 to $2H_2O$ is a bimetallic O_2-reducing site. This is initially reduced by two electrons transferred intramolecularly from the two primary electron acceptors, types 1 and 2 Cu^{2+} or cytochrome a and Cu_A, respectively. In this way O_2 is rapidly reduced to O_2^{2-} without the intermediate formation of O_2^-. This is a key feature in the ability of these enzymes to overcome the kinetic inertness of O_2, as the $1e^-$-reduction of O_2 by an electron donor with a high reduction potential is thermodynamically unfavourable.[2]

The peroxy intermediates in both laccase and cyto-chrome oxidase are further reduced by $1e^-$ to yield para-magnetic intermediates, which, however, have different structures in the two enzymes. Cytochrome oxidase utilizes the ability of a heme group to form a ferryl ion ($Fe^{4+}=O$), whereas laccase probably has an O^- radical bound to the two type 3 copper ions. In both enzymes the last step in the reduction of O_2 involves the addi-tion of one more electron to the paramagnetic inter-mediate.

REFERENCES

1.   T. Keevil and H. S. Mason, Molecular Oxygen in Bio-logical Oxidations—An Overview, Methods Enzymol. 52:3 (1978).

2.   B. G. Malmström, Enzymology of Oxygen, Ann. Rev. Biochem. 51:in press (1982).

3.   B. G. Malmström, A. Finazzi Agrò, and E. Antonini, The Mechanism of Laccase-Catalyzed Oxidations: Kinetic Evidence for the Involvement of Several Electron-Accepting Sites in the Enzyme, Eur. J. Biochem. 9:383 (1969).

4.   E. Antonini, M. Brunori, C. Greenwood, and B. G. Malmström, Catalytic Mechanism of Cytochrome Oxidase, Nature 228:936 (1970).

5.   B. G. Malmström, The asymmetric nature of oxidases containing several copper atoms per molecule, in "Symmetry and Function of Biological Systems at the Macromolecular Level," A. Engström and B. Strandberg, eds., John Wiley & Sons, Inc., New York (1969).

6.   B. Reinhammar and B. G. Malmström, "Blue" Copper-Containing Oxidases, in "Copper Proteins," T. G. Spiro, ed., John Wiley & Sons, Inc., New York (1981).

7.   D. M. Dooley, J. Rawlings, J. H. Dawson, P. J. Stephens, L.-E. Andréasson, B. G. Malmström, and H. B. Gray, Spectroscopic Studies of *Rhus verni-cifera* and *Polyporus versicolor* Laccase. Elec-tronic Structures of the Copper Sites, J. Am. Chem. Soc. 101:5038 (1979).

8.   B. G. Malmström, Some Aspects of Structure and Func-tion in Copper Containing Oxidases, in "New Trends in Bio-Inorganic Chemistry," R. J. P. Williams and J. R. R. F. Da Silva, eds., Academic Press, London (1978).

9. B. Reinhammar, R. Malkin, P. Jensen, B. Karlsson, L.-E. Andréasson, R. Aasa, T. Vänngård, and B. G. Malmström, A New Copper(II) Electron Paramagnetic Resonance Signal in Two Laccases and in Cytochrome c Oxidase, J. Biol. Chem. 255:5000 (1980).

10. C. T. Martin, R. H. Morse, R. M. Kanne, H. B. Gray, B. G. Malmström, and S. I. Chan, Reactions of Nitric Oxide with Tree and Fungal Laccase, Biochemistry, in press (1981).

11. R. S. Himmelwright, N. C. Eickman, C. D. LuBien, K. Lerch, and E. I. Solomon, Chemical and Spectroscopic Studies of the Binuclear Copper Active Site of Neurospora Tyrosinase: Comparision to Hemocyanins, J. Am. Chem. Soc. 102:7339 (1980).

12. L.-E. Andréasson, R. Brändén, B. G. Malmström, and T. Vänngård, An Intermediate in the Reaction of Reduced Laccase with Oxygen, FEBS Lett. 32:187 (1973).

13. R. Aasa, R. Brändén, J. Deinum, B. G. Malmström, B. Reinhammar, and T. Vänngård, A ^{17}O-Effect on the EPR Spectrum of the Intermediate in the Dioxygen-Laccase Reaction, Biochem. Biophys. Res. Commun. 70:1204 (1976).

14. B. Reinhammar and Y. Oda, Spectroscopic and Catalytic Properties of Rhus vernicifera Laccase Depleted in Type 2 Copper, J. Inorg. Biochem. 11:115 (1979).

15. L.-E. Andréasson and B. Reinhammar, The Mechanism of Electron Transfer in Laccase-Catalyzed Reactions, Biochim. Biophys. Acta 568:145 (1979).

16. M. Wikström, K. Krab, and M. Saraste, Proton-Translocating Cytochrome Complexes, Ann. Rev. Biochem. 50:623 (1981).

17. B. G. Malmström, Cytochrome c Oxidase. Structure and Catalytic Activity, Biochim. Biophys. Acta 549:281 (1979).

18. R. Aasa, S. P. J. Albracht, K.-E. Falk, B. Lanne, and T. Vänngård, EPR Signals from Cytochrome c Oxidase, Biochim. Biophys. Acta 422:260 (1976).

19. H. Beinert and R. W. Shaw, On the Identity of the High Spin Heme Components of Cytochrome c Oxidase, Biochim. Biophys. Acta 462:121 (1977).

20. B. Karlsson and L.-E. Andréasson, The Identity of a New Copper(II) Electron Paramagnetic Resonance Signal in Cytochrome c Oxidase, Biochim. Biophys. Acta 635:73 (1981).

21. G. W. Brudvig, T. H. Stevens, and S. I. Chan, Reactions of Nitric Oxide with Cytochrome c Oxidase, Biochemistry 19:5275 (1980).

22. B. G. Malmström, Cytochrome c Oxidase, in "Metal Ion Activation of Dioxygen," T. G. Spiro, ed., John

Wiley & Sons, Inc., New York (1980).

23. C. Greenwood and Q. H. Gibson, The Reaction of Reduced Cytochrome c Oxidase with Oxygen, J. Biol. Chem. 242:1782 (1967).

24. B. Chance, C. Saronio, and J. S. Leigh, Jr., Functional Intermediates in the Reaction of Membrane-bound Cytochrome Oxidase with Oxygen, J. Biol. Chem. 250:9226 (1975).

25. B. Chance, C. Saronio, J. S. Leigh, Jr., W. J. Ingledew, and T. E. King, Low-Temperature Kinetics of the Reaction of Oxygen and Solubilized Cytochrome Oxidase, Biochem. J. 171:787 (1978).

26. G. M. Clore, L.-E. Andréasson, B. Karlsson, R. Aasa, and B. G. Malmström, Characterization of the Low-Temperature Intermediates of the Reaction of Fully Reduced Soluble Cytochrome Oxidase with Oxygen by Electron-Paramagnetic-Resonance and Optical Spectroscopy, Biochem. J. 185:139 (1980).

27. G. M. Clore, L.-E. Andréasson, B. Karlsson, R. Aasa, and B. G. Malmström, Characterization of the Intermediates in the Reaction of Mixed-Valence-State Soluble Cytochrome Oxidase with Oxygen at Low Temperatures by Optical and Electron-Paramagnetic-Resonance Spectroscopy, Biochem. J. 185:155 (1980).

28. B. Karlsson, R. Aasa, T. Vänngård, and B. G. Malmström, An EPR-Detectable Intermediate in the Cytochrome Oxidase-Dioxygen Reaction, FEBS Lett. 131:186 (1981).

STRUCTURE AND FUNCTION OF THE REDOX SITE OF CYTOCHROME OXIDASE

Britton Chance

Johnson Research Foundation, University of Pennsylvania
Philadelphia, PA 19104   USA

L. Powers and Y. Ching

Bell Laboratories, Murray Hill, NJ 07974   USA

INTRODUCTION

The respiratory enzyme of Warburg and Keilin[1,2] has remained an
enigma since its discovery by these two pioneers in the field of
cell respiration. Warburg identified the heme nature of the enzyme
while Keilin perceived the chain of electron transfer components as
an integral factor in cell respiration. Aside from certain further
elaborations of the function of cytochromes in energy coupling and
in respiratory control, the molecular mechanisms for oxygen reduction and energy conservation have remained largely in "the black
box" status,[3] where a number of structural and functional properties
were ascribed to the system in order to be consistent with the
ingenious hypothesis of transmembrane charge transfer according to
Lundegårdh and Mitchell.[3] There was one good reason for the popu-
larity of "the black box" approach--the lack of protein crystals and
appropriate derivatives for high resolution X-ray crystallography of
electron transfer components. Thus, the dearth of knowledge on the
nature of the iron and copper atoms, their ligands, and interatomic
distances has rendered incomplete, detailed explanations of electron
transfer mechanisms and energy conservation.

The recent advent of synchrotron radiation techniques for
determining the charge density of metal atoms by edge absorption
techniques and the interatomic distance of binuclear metallic clusters by extended absorption fine structure (EXAFS) has radically
altered the possibilities of structural studies of non-crystalline
metalloenzymes and proteins, particularly membrane proteins. X-ray

synchrotron studies have afforded scalar distances of the ligands
of metal atoms, initially in hemoglobin[4] and rubredoxin,[5] and more
recently in a variety of structural states[6] of important transport
pigments[7] and metal enzymes[8,9,10]. Cytochrome oxidase has, however,
presented a problem of the greatest difficulty because of its large
molecular weight permitting only relatively low concentrations of
the enzyme available for study. This has been overcome due to the
dedicated electron synchrotron operation, increased efficiency of
fluorescent photon detection,[11,12] online optical and offline epr
monitoring of sample integrity,[13] and improvements of sample biochemistry[9,10]. More details of the following are found in Ref. 9.

     X-ray absorption spectroscopy (Fig. 1) can be used to determine
distances of nearby atoms ($\lesssim 4.5$ Å) surrounding the metal atoms of
cytochrome oxidase (two iron and two copper atoms in this case). In
addition, the edge structure, which measures the charge density on
the absorbing atom contains additional information about the number,
chemical type, and geometry of the ligands. Absorption of x-rays by
the metal atoms of cytochrome oxidase at energies near the K_α edge
(Fe, ~ 7 and Cu, ~ 9 KeV) causes transitions from the 1s shell to
higher bound states. At energies above the edge, the ejected photo-
electrons are backscattered from the nearby ligands. As the energy
of excitation is increased, oscillations in the absorption are
observed (Figs. 2 and 3), representing this interference (EXAFS):
in this case, the contributions of the two iron or copper atoms are
summed. Figure 2 illustrates the X-ray absorption spectra for iron
and copper in 1 mM cytochrome oxidase determined at Stanford
Synchrotron Radiation Laboratory using beam lines II-3 with ~ 3 eV
resolution and I-5 having ~ 1 eV resolution, respectively. In order
to better discern the oscillations of intensity, a linear background
is subtracted, and thereby the absorption below the edge is set at

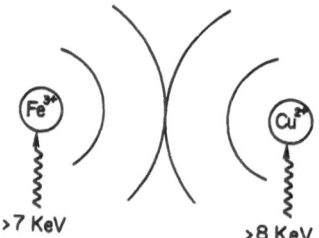

Fig. 1.   Illustration of the principle of the EXAFS study of the
          Fe → Cu, Cu → Fe distances in cytochrome oxidase
          (EXAFS = Extended X-ray Absorption Fine Structure) by
          selective excitation of Fe and Cu at energies above their
          respective absorption edges and observation of the back-
          scatter from the paired atom.

zero. The data of Fig. 2 comprise an average of 27 7-minute scans at an intensity measured at the sample (3.0 GeV and 50 ma) of 10^{10} and 10^{11} photons/sec, respectively, for the two beam lines.

The signal is given by the expression[9]

$$\chi(k) - \sum_{0} \frac{r_i - N_i|f_i(k,\pi)|}{kr_1^2} e^{-r_i/\lambda(k)} e^{-2\sigma_i(k)k^2} \sin(2[kr_i+\alpha_i(k)]) \quad (1)$$

where the summation is over the distances r_i from the absorbing atom, N_i is the number of the same type of backscattering atoms at r_i, $f_i(k,\pi)$ is the backscattering amplitude of the ith atom which is Z/k^2, for $k >\sim 4$ $\overset{\circ}{A}^{-1}$, Z is the atomic number, and $\lambda(k)$ is the photoelectron mean free path. $\sigma^2(k)$ is the Debye-Waller factor describing the mean square displacement in r_i (from the thermal and lattice disorder), $\phi_i(k)$ is the energy-dependent phase shift of the photoelectron caused by the potentials of both the absorbing and

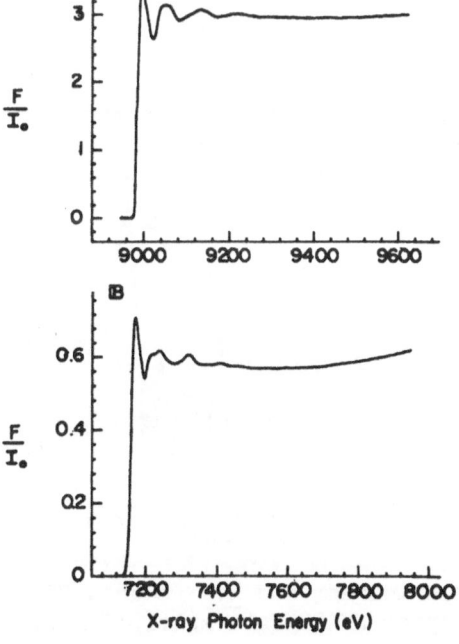

Fig. 2. A. Cu EXAFS, and B. Fe EXAFS of 1 mM cytochrome oxidase in the fully oxidized or resting state. F is fluorescent signal and I_0 is beam intensity. (Courtesy Biophys. J.[9]).

the backscattering atom. k is the magnitude of the photoelectron
wave vector given by

$$k = [2m_e \ (E-E_o)]^{\frac{1}{2}}\hbar \tag{2}$$

where m_e is the electron mass, E is the X-ray energy and E_o is the
edge energy or threshold.

The EXAFS signal is isolated by a cubic B-spline fit that
removes the "isolated atom" contribution, multiplied by k^3 to correct for the approximate k^{-3} dependence of the EXAFS, and to
equalize the oscillations over the observed range of k. The function is then normalized to one absorbing atom by the magnitude of
the edge (which represents two atoms). The signal-to-noise ratio
of the corrected EXAFS of cytochrome oxidase (Fig. 3) is 3 times

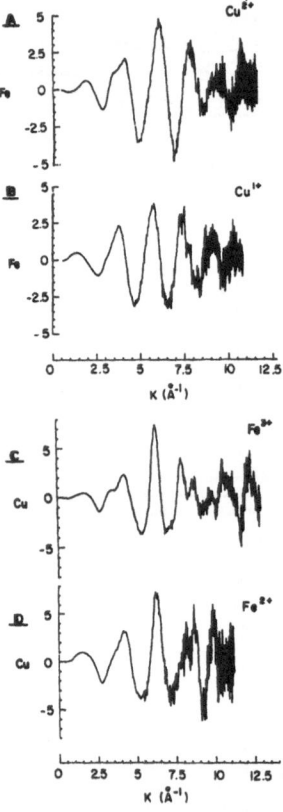

Fig. 3.   <u>A</u>. Fully oxidized; and <u>B</u>. reduced CO states: Background
          removed copper EXAFS data multiplied by k^3 and normalized
          to one copper atom.
          <u>C</u>. Fully oxidized; and <u>D</u>. reduced CO states: Background
          removed iron EXAFS data multiplied by k^3 and normalized to
          one iron atom. (See also Ref. 9.)

that of the classic experiment on 20 mM hemoglobin used to obtain
the Fe-N (pyrrole) bond lengths,[4] and is adequate without smoothing.
However, much more beam time at higher intensities was required for
these studies.

Fourier transformation of the EXAFS gives an intensity-distance
function that is very useful in identifying structural changes. It
is a relative radial distribution function (RRDF). The abscissa of
Fig. 4 contains the absorber-scatter phase shift $(\alpha(k) \approx 0.20$ to $0.40\text{Å})$.

Isolation of contributions from a single shell is accomplished
by a Fourier filter of the shell in question[9]. The data for each
shell were then decomposed into a unique solution of an amplitude
and phase. The phase is unique to the absorber-scatterer pair and
has been experimentally as well as theoretically proven to be
"chemically transferrable" from the chemical environment of one
compound to another[14]. The amplitude, which contains N_i, $\sigma(k)$ and

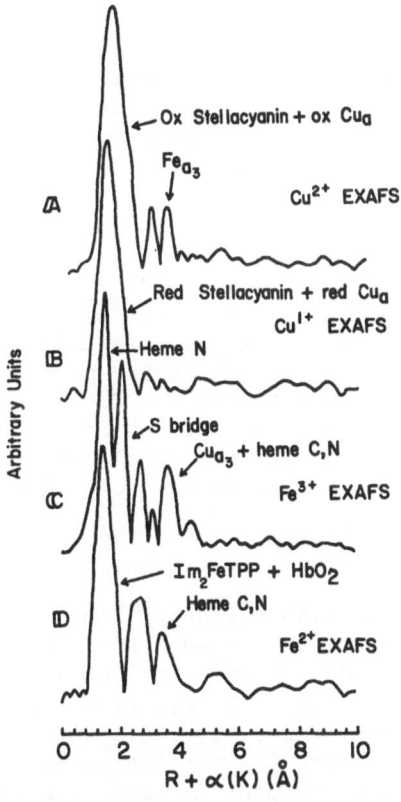

Fig. 4. A. Fully oxidized; and B. reduced CO states: Fourier-
transformed copper EXAFS data.
C. Fully oxidized; and D. reduced CO states: Fourier-
transformed iron EXAFS data. (See also Ref. 9.)

$\alpha(k)$, on the other hand, has proved to vary as much as 50% due to the chemical environment and inelastic scattering effects.[15] For this reason, only phases were used to identify the chemical type of a scattering atom (one type per shell). Shells containing multiple types of atoms and distances were identified by direct comparison of filtered data with that of appropriate model compounds with no fitting procedures or variables.[16] Thus, one can precisely compare the various shells of cytochrome oxidase Fe and Cu with those of models.[9]

The analysis of the EXAFS data is difficult because the contributions of the two iron and two copper atoms to the filtered data of each shell need to be separated. This is accomplished by an appropriate model compound. In the case of Cu, the models used to reproduce the filtered data were suggested from previous edge investigations[9] which were used in both the oxidized and the reduced states to represent the redox site in the four redox combinations that can be obtained experimentally. Fe_a and Cu_a changed only slightly but the contributions of Fe_{a3} and Cu_{a3} to the various peaks clearly vary with the redox state. The third shell of Cu^{2+} EXAFS peak is attributed to the spin-paired Fe atom and to the spin-paired Cu atom plus heme contributions in the Fe^{3+} data. The bridging S atom causes a split on the first shell Fe^{3+}, but not in the Fe^{2+} RRDF.

Appropriate models are depicted in Fig. 5 and show the structures and bond lengths for the four centers and the dashed lines indicate ligands that are possible but are too long to form real bonds. In several cases, possible alternatives for the ligand are given in parentheses. Error bars are omitted for simplicity (± 0.03 Å except for Fe^{3+} - Cu^{2+} distance, which is ± 0.05 Å), and a line over the distance indicates an average for those of the same ligands in that metal center. Only formal valences are indicated, since charge delocalization is likely in such a complex system.

The sulfur-ligated 6-coordinate high spin structure Fe_{a3}^{3+} (Fig. 5) is consistent with resonance Raman studies[17] and with the lack of reactivity of the oxidized state with characteristic ligands such as fluoride, cyanide, azide, and sulfide,[18] and with H_2O_2 itself, which, to the contrary, react rapidly with peroxidase, metmyoglobin, and methemoglobin.[2] This unreactivity is similar to that of cytochrome c where the methionine is in the 6-coordination place[2] but reduction of cytochrome oxidase breaks the Fe-S-Cu bond and rapid and effective reaction with oxygen or the inhibitory reaction with CO is observed.

Figure 6 traces structural changes of cytochrome oxidase to electron acceptance for reduction and electron donation for oxidation. A cycle of bond breaking and bond making is suggested by the EXAFS structure changes and conformation changes occur following stepwise acceptance of one or more electrons. In a one-electron mechanism, the S bridge between iron and copper would be pivotal, one of the electron pair being transmitted to the Cu^{2+} and the other to the Fe^{3+} and thereby utilizing the $Fe^{3+}_{a3} - S - Cu^{2+}_{a3}$ bridge as a pathway for transfer of electrons between the two metal atoms.

The Fe-S bond rupture is necessary so that 5-coordinated Fe^{2+}_{a3} may react with oxygen. Absence of steric hindrance to the Fe from the Cu-S ligand in the reduced state can be inferred from the velocity of the reaction with oxygen--10^8 M^{-1} sec^{-1} (as rapid as the most rapid of the oxyhemoglobin or oxymyoglobin reactions). Occupancy of the active site in cytochrome oxidase by gaseous competitors such as acetylene and nitrous oxide further emphasizes the availability of an adequate space for an effective "collision-limited" reaction rate. Thus, the conformation change from the

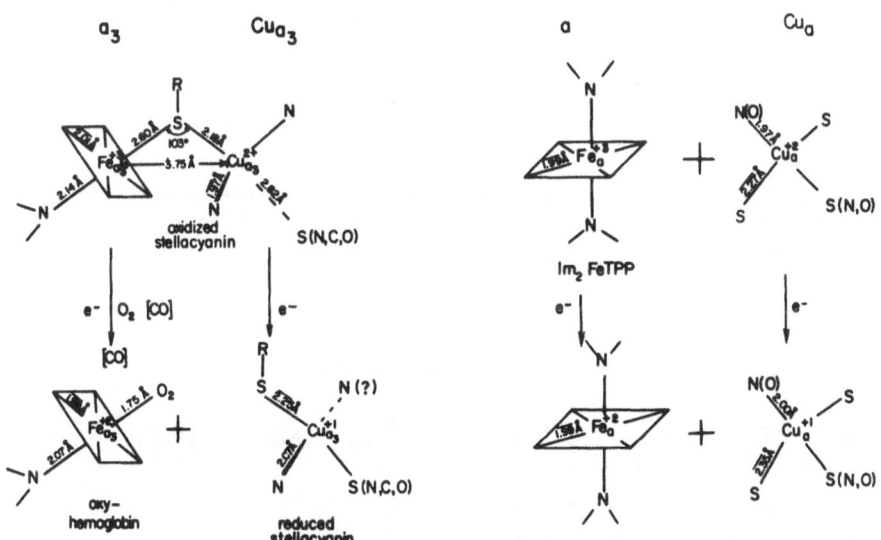

Fig. 5. Pictorial representation of X-ray absorption results for the redox centers in the fully oxidized and reduced CO states. Other ligand possibilities are given in parentheses and (?) indicates ligands not observed but postulated from crystallographic data of similar models. Model compounds are identified where appropriate, error bars are omitted but given in text, and bars over distances indicate average distance for all the same type of ligands in that center. (Courtesy of Biophys. J., Ref. 9.)

Fe_{a3}^{3+} – S – Cu_{a3}^{2+} oxidized state to the reduced state renders Fe_{a3}^{2+}
structurally similar to the oxygen transport pigments and equally
reactive to dioxygen.

The Cu–S moiety does not impede the oxygen reaction even
though it appears to be a binding site for CO,[19] and furthermore,
seems pivotal in securing a two-electron reduction of oxygen as
observed experimentally concomitant with the low temperature
oxidation of Fe_{a3}^{2+} and Cu_{a3}^{1+} at –120°.[19] Thus, any increased
separation of Fe_a and Cu_a in the reduced conformation is not so
large that rapid electron transfer from Cu_{a3}^{1+} to an Fe_{a3}^{2+} bound
oxygen is impeded. An Fe–Cu peroxide bridge in Compound B[9,21] is
consistent with a distance between the two metal Fe_{a3} and Cu_{a3}
centers of 3.75 Å observed in the oxidized state.

Further electron donation to oxygen can occur by electrons
transmitted from Fe_a^{2+} and Cu_a^{1+} or from cytochromes c^{2+}, c_1^{2+}, the
Fe–S/protein, etc., as indicated by Fig. 6, completing the reduc-
tion of oxygen to water and allowing reformation of the Fe–S bond
in catalytic activity as is confirmed by recent EXAFS studies.

Obviously, the key compound of this oxygen reaction is the
intermediate Compound B. Our optical data are consistent with the
"half-oxidized" structure (Fe_{a3} and Cu_{a3} oxidized and Fe_a, Cu_a
remaining reduced at low temperatures). Preliminary data on
Compound B (Chance and Powers, in preparation) show an X-ray edge

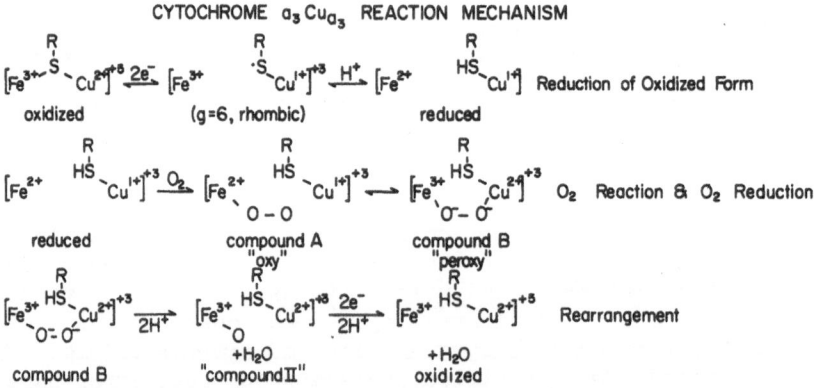

CYTOCHROME $a_3 Cu_{a_3}$ REACTION MECHANISM

Fig. 6.   Sequence of reactions in cyclic reduction and oxidation
          of cytochrome oxidase. (Courtesy of Biophys. J., Ref. 9.)

absorption that confirms these valence assignments. Figure 7B (dashed trace) shows a superposition of the mixed valence formate and Compound B edges in which $Fe_{\underline{a}}$ and $Cu_{\underline{a}}$ are reduced and are distinctly different from the fully oxidized state (Fig. 7A, solid trace).[22]

Preliminary Fe and Cu EXAFS studies of Compound B show the higher shells to be similar to those of the oxidized state; i.e., to contain respectively $Cu_{\underline{a}}^{2+}$ and $Fe_{\underline{a}}^{3+}$ at about 3.75 Å spacing

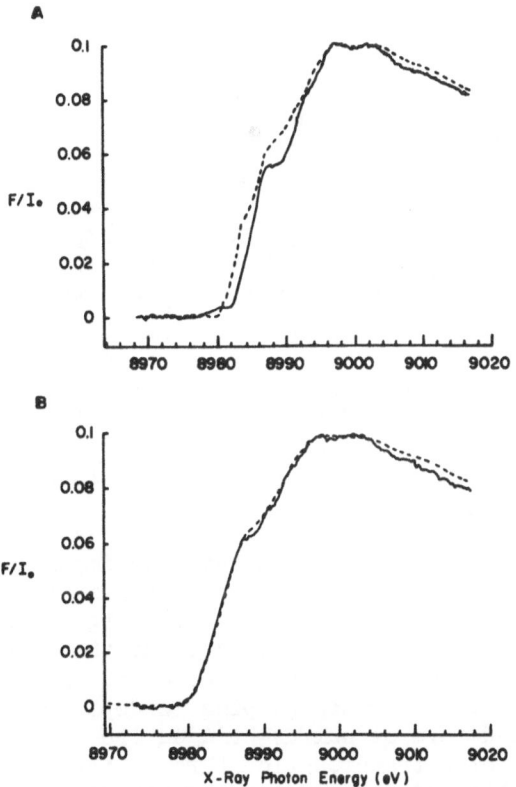

Fig. 7. <u>A</u>. A comparison of the Cu edge absorption of fully oxidized cytochrome oxidase (———) and Compound B (----) as trapped at -120°.
<u>B</u>. A similar comparison of the mixed valence formate compound of cytochrome oxidase ($\underline{a}^{2+}$ $Cu_{\underline{a}}^{1+}$, $\underline{a}_3^{3+}$ $Cu_{\underline{a}_3}^{2+}$)(———) with Compound B (----) as trapped at -120°.
In both cases, the abscissa is incident energy and the ordinate is the ratio of fluorescence of the Cu atoms (measured in counts/sec) to the incident intensity measured in millivolts).

of the mixed valence formate and fully oxidized states.[9] The first shell of Compound B lacks an S atom at 2.6 Å in the Fe data and appropriate models for a peroxide bridge structure are under analysis. Thus, three important chemical and structural features of Compound B are indicated: its mixed valence state, its approximate Fe-Cu distance (~3.8 Å) and the lack of a first shell sulfur, all being consistent with Fig. 6.

Structural explanations for altered reactivities of cytochrome oxidase in the seemingly mysterious "oxygenated" and "pulsed" states described by Orii et al[23] and by Antonini et al[24] may be based upon a) the variation of distance between Fe_{a_3} and Cu_{a_3}, and/or b) the presence and absence of a bridging ligand(s), and c) factors unrelated to the redox center. The distances of Fe and Cu may range from ~3.75 Å to outside the range of EXAFS detectability (> 4 Å) and the ligands may range from the identified S bridge to the proposed μperoxo bridge. Some data suggest that the recently oxidized redox center lacks the S bridge, as based on measurements of its high reactivity towards cyanide as would be expected for a five coordinate ferric hemoprotein.[25] Other data suggest, however, that Cu_{a_3} may be difficult to reduce unless a_3 and the S bridge are present (unpublished observations). In addition, the oxidation product of Compound B has the sulfur bridge.[22] These observations, together with the lack of a published report on any significant concentration of the "g = 5"[26] oxygenated species in the steady-state turnover of cytochrome oxidase, indicates that the S-bridge species may be predominant over the "pulsed" form during catalytic function.

These structures have an immediate and important impact upon the function of cytochrome oxidase, not only in the above-mentioned electron transfer reactions, but also in charge separation and energy conservation[9,27]. Considering first that a proton gradient intervenes between the cytochrome oxidase and the ATPase, the role of the active center in proton pumping is depicted by Figs. 8, where the oxidized form of sulfur-bridged iron copper binuclear complex blocks proton translocation through the active center from matrix to cytosolic side by creating a water-free space (Fig. 8A). Reduction of the redox site involves rupture of the sulfur bridge and the entry of water between iron and copper with the possibility of proton translocation (Fig. 8B).

A mechanism based upon Wyman's linked function[28] as exemplified by the Bohr effect in hemoglobin[29] affords transmembrane charge separation. One requirement for charge separation is that there be an unsymmetrical change on each side of the membrane.[30] Thus, the coupling between the redox center and a cytosolic space carboxyl group causes deprotonation and the entry of a proton into the channel from matrix to cytosolic side. In Fig. 8C, the

A Hypothesis for Energy Coupling through Cytochrome
Oxidase in Mitochondrial Membranes

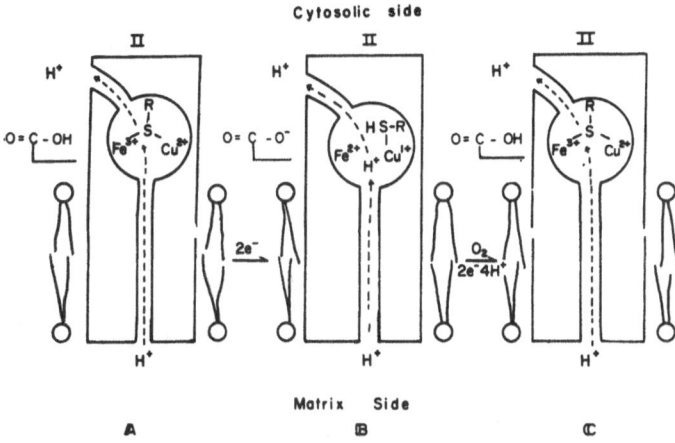

Fig. 8.   An illustration of the redox center's function in
          charge separation by the Membrane Bohr Effect.

combination with oxygen, its reduction eventually to water by
protonation and further electron transfer, reestablishes closure
of the gate and reversal of the asymmetric structural change. This
causes reprotonation of the carboxyl group and formation of water
and a hydroxyl ion on the matrix side and a hydrogen ion on the
cytosolic side. The net effect is to withdraw a proton from the
matrix space. Thus, the mechanism has many features of transmembrane ion pumps.[31] The number of protons generated in this
cycle[32,33] will depend upon the asymmetry of the charge changes on
each side as indeed is the case with the purple membrane protein.[34]
Thus, the mechanism differs from Mitchell's electron/proton loop[35,36]
in which the cytochrome oxidase redox center is required for transmembrane electron flow and is proposed by Mitchell to be located in
the matrix side of the membrane. Thus, all the redox components,
plus the coupling to the protolytic reaction, may be located on the
cytosolic side of the mitochondrial membrane.[22]

An alternate energy coupling mechanism is one involving
juxtaposition of the ATPase and the cytochrome oxidase in which
the conformation change[3] is propagated laterally or peripherally
from one transmembrane molecule to the other and would couple the
microenvironments of the two molecules so that localized charge
separations would be effective in energy conservation. In this
case, microscopic rather than macroscopic charge separations would
be effective in the energy coupling process presumably translocated

by special subunits of the two transmembrane macromolecules which most likely would be subunit III of the cytochrome oxidase and its corresponding subunit of the F_1 ATPase.[37]

In summary, the generation of a large and significant conformation change in the oxidase makes possible the employment of both macroscopic and microscopic charge separation mechanisms in energy transfer.

ACKNOWLEDGEMENTS

This research was supported by NIH grants GM-27308, HL-SCOR-15061, GM-27476 & 28385. The work was done partially at the Stanford Synchrotron Radiation Laboratory (Project 423B), which is supported by the NSF through the Division of Materials Research and the NIH through the Biotechnology Resource Program in the Division of Research Resources in cooperation with the Department of Energy.

REFERENCES

1. O. Warburg, Wasserstoffübertragende Enzyme, Deutsche Zentraldruckerei, Berlin (1948).
2. D. Keilin, The History of Cell Respiration and Cytochrome, Cambridge University Press, Cambridge (1966).
3. P. Boyer, B. Chance, L. Ernster, P. Mitchell, E. Racker and E.C. Slater, "Oxidative Phosphorylation and Photophosphorylation," Ann. Rev. Biochem. 46:955-1026 (1977).
4. P. Eisenberger, R. Shulman, G. Brown and S. Ogawa, "Structure-function Relations in Hemoglobin as Determined by X-ray Absorption Spectroscopy," Proc. Natl. Acad. Sci. USA 73:491-495 (1976).
5. R. Shulman, P. Eisenberger, K. Teo, B. Kincaid and G. Brown, "Fluorescence X-ray Absorption Studies in Rubredoxin and its Model Compounds," J. Mol. Biol. 124:305-321 (1978).
6. S. Doniach, P. Eisenberger and K. Hodgson, "X-ray Absorption Spectroscopy of Biological Molecules," in Synchrotron Radiation Research, H. Winick and S. Doniach, eds., Plenum Press, New York, 425-458 (1980).
7. J. Brown, L. Powers, B. Kincaid, J. Larrabee and T. Spiro, "Structural Studies of the Hemocyanin Active Site: I. EXAFS (Extended X-ray Absorption Fine Structure) Analysis," J. Am. Chem. Soc. 102(12):4210-4216 (1980).
8. R.A. Scott, S.P. Cramer, R.W. Shaw, H. Beinert and H.B. Gray, "Extended X-ray Absorption Fine Structure of Copper in Cytochrome c Oxidase: Direct Evidence for Copper-Sulfur Ligation," Proc. Natl. Acad. Sci. USA 78(2):664-667 (1981).

9. L. Powers, B. Chance, Y. Ching and P. Angiolillo, "Structural Features and the Reaction Mechanism of Cytochrome Oxidase," Biophys. J. 34(3):465-498 (1981).

10. L. Powers, W. Blumberg, B. Chance, C. Barlow, J.S. Leigh, Jr., J. Smith, T. Yonetani, S. Vik and J. Peisach, "The Nature of the Copper Atoms of Cytochrome c Oxidase as Studied by Optical and X-ray Absorption Edge Spectroscopy," Biochim. Biophys. Acta 546:520-538 (1979).

11. E. Stern and S.M. Heald, "X-ray Filter Assembly for Fluorescence Measurements of X-ray Absorption Fine Structure," Rev. Sci. Instr. 50:1579-1582 (1979).

12. M. Marcus, L. Powers, A. Storm, B. Kincaid and B. Chance, "Curved-crystal (LiF) X-ray Focussing Array for Fluorescence EXAFS in Dilute Samples," Rev. Sci. Instr. 51(8):1023-1029 (1980).

13. B. Chance, P. Angiolillo, E. Yang and L. Powers, "Identification and Assay of Synchrotron Radiation Induced Alterations on Metalloenzymes and Proteins," FEBS Lett. 112(2):178-182 (1980).

14. P.H. Citrin, P. Eisenberger and B. Kincaid, "Transferability of Phase Shifts in Extended X-ray Absorption Fine Structure," Phys. Rev. Lett. 36:1346-1349 (1976).

15. P. Eisenberger and B. Lengler, "Extended X-ray Absorption Fine Structure Determinations of Coordination Numbers Limitations," Physiol. Rev. B. 22:3551-3562 (1980).

16. T.D. Tullius, P. Frank and K.O. Hodgson, "Characterization of the Blue Copper Site in Oxidized Azurin by Extended X-ray Absorption Fine Structure: Determination of a Cu-S Distance," Proc. Natl. Acad. Sci. USA 75:4069-4073 (1978).

17. G. Babcock and I. Salmeen, "Resonance Raman Spectra and Optical Properties of Oxidized Cytochrome Oxidase," Biochem. J. 18: 2493-2498 (1979).

18. M. Erecinska and D. Wilson, "Inhibitors of Cytochrome c Oxidase," J. Pharmacology and Therapeutics 8:1-20 (1980).

19. J.O. Alben, F. Fiamingo and R.A. Altschuld, "The Heme Pocket of Hemoglobin, Myoglobin, and Cytochrome c Oxidase, Probed by FTIR Spectroscopy and Low Temperature Photolysis," VII Intl. Biophys. Congr. and III Pam-Am Biochem. Congr. Abstr. No. M-C12, p. 64 (1981).

20. B. Chance and J.S. Leigh, Jr., "Oxygen Intermediates and Mixed Valence States of Cytochrome Oxidase: Infrared Absorption Difference Spectra of Compounds A, B, C of Cytochrome Oxidase and Oxygen," Proc. Natl. Acad. Sci. USA 74(11):4777-4780 (1977).

21. W. Blumberg and J. Peisach, "Some Possible Chemical and Electronic States of Cytochrome Oxidase and its Intermediate Redox States," in Cytochrome Oxidase, T. King, Y. Orii, B. Chance and K. Okunuki, eds., Elsevier/North-Holland Biomedical Press, Amsterdam (1979), pp. 153-160.

22. B. Chance, L. Powers, Y. Ching and B. Muhoberac, "The Function of Cytochrome Oxidase Based on Structure Determinations by X-ray Absorption Fine Structure," (1981), In press.

23. L. Powers, W. Blumberg, B. Chance, C. Barlow, J.S. Leigh, Jr. J. Smith, T. Yonetani, S. Vik and J. Peisach, "The Nature and Function of Copper in Cytochrome Oxidase," in Cytochrome Oxidase, T. King, Y. Orii, B. Chance and K. Okunuki, eds., Elsevier/North-Holland Biomedical Press, Amsterdam (1979), pp. 189-195.

24. E. Antonini, M. Brunori, A. Colosimo, C. Greenwood and M. Wilson, "Oxygen "Pulsed' Cytochrome c Oxidase: Functional Properties and Catalytic Relevance," Proc. Natl. Acad. Sci. USA 74(8): 3128-3132 (1977).

25. T. Brittain and C. Greenwood, "Kinetic Studies on Binding of Cyanide to Oxygenated Cytochrome c Oxidase," Biochem. J. 155:453-455 (1976).

26. R. Shaw, R. Hansen and H. Beinert, "A Novel Electron Paramagnetic Resonance Signal of 'Oxygenated' Cytochrome c Oxidase," J. Biol. Chem. 253(19):6637-6640 (1978).

27. B. Chance, L. Powers and Y. Ching, "Structure, Function, and Charge Separation in Cytochrome Oxidase, and EXAFS Study," in Mitochondria and Microsomes: In Honor of Lars Ernster, C.P. Lee, G. Schatz and G. Dallner, eds., Addison Wesley, Reading, MA (1981), pp. 271-292.

28. J. Wyman, "Relation of Physiological Function and Molecular Studies in Hemoglobin," Fed. Proc. 7(3):502-508 (1948).

29. C. Bohr, K.A. Hasselbach and A. Krogh, "Ueber Einen Biologischer Beziehung Wictigen Einfluss Dendie Kohlensaurespannung des Blutes Auf Deffen Saurstoffbindung Ubt," Skan. Arch. Physiol. 16:402 (1904).

30. B. Chance, A. Crofts, N. Nishimura and B. Price, "Fast Membrane H^+ Binding in the Light-activated State of Chromotium Chromatophores," Eur. J. Biochem. 13:364-374 (1970).

31. R.E. Davies and R.D. Keynes, "A Coupled Sodium-potassium Pump," in Membrane Transport Metabolism, A. Kleinzeller and A. Kotyk, eds., Czech. Acad. Sci., Prague (1961) pp. 336-340.

32. K. Krab and M. Wikström, "Proton-translocation Cytochrome c Oxidase in Artificial Phospholipid Vesicles," Biochim. Biophys. Acta 504:200-214 (1978).

33. A. Lehninger, "Proton Ejection Coupled to Ferrocytochrome c Oxidase by Rat Liver Mitoplasts," First Eur. Bioenergetics Conf., Patrone Editore, Bologna, Italy (1980).

34. B. Hess, and D. Kürschmitz, "Establishments of Electrochemical Gradients: General Views and Experiments on Purple Membrane," in Frontiers of Biological Energetics, P.L. Dutton, J.S. Leigh, Jr. and A. Scarpa, eds., Academic Press, New York (1978), pp. 257-264.

35. P. Mitchell, Chemiosmotic Coupling and Photosynthetic Phosphorylation, Glynn Res., Bodmin, Cornwall, England, 1966.

36. B. Chance, "Structural and Kinetic Approaches to Electron
     Transfer Oxygen Reaction and Energy Conservation in Cyto-
     chrome Oxidase," in The Proton Cycle: In Honor of Peter
     Mitchell, P. Hinkle and J. Delatz, eds., Addison-Wesley,
     New York (1981), In press.
37. P.L. Pedersen and J. Hullehen, "Adenosine Triphosphatase of
     Rat Liver Mitochondria," J. Biol. Chem. 253(7):2176-2183
     (1978).

KINETIC STUDIES OF CYTOCHROME-*c*-OXIDASE: SIGNIFICANCE OF

DIFFERENT FUNCTIONAL STATES OF THE ENZYME

Maurizio Brunori, Alfredo Colosimo and Michael T. Wilson[o]

Institutes of Chemistry and Biochemistry and C.N.R. Center
of Molecular Biology, Faculty of Medicine, University of
Rome, Rome, Italy
[o]Department of Chemistry, University of Essex, Colchester,
Essex, U.K.

INTRODUCTION

Cytochrome-*c*-oxidase, the terminal electron acceptor of the
mitochondrial respiratory chain, is the fundamental enzyme of aerobic
life. In the last few years, the structural features of this enzyme
have been actively investigated with considerable success. The
number of polypeptide chains, their amino acid sequences, their
mutual contacts have been studied and an approximate three-dimensional model has been proposed.[1,2] The structure of the metal centers,
the hemes and the copper atoms, has been partially elucidated by
chemical and spectroscopic techniques, in different oxidation
states.[3] Thus the prospect of quantitative correlations between
structure and function for this important enzyme is indeed very
good at this stage.

Cytochrome oxidase is a polyfunctional macromolecular complex
which displays several types of interaction phenomena. Its two
basic functions may be summarized in the following way:

(i) To catalyze the electron transfer from reduced cytochrome *c* to
dioxygen, according to the reaction

$$4 \text{ cyt } \underline{c}^{2+} + 4 \text{ H}^+ + O_2 \longrightarrow 4 \text{ cyt } \underline{c}^{3+} + 2H_2O \qquad \text{(Eq. 1)}$$

This is the basic function which demands coupling between the
one electron donor, cytochrome *c*, and the four electron acceptor,
dioxygen.

(ii) To bind other solvent components, such as protons and calcium,
and possibly to act as a pump in the build up of the proton
gradient involved in oxidative phosphorylation. [4]

In this paper we shall describe some of the kinetic work carried out in the Institute of Biochemistry at Rome during the
last 10 years, and outline some of the possible physiological
implications of different functional states of cytochrome oxidase.

KINETICS OF THE REACTION OF SOLUBLE OXIDASE

A schematic view of the basic functional unit of cytochrome
oxidase is shown in Fig. 1. The spectroscopic properties of the
four metals have been the subject of very extensive investigations
by a number of workers (see 5) since the preparation of solubilized
purified cytochrome oxidase was initially reported.[6] In Fig. 1, the
commonly accepted nomenclature for the two hemes (cytochromes a and
a_3) and the two copper atoms (Cu_A and Cu_B) is indicated.

Transient studies have been carried out at high concentrations
of the enzyme with both substrates, i.e. oxygen and cytochrome c.
These investigations have made use of rapid mixing methods, as well
as flash photolysis and temperature jump ones, in view of the very
short times involved.

The reaction of fully or partially reduced oxidase with O_2 has
been investigated by flow-flash starting from the CO derivative.[7]
With the enzyme in solution, the overall process is indeed very
rapid, being complete in a few msec at 20°C. Spectrally distinct
intermediate phases have been observed over the time range from
μsec to msec: the slowest event, corresponding to oxidation of
cytochrome a, has a first order rate constant of 700 s^{-1} at 20°C;
electron donation from Cu_A occurs at 7000 s^{-1} with the oxygen bound
species, while the fastest event is O_2 concentration dependent only
in the lower concentration range (i.e. < 50 μM O_2).

The problem of the identification of the intermediates in the
reduction of O_2 by the reduced enzyme has been a very active field
of investigation by the low-temperature freeze-trapping method introduced by Chance and coworkers.[8] The characterization of species
in which the binuclear center, cyt a_3–Cu_B, becomes EPR detectable
has been successfully achieved by the Swedish group.[9]

The mechanism of electron addition has been largely elucidated.
Using cytochrome c^{2+} as the electron donor, it was shown that one
of the sites, cytochrome a, is involved in an electron transfer
bimolecular process.[10,11] The sequential pathway for electron entry
into the enzyme is schematically shown in Fig. 1; the detailed
kinetic scheme which we have used for computational analysis of the

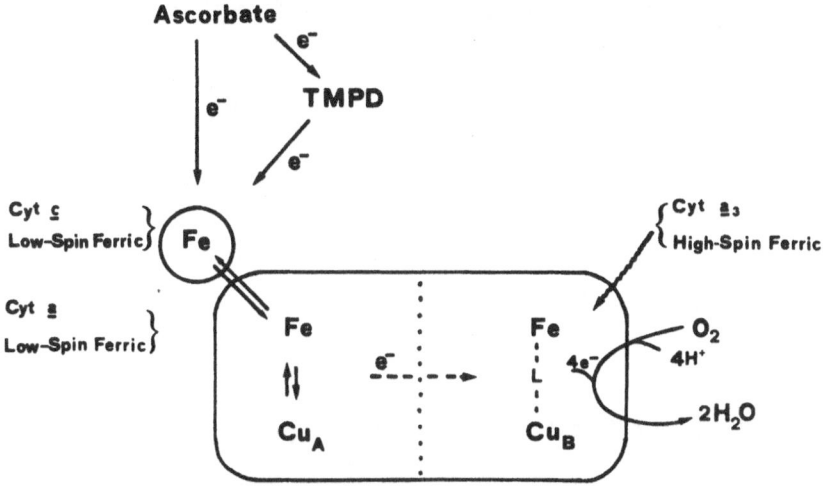

Fig. 1.   Schematic view of the functional relationships between the
         four metal centers in cytochrome c oxidase.
         Electrons are transferred from the first two sites
         (Fe_a – Cu_A) to the other two (Fe_{a3} – Cu_B), by an intramo-
         lecular reaction which is the slowest step in the overall
         process catalyzed by the enzyme.
         L indicates the still unknown ligand bridging cytochrome
         a and Cu_B, which are antiferromagnetically coupled in the
         oxidized resting enzyme.

transient kinetics is given in Fig. 2 (together with the parameters
which apply under the conditions of the experiments with the solubilized enzyme).

      The assumptions on which this minimal scheme is based are as
follows (see also 12):

   i) The "basic functional unit" of oxidase is the monomer contai-
      ning four metal centers, as also apparent from the overall
      stoichiometry with cytochrome c and dioxygen given in eq. 1.
  ii) Electrons enter the unit through cytochrome a in a bimolecular

A   MINIMAL SCHEME

$$c^{2+} + \left[a^{3+} Cu_A^{2+}\ Cu_B^{2+}\ a_3^{3+}\right] \underset{k_{-1}}{\overset{k_1}{\rightleftharpoons}} c^{3+} + \left[a^{2+}\ Cu_A^{2+}\ Cu_B^{2+}\ a_3^{3+}\right]$$

$$\left[a^{2+}\ Cu_A^{2+}\ Cu_B^{2+}\ a_3^{3+}\right] \underset{k_{-2}}{\overset{k_2}{\rightleftharpoons}} \left[a^{3+}\ Cu_A^{+}\ Cu_B^{2+}\ a_3^{3+}\right]$$

$$c^{2+} + \left[a^{3+}\ Cu_A^{+}\ Cu_B^{2+}\ a_3^{3+}\right] \underset{k_{-3}}{\overset{k_3}{\rightleftharpoons}} c^{3+} + \left[a^{2+}\ Cu_A^{+}\ Cu_B^{2+}\ a_3^{3+}\right]$$

$$\left[a^{2+}\ Cu_A^{+}\ Cu_B^{2+}\ a_3^{3+}\right] \overset{k_4}{\longrightarrow} \left[a^{3+}\ Cu_A^{2+}\ Cu_B^{+}\ a_3^{2+}\right]$$

$$2c^{2+} + O_2 + \left[a^{3+}\ Cu_A^{2+}\ Cu_B^{+}\ a_3^{2+}\right] \overset{k_5}{\longrightarrow} 2c^{3+} + \left[a^{3+}\ Cu_A^{2+}\ Cu_B^{2+}\ a_3^{3+}\right] + 2H_2O$$

$$c^{3+} + \text{Ascorbate} \overset{k_6}{\longrightarrow} c^{2+} + \text{Products}$$

B   PARAMETERS : $k_1 \approx k_{-1} \approx k_3 \approx k_{-3} = 5 \times 10^6\ M^{-1}\ s^{-1}$

$k_2 \approx k_{-2} \approx 50\ s^{-1}$

$k_4 \approx 0.5 \div 40\ s^{-1}$; $k_5 = $ FAST

Fig. 2.   A) Minimal scheme representing the various steps of the
electron transfer from cyt $\underline{c}^{2+}$ to O_2, catalyzed by
cytochrome $\underline{c}$ oxidase

B) Numerical values of the basic rate constants used for
simulation (see for example fig. 4).

mode, and the value of the bimolecular rate constant as obtained from stopped-flow and T-jump experiments[10,11,13] is
known to approach at low ionic strength (0.01 M) the diffusion
controlled limit of ~ $10^8 M^{-1} s^{-1}$.[14]

iii) Cytochrome $\underline{a}$ is in reversible redox equilibrium with another
site in the functional unit, most probably Cu_A. Since this
equilibrium is not far from unity,we have taken the rate
constants in both directions to be the same and identical to
that measured by T-jump experiments on mixed-valence CO
oxidase.[13]

iv) Cytochrome a_3 (and Cu_B) are reduced intramolecularly by an internal electron transfer process which is the rate limiting step in the turnover of the enzyme. Its rate can be measured from the associated reconstitution of the CO binding capacity[10] and has been shown to depend largely on the experimental set-up.[15]

v) Fully or partially reduced cytochrome oxidase reacts very rapidly with molecular oxygen, and cytochrome a_3 must be reduced for the oxygen reaction to occur. The reaction with O_2 is written as a single-step irreversible process because the intermediates are known to be extremely short-lived at room temperature.[7]

FUNCTIONAL STATES OF CYTOCHROME OXIDASES

It is recognized that the fully oxidized form of cytochrome oxidase may exist in different spectroscopic forms, as shown by optical absorption, and more recently under controlled conditions by EPR spectroscopy of oxidase or its complexe(s) with NO (see 16). Among the different forms, it was shown by Antonini et al.[15] that the product of the reaction of the fully reduced enzyme with O_2 is functionally and spectrally different from the fully oxidized enzyme as obtained after preparation or storage. These species were given the name of "pulsed" and "resting" oxidases.

The reaction of reduced oxidase with O_2 is largely complete within the "dead time" ($\sim$ 4 ms) of the stopped-flow instrument; thus pulsed and resting oxidase can be compared relative to phenomena reflected by the various kinetically discernible phases in the time course of the reaction with cyt c.[17]

In this type of kinetic experiments, it was found that the behaviour of pulsed and resting oxidases is not very different in the first phase (electron transfer between cytochrome c and cytochrome a); however the two forms are obviously different in the electron transfer to oxidized cytochrome a_3. Following the development of the CO binding capacity by cytochrome a_3 as a probe of its reduction, this process was estimated to be 4-5 times faster in the pulsed than in the resting form. As a result of the more efficient intramolecular electron transfer, the oxidation level of cytochrome c at steady state is higher for the pulsed enzyme.

The amount of pulsed oxidase is strictly related to the amount of available oxygen. A stoichiometric ratio of 1:1 between oxygen and the functional unit of oxidase could be demonstrated by kinetic titrations in which the reduced enzyme was mixed with different O_2 concentrations.[17] This result, taken together with the finding that 4 moles of reduced cytochrome c were oxidized per mole of pulsed oxidase formed, indicates that the pulsed enzyme contains four oxidizing equivalents.

The absorption spectrum of the pulsed enzyme, though somewhat similar to the oxidized derivative, shows distinct features, with characteristic differences both in the absorption maximum and in the extinction coefficient. These differences are observed both in the α-band and in the Soret region, where, similarly to the oxygenated species described by Okunuki and its school[18] and by Orii and King,[19] the maximum of pulsed oxidase is at 428 nm (while the resting one absorbs at 418 nm).

Using a "rapid freezing" technique, Beinert and his associates[20] reported the identification of a new intermediate trapped after mixing reduced oxidase and O_2. This species shows an intense EPR signal with g=5, 1.78 and 1.69. The intensity of the signal is dependent on O_2 concentration at substoichiometric O_2, and the EPR of oxidized Cu_A and cytochrome a are both observed. These EPR features are crucial to the assignment of the oxidation state of the metal centers in pulsed oxidase. After substitution of $^{17}O_2$ for $^{16}O_2$ in the reaction mixture, no evidence could be found for a broadening of the new signal, and thus for the possible role of O_2 in the electronic arrangement of the binding site (although this conclusion is not necessarily definitive).

Pulsed oxidase was initially obtained using the beef heart enzyme in detergent, but it has now been described using oxidase from a number of other species (i.e. camel, rat, shark and Neurospora oxidases). It has been found that not only solubilized oxidase, but also the Keilin-Hartree particles as well as the enzyme reconstituted into proteoliposomes can be "pulsed".[21,22] Moreover its functional properties, at least to a first approximation, are the same irrespective of the nature of the reductant used (i.e. whether dithionite, ascorbate and cytochrome c, NADH and phenazine methosulphate or endogenous reductants). Finally Brunori et al.[23] have shown that O_2 is not unique in producing pulsed oxidase, but also ferricyanide can be used with functionally similar results. This finding indicates, together with other results, that the difference between resting and pulsed oxidase is of conformational origin.

A TWO STATE MODEL

According to our interpretation,[12] both resting and pulsed oxidase are catalytically competent, i.e. both forms of the enzyme can catalyze the oxidation of cytochrome c by molecular oxygen. The different catalytic efficiency of these two states has been related to the different value of the rate limiting step, i.e. the electron transfer to the binuclear coupled metal center where O_2 is bound (see Fig. 1).

The notion of two functionally distinct states of the enzyme has been incorporated into a scheme, which is depicted in Figure 3. This model was inspired by the classical two state allosteric scheme

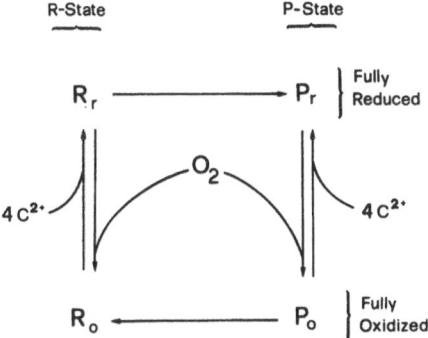

Fig. 3. "Square" diagram of the two-state model for cytochrome $\underline{c}$
oxidase (from 12, modified). The "circulation" of the
enzyme through the four corners of the square is induced
by the reactions with the reducing (cyt $\underline{c}^{2+}$) and the oxidizing (O_2) substrates. The transition $R_r \longrightarrow P_r$ and
$P_o \longrightarrow R_o$ are described as quasi-irreversible, to represent the higher stability of the pulsed form in the reduced
state and of the resting form in the oxidized state.

proposed by Monod, Wyman and Changeux[24] for the interpretation of
interaction phenomena in enzymes and respiratory proteins.

The main features of the scheme are summarized in the legend
to Figure 3. Each state of oxidase undergoes a full redox cycle,
according to Eq. (1) and Fig. 2, and the two states are interconvertible. The following comments to the scheme are in order:
(a) This is admittedly an oversimplified model for such a complex

molecule, but it is still realistic, as shown by its capability to describe some of the kinetic features of cytochrome oxidase in solution.

(b) It demands that also the fully reduced enzyme (as well as the various oxidation intermediates, not shown) should exist in a resting (R) and pulsed (P) configuration, as indicated.

(c) The difference between the two states is of conformational origin, and this contention is supported by several experimental results (such as the possibility to freeze the molecule in the resting state by covalent cross-linking).

(d) The population of the enzyme is distributed among the two states following dynamic as well as equilibrium considerations. The only information at present available indicates that the pulsed state prevails in the fully reduced enzyme, while the resting one is the stable state in the fully oxidized species.

A detailed presentation of this scheme has been published by Wilson et al.[12] Extensive computer simulation has made it possible to describe the kinetics of the oxidase starting from the resting and pulsed conditions. An example of the comparison between computed and experimental time course of reaction is presented in Fig. 4; the quantitative agreement is more than satisfactory.

The rate of interconversion has been estimated only for the fully reduced and fully oxidized species. The rate constant for the $R_r \longrightarrow P_r$ transition has been calculated from the computation; owing to its competition with the binding of O_2, which is a very fast process, activation to the P state is a slow process during turnover, in spite of the relatively high value of the relevant rate constant $(k(R_r \longrightarrow P_r) \simeq 10^2 - 10^3 \ s^{-1})$. The opposite process, $P_o \longrightarrow R_o$, corresponds spectrally to the transition of the so-called "oxygenated" oxidase discovered by the Japanese School.[18] It is a very slow transition which occurs through more than one spectral intermediate,[19] and whose half time may be varied from $\sim$ 1 min to over 1 hour (depending on the experimental conditions) (unpublished work). The understanding of the parameters which control these decay processes are crucial to the realistic evaluation of the general physiological significance of the two state model illustrated above.

CONCLUDING REMARKS

The picture of cytochrome oxidase introduced here is that of a macromolecular complex existing in (at least) two different and interconvertible states, the resting and the pulsed form, both catalytically competent. A strong element in favour of this interpretation is that the kinetic behaviour of the pulsed form has been observed in all the properly devised experiments of oxygen combination under very different experimental situations (see above).

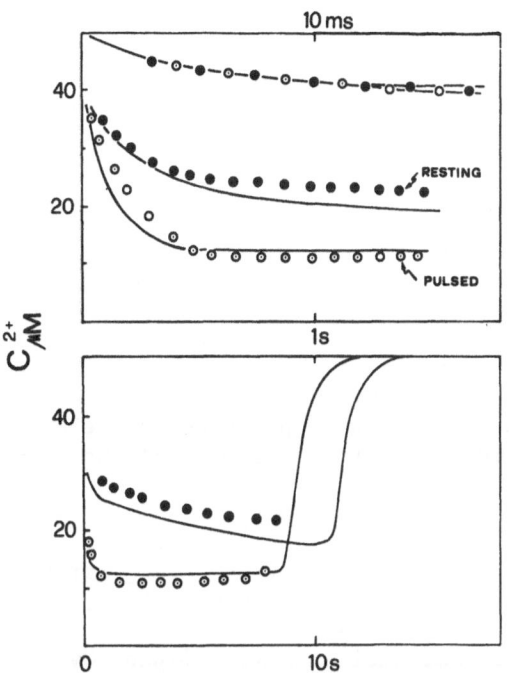

Fig. 4. Comparison between experimental data and simulated time
courses for the reaction with cytochrome c^{2+} (50 μM) of
resting (●) and pulsed (O) cytochrome oxidase (10 μM) in
the presence of O_2 (135 μM), ascorbate (5 mM) and TMPD.
The solid lines have been calculated on the basis of the
minimum scheme and the kinetic constants depicted in Fig.2,
assuming that the difference between resting and pulsed
oxidase resides in the rate constant for the intramolecular
electron transfer, k_4 (which is 5s^{-1} for resting and 30 s^{-1}
for pulsed respectively). Notice, in the upper panel, the
two different abscissas, the millisecond time scale referring to the uppermost curve.

The involvement of pulsed oxidase in the catalytic cycle is
demanded for in any future attempt to elucidate the complex physiological role of cytochrome oxidase. Moreover, the bulk of functional
data in the literature, which are often interpreted in terms of a
single molecular population, should be properly reinterpreted taking
into account the presence of, at least, two active conformers. The
proposed model (Fig. 3), in which the dynamic distribution of oxidase
between two catalytic states is coupled to the electron transfer
process, may have obvious implications as a regulatory mechanism.
Its in vivo significance needs to be assessed.

A natural extension of the studies reported in this paper will
be directed to assess the role of possible effectors on the distribution of oxidase between states. Among possible effectors the role of
protons is fascinating because it has been demonstrated that pH has
a definite influence on the catalytic activity of the soluble enzyme.
Since our ultimate interest in the enzyme lies in its role in energy
transduction, and accepting the view that cytochrome oxidase acts as
a proton pump through the mitochondrial membrane, the transduction
ought to be shown by proving coupling between electron and vectorial
proton transport in the different states of the enzyme.

REFERENCES

1. R. A.Capaldi, Structure of Cytochrome c oxidase, in "Membrane
    Proteins in Energy Transduction", R. A. Capaldi, ed., Dekker
    Inc., New York (1979).
2. M. Brunori, E. Antonini and M.T. Wilson, Cytochrome Oxidase:
    an overview of recent work, in: "Metal ions in biological
    systems", vol. 13, H. Sigel ed., Dekker, New York (1981).
3. B. G. Malmström, Cytochrome c Oxidase: structure and catalytic
    activity, Biochim. Biophys. Acta, 549: 281 (1979).
4. M. Wikström, K. Kraab and H. Saraste, Proton-translocating
    cytochrome complexes, Ann. Rev. Biochem. 50: 623 (1981).
5. W. E. Blumberg and J. Peisach, Some possible chemical and
    electronic states of cytochrome c oxidase and its interme-
    diate redox states, in: "Cytochrome oxidase", T. E. King,
    Y. Orii, B. Chance and K. Okunuki eds., Elsevier/North
    Holland, Amsterdam (1979).
I. T. Yonetani, Studies on Cytochrome Oxidase I. Absolute and
    difference absorption spectra, J. Biol. Chem. 235: 845 (1960).
7. C. Greenwood and Q. H. Gibson, The reaction of cytochrome c
    oxidase with oxygen, J. Biol. Chem. 242: 1782 (1967).
8. B. Chance, C. Saronio and J. S. Leigh Jr., Functional interme-
    diates in the reaction of membrane-bound cytochrome oxidase
    with oxygen, J. Biol. Chem., 250: 9226, (1975).
9. B. Reinhammar, R. Malkin, P. Jensen, B. Karlsson, L. Andréasson,
    R. Aasa, T. Vänngard and B. G. Malmström, A new copper (II)
    electron paramagnetic resonance signal in two laccases and
    in cytochrome c oxidase, J. Biol. Chem. 255: 5000 (1980).

10. Q. H. Gibson, C. Greenwood, D. C. Wharton and G. Palmer, The reaction of cytochrome oxidase with cytochrome c, J. Biol. Chem., 240: 888 (1965).

11. M. T. Wilson, C. Greenwood, M. Brunori and E. Antonini, Kinetic studies on the reaction between cytochrome c oxidase and ferrocytochrome c, Biochem. J., 147: 145 (1975).

12. M. T. Wilson, J. Peterson, E. Antonini, M. Brunori, A. Colosimo and J. Wyman, A plausible two-state model for cytochrome c oxidase, Proc. Natl. Acad. Sci. U.S.A., 78: 7115 (1981).

13. C. Greenwood, T. Brittain, M. Wilson and M. Brunori, Studies on partially reduced mammalian cytochrome oxidase reactions with ferrocytochrome c, Biochem. J., 157: 591 (1976).

14. K. J. H. Van Buuren, B. F. Van Gelder, J. Wilting and R. Braams, Biochemical and biophysical studies on cytochrome c oxidase. The reaction with cytochrome c as studied by Pulse Radiolysis, Biochim. Biophys. Acta, 333: 421 (1974).

15. E. Antonini, M. Brunori, A. Colosimo, C. Greenwood and M. T. Wilson, Oxygen "pulsed" cytochrome c oxidase: Functional properties and catalytic relevance, Proc. Natl. Acad. Sci. U.S.A., 74: 3128, (1977).

16. C. T. Martin, R.H. Morse, R. M. Kanne, N. B. Gray, B. G. Malmström and S.I. Chan, Reactions of nitric oxide with tree and fungal laccase, Biochemistry, 20: 5147 (1981).

17. M. Brunori, A. Colosimo, G. Rainoni, M. T. Wilson and E. Antonini, Functional intermediates of cytochrome oxidase. Role of "pulsed" oxidase in the pre steady state and steady state reactions of the beef enzyme, J. Biol. Chem., 254: 10769 (1979).

18. K. Okunuki, Cytochrome oxidase, in: "Oxygenases", O. Hayaishi, ed., Academic Press, New York (1962).

19. Y. Orii and T. E. King, On the nature of the three intermediate species formed after reaction of reduced cytochrome oxidase with oxygen, J. Biol. Chem., 251: 7487 (1976).

20. R. W. Shaw, R. E. Hansen and H. Beinert, The oxygen reactions of reduced cytochrome c oxidase. Position of a form with an unusual EPR signal in the sequence of early intermediates, Biochim. Biophys. Acta, 548: 386 (1979).

21. C. Bonaventura, J. Bonaventura, M. Brunori and M. T. Wilson, Functional studies on crosslinked bovine cytochrome c oxidase, FEBS Letters, 85: 30 (1978).

22. M. Brunori, A. Colosimo, P. Sarti, E. Antonini, W. Lalla-Maharajh and M. T. Wilson, Activation of cytochrome c oxidase, in "Proc. of the third International Symposium on oxidases and related oxidation-reduction systems", Albany New York, (1979).

23. M. Brunori, A. Colosimo, P. Sarti, E. Antonini and M. T.Wilson, "Pulsed" cytochrome oxidase may be produced without the advent of dioxygen, FEBS Letters, 126: 195 (1981).

24. J. Monod, J. Wyman and J.P. Changeux, On the nature of allosteric transitions: a plausible model, J. Mol. Biol., 12: 88 (1965).

NON-HEME COMPONENTS OF ELECTRON TRANSFER SYSTEMS:

INTERACTIONS AND CONFORMATIONAL EFFECTS

Helmut Beinert

Institute of Enzyme Research
University of Wisconsin
Madison, 53706 USA

INTRODUCTION

Traditionally, in our thinking of electron transfer systems, cytochromes have had dominant attention. By contrast, at this meeting, in which heme proteins will duly receive prime consideration, it may be useful also to discuss features of some of the biological electron transfer agents that were more recently discovered, such as Fe-S proteins, ubiquinone, and copper. Their very properties have forced us to apply approaches different from those suitable for heme proteins; the EPR technique has largely replaced spectrophotometry in their study; in turn, EPR spectroscopy then also applied to cytochromes has greatly enriched our knowledge of their properties and behaviour. In addition, EPR spectroscopy, which allows us to study molecules in what is essentially their ground state, is more suited than spectrophotometry for detecting subtle interactions between neighboring groups or sites. This aspect will form an essential part of my presentation.

SPIN-SPIN INTERACTIONS

Iron Sulfur Proteins

There seems to be no end in sight yet to the discovery of new Fe-S proteins, new types of Fe-S clusters and to the extent and scope of their involvement in biological functions; but even a look at a more limited area, such as the well studied and publicized respiratory chain of mitochondria, can give us an introductory lesson as to the variety of components and interactions between them that nature has evolved. In heart mitochondria, which do not possess the number of auxiliary metabolic systems as does liver or

123

kidney, we count 8 to 11 individual Fe-S components and in addition
two different mitochondrial Fe-S clusters, which are not part of
the respiratory chain, not to mention some subspecies which have
been resolved recently (Table I) (1, 2). A decision between the
numbers 8 and 11 is not possible at this time. Of these, say, 8 Fe-S
components of the respiratory chain only one is associated with
cytochromes, namely the Fe-S protein of Complex III (the "Rieske"
Fe-S protein) (3) (midpoint oxidation-reduction potential, E_0' pH
7.2: + 280 mV), the remaining seven which are of lower midpoint
potential ($\leq$ 40mV), are associated with the flavoproteins that
channel electrons from the principal substrates, namely succinate,
NADH and fatty acids, into the ubiquinone (UQ)-cytochrome system.
Little is known about one (or two) additional Fe-S cluster(s)
which is (are) located in the outer mitochondrial membrane, and
curiously, one Fe-S cluster in the mitochondrial matrix is part
of a non-oxidative enzyme, namely aconitase (4). Of all these
clusters roughly one half to two thirds seem to be [2Fe-4S] clusters,
and one third to one half [4Fe-4S] clusters and one, namely that of
aconitase is a [3Fe-3S] cluster, at least in the enzyme, as purified
(5). Among the [4Fe-4S] clusters there is one that utilizes the
upper two oxidation levels (+3 and +2) (6, 7), typical for Chromatium vinosum "High Potential" Fe-S protein (HiPIP) (8), while
others use the lower oxidation state (+2 and +1) typical for ferredoxins (Fds). Thus far the individual Fe-S proteins or clusters
referred to above have not been observed to use all three oxidation
states in their biological function. As we can see, there is a
variety of structures and behaviour.

One of the reasons why we are not certain about the total
number of clusters present in mitochondria is that there seem to be
Fe-S clusters, which are not at all or not under all circumstances
detectable by EPR (9 - 11). Thus there is some controversy as to
whether succinate dehydrogenase contains two or three Fe-S clusters.
If there are three, one of them would have to be EPR silent in the
more intact preparations. The alternative view is that the third
cluster, which is only observed in purified preparations, has
arisen through some artefact (12). This third and disputed cluster
(named cluster 2) is not reducible by substrate, but can be reduced by dithionite or other strong reducing agents. When this is
done, the spin relaxation of the reduced cluster 1 - a cluster
readily seen in all preparations on reduction with succinate -
is drastically increased (9, 13 - 15) (Table II). Thus it is obvious
that either cluster 2, when reduced, interacts with cluster 1, which
leads to increased spin relaxation, or, if cluster 2 does in fact
not exist, a conformational rearrangement in the enzyme would have
to bring about the increased spin relaxation of cluster 1 (12).
Other evidence has been brought forth by Ohnishi and her colleagues
(13, 15) that there is indeed interaction between the two Fe-S
clusters 1 and 2. A so called half-field signal was detected at

Table 1. Known Iron-Sulfur Clusters of Beef Heart Mitochondria

	NADH - UQ	Succinate - UQ	ETF - UQ	UQ - c_1	outer membrane	citric acid cycle/mito-chondrial matrix
Segment of respiratory chain in which Fe-S cluster operates a) or location						
"Complex" of respiratory chain b)	I	II	-----	III	-----	1
Number of individual clusters	4 - 6	2 - 3	1	1 - 2	2	1
Type of cluster	[2Fe-2S]c)(2+; 1+) d) [4Fe-4S]	1-2 [2Fe-2S] (3+; 2+) d) 1 [4Fe-4S]	[4Fe-4S](2+; 1+) d)	[2Fe-2S]	[2Fe-2S]	[3Fe-3S]
Midpoint oxidation-reduction potential at pH 7.2 e)	-20 to -400 mV	+30 to -400 mV (~+100 mV for HiPIP)	+ 40 mV	+280 mV	-----	~+100 mV

a) This does not imply that the compounds tested are the direct electron donors or acceptors of the cluster in question.

b) Refers to the usual trivial nomenclature.

c) Contradictory information exists on the number of each cluster type; therefore no numbers are given.

d) According to the accepted nomenclature of Fe-S proteins the oxidation state in which a cluster can occur is indicated by the upper numbers in parethesis following the brackets. Only 4Fe-4S clusters have been observed to assume either the (2+; 1+), i.e. the Fd type or the (3+; 2+), i.e. the HiPIP oxidation levels.

e) These values give only an approximate range. The values depend on the type of preparation used. The original literature (or Ref. 9) has to be consulted.

Table 2. Saturation with Microwave Power of Iron-Sulfur Centers of Succinate Dehydrogenase after Reduction with Succinate Dithionate

Preparation	condition	Temperature	Microwave power	Intensity of Fe-S signal at g = 1.94		
				reduced with succinate, A (center 1)	reduced with dithionite, B (center 1 + 2)	difference B - A
		°K	μW	% of concentration of bound flavin		
1	a	90	2700	73	103.5	30.5
	b	33	270	78	106.5	28.5
	c	9.5	9	81	118	37
	d	9.5	900	16	64.5	48.5 *
2	a	33	270	115	156	41
	b	13	27	108.5	164	55.5
	c	9.5	27	95	137	42
	d	9.5	900	23	90.5	67.5 *

* Note that at high power and low temperature the apparent contribution of B - A rises to values ~1.5 times higher than those expected from measurements at non-saturating conditions (A, condition a, b, c), showing that with both centers reduced, the contribution of center 1 under saturating conditions must be higher than shown under A, condition d, when only center 1 is reduced.

g=3.9 which is typical for spin-spin interaction in a triplet state. Similarly, evidence was obtained for interaction between flavin-semi-quinone and a Fe-S cluster of NADH dehydrogenase (16). While these indications of spin-spin interactions are unambiguous, the interacting partners are not clearly defined at this time.

If I may take the liberty of digressing from the respiratory chain for a moment: We have, in a bacterial dehydrogenase system, encountered a most striking example of spin-spin interaction which also furnishes information on a conformational change in the enzyme as well as on the path of electron transfer between the electron carriers of that enzyme. The enzyme in question is trimethylamine (TMA) dehydrogenase from a methylotrophic bacterium (17,18), a Fe-S flavoprotein, containing a single FMN and Fd type [4Fe-4S] (+2; +1) cluster. On reduction with TMA the FMN in this enzyme is immediately ($t_{1/2}$ <2 msec) reduced to $FMNH_2$ followed by a slow ($\leq$ 100 msec) transfer of one electron to the Fe-S cluster. This can readily be observed by EPR in that a strong triplet signal appears at g=2 and at half-field, while signals of the flavin semiquinone and the reduced Fe-S cluster, which one would have expected, are hardly detectable (Fig. 1). These observations today constitute one of the most explicit examples for a plausible mode of electron transfer in Fe-S flavoproteins. However, the triplet EPR signal is able to provide additional information in that the shape of this EPR signal changes with time as the enzyme is exposed to substrate. The zero-field splitting parameter D increases, which means an increase in the strength of the interaction (19).

We can conclude, therefore, that either the mutual orientation of the interacting partners or their distance changes, or both. Simulation of the spectra and calculations indicate that the center to center distance of the partners must be $\leq$10 Å with the edge to edge distance being <5 Å . We interpret these results to mean that additional TMA is bound to the enzyme as effector, which brings about the conformational change reflected in the increased splitting of the EPR lines (cf. Fig. 1). This idea is supported in that tetramethylammonium chloride, which is not a substrate but an inhibitor, is able to produce an analogous effect on the EPR spectrum as does TMA (18).

Ubiquinone

Returning to the mitochondrial electron transfer system I would like to mention the most impressive example of spin-spin interaction which we have observed there. Those not intimately concerned with the respiratory chain may be reminded that the most abundant electron carrier in this system, the one with the highest total capacity for electron uptake, is ubiquinone (UQ), if we omit NAD from consideration here. Interest has always been great, and in fact still is, as to whether and when UQ carries out one- or two-electron

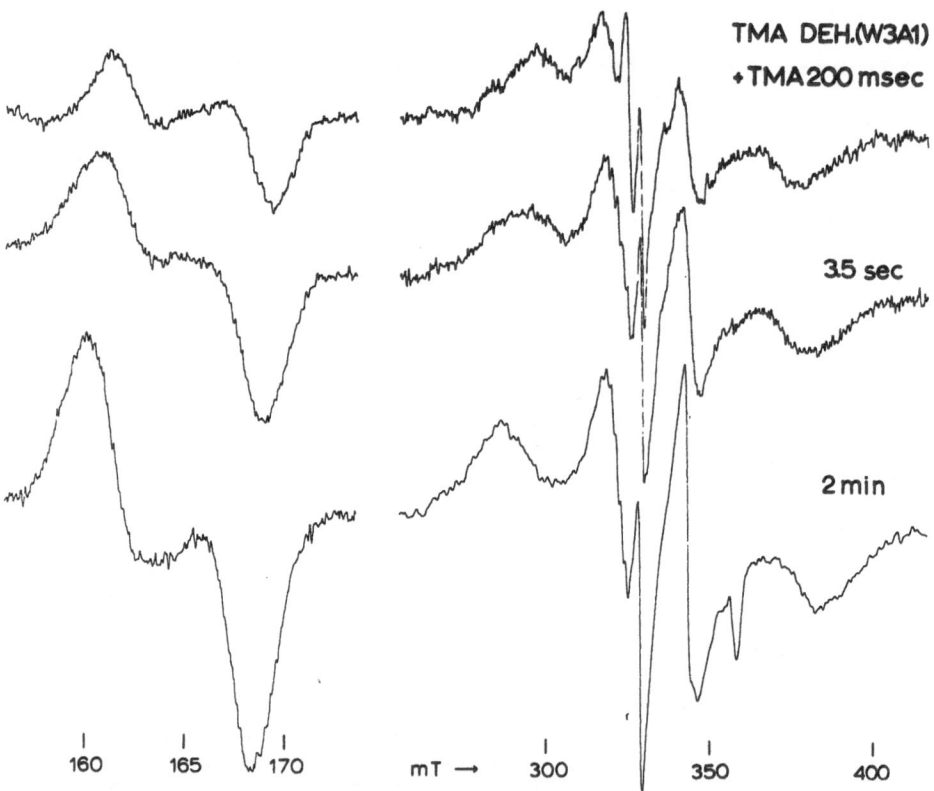

Fig. 1 Comparison of EPR signals of trimethylamine dehydrogenase
        of bacterium W3A1 when exposed to TMA for various lengths
        of time at 17°. Top and center: with 0.1 mM TMA for 0.2 sec
        and 3.5 sec, respectively; bottom: with 2 mM TMA for 2 min.
        The EPR spectra were recorded at 9.2 GHz, 0.8 mT modulation
        amplitude and 12 K and with 9 mWatt power at low field
        (left) and 2.7 mW at high field (right). Up to 9 scans
        were averaged by a signal averager. Note that the ampli-
        tudes of the signals at high field would be 1.5 fold higher,
        if microwave power and gain had been identical for low and
        high field recording.

transfer (20). EPR signals for UQ have been observed (21), but they
were weak and measurements on free quinones in solution indicated
that dispropotionation of the semiquinone (SQ) to quinone and quinol
is very high so that formation of stable SQ could not be expected
(22). Evidence has been presented in recent years that there are
UQ binding proteins associated with the complexes of the respiratory
chain which are capable of binding UQ and stabilizing SQ formation (23).
The question as to the specificity of this binding, however, is
still open. An unexpected finding, therefore, was that of a USQ
pair connected with the electron transfer system of mitochondria
(24). This was detected by its unusual EPR signal (Fig. 2). This
signal is prominent in partly reduced tissue samples, mitochondria
and submitochondrial particles (SMP) and has long resisted a plausible interpretation. One of the obstacles was the overlap and apparent association of the typical EPR signal with that due to the
HiPIP-type $[4Fe-4S]^{(+3; +2)}$ cluster of succinate dehydrogenase. The
unknown EPR signal could be shown to involve USQ by extraction of
UQ from, and restoration into SMP (24). Analysis and simulation of
the EPR spectrum indicated that it is due to a special pair of USQ
molecules, which apparently are close to the HiPIP-type Fe-S cluster
of succinate dehydrogenase (25). This was concluded from the spin
relaxation behaviour of the EPR signal of the USQ pair (24, 25),
which is that of a transition metal ion rather than of an organic
radical, and from the measured midpoint potentials. On the assumption
that the interaction between the USQ molecules of the pair is essentially dipolar in nature - as indicated by the simulations - a
distance of 7 Å between partners was calculated (24). This USQ pair
seems to be specific for the pathway of electrons from succinate
via succinate dehydrogenase to UQ; it is not seen with isolated
soluble succinate dehydrogenase or succinate UQ reductase (Complex
II); however, oxidation of reduced Complex II with UQ produces
strong signals of the pair (26). The concentration of the USQ pair
comes close to that of succinate dehydrogenase in mitochondria,
indicating that a considerable stabilization of USQ is reached in
this specific pair.

CONFORMATIONAL EFFECTS; REDUCTIVE ACTIVATION

Iron-Sulfur Proteins

There are two fairly long known samples of interactions among
redox-active sites in components of the respiratory chain. The Fe-S
protein of the UQH_2-cytochrome c reductase (Complex III), or better
one of the subspecies of this Fe-S protein undergoes an EPR detectable change in state during reduction of Complex III (1,27). This
occurs when the midpoint potential is lowered toward 40-60 mV
(pH 7.2). The range of potential when the change of state of the
Fe-S protein occurs is pH dependent in the same way as the potential
of the fumarate-succinate couple. It was suggested, therefore, that
the redox state of UQ might be involved in the change observed with

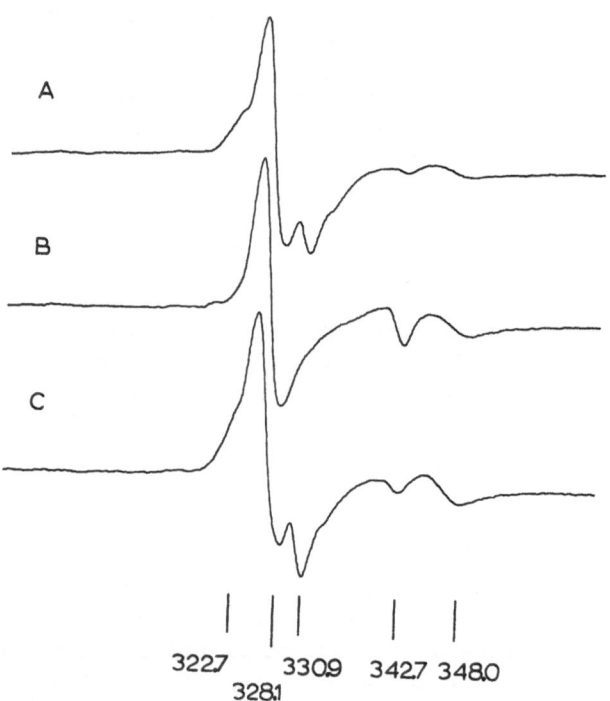

Fig. 2   EPR signals in the g=2 region of samples of lyophilized,
         extracted and reconstituted submitochondrial particles (30
         mg protein per ml), partly reduced with dithionite. A,
         lyophilized ETP; B, lyophilized, pentane extracted ETP; C,
         lyophilized, pentane extracted ETP after reconstitution with
         UQ-10. The conditions of EPR spectroscopy were those of
         Fig. 1 (2.7 mwatt). The field positions of prominent peaks
         in the spectra are indicated in mT.

the Fe-S component. Most recently this notion found independent support from work in which single electron transfers could be induced in mitochondrial complex III by coupling to photochemical centers of <u>Rhodospseudomonas sphaeroides</u>. It was concluded from these experiments that a unique ubiquinone, Q_Z, not part of the UQ pool, specifically interacts with the Fe-S cluster, thus producing the described effect and controlling the redox behaviour of the Fe-S cluster (28). This interaction may assume particular significance with respect to the question of one- versus two -electron transfer by UQ in the cytochrome $\underline{b}-\underline{c}_1$ region of the electron transfer system.

It is also well known that succinate dehydrogenase requires activation to reach maximal catalytic activity. A number of agents, such as succinate, malonate, nitrate or phosphate are able to convert the enzyme into the active conformation. It has been shown that on titration with dithionite activation ensues when the bound flavin of the enzyme is reduced (29). This is one of the longest known examples of "reductive activation", a phenomenon which we will encounter in two other examples in what follows below.

Yet another EPR signal in mitochondria has led us to recognize one of these additional examples of reductive activation and has also revealed conformational effects. In EPR spectra of oxidized mitochondria there is, associated with the HiPIP type signal of succinate dehydrogenase (6, 7), a similar, overlapping EPR signal, which belongs to a Fe-S component of electron spin relaxation slower than that of the succinate dehydrogenase cluster (30). This component could be solubilized from mitochondria by sound treatment (31) and was shown, eventually, to be the Fe-S cluster of the long known citric acid cycle enzyme aconitase (4). While aconitase has been found to contain iron and labile sulfide a few years ago (32, 33), the properties and function of its Fe-S cluster had not been investigated. It is now known that aconitase, as obtained on routine purification, contains a [3Fe-3S] cluster (5). The question emerged: What is the meaning and function of a Fe-S cluster in an enzyme, which is not involved in oxidation-reduction? Purified aconitase is inactive and requires exposure to some activating agent(s) to regain activity (34, 35). It is characteristic of all activating mixtures that they contain one or more reducing agents such as Fe^{2+}, ascorbate, or thiols. We could show by EPR spectroscopy that any of the activating mixtures reduced the Fe-S cluster of the enzyme and that reduction of the cluster was necessary for reactivation of inactive enzyme (36). However, mere reduction does not bring about activity of the inactivated enzyme. A certain period, the length of which highly depends on temperature, has to elapse after reduction or reoxidation, until full activity is reached or until all activity has ceased, respectively. Thus, it became obvious that the state of the Fe-S cluster in this enzyme determines whether the enzyme can assume an active or an inactive conformation. I say "an active" or "an inactive" form, because we have observed by EPR and Mössbauer spectroscopy

that   the Fe-S cluster may occur in several states that allow or
disallow activity.* We would then assume that ligand groups supplied
by the protein must be involved in stabilizing the various states of
the cluster that can occur, and that these ligand groups must be
organized in different ways for different states of the cluster.
We should add the caveat here, that under physiological conditions
inactive forms may not readily or possibly never arise.

Cytochrome Oxidase

One of the prime examples of interactions between electron
carriers and the involvement of conformational changes among components of the respiratory chain remains cytochrome c oxidase. My
mentioning of this enzyme in a lecture of non-heme components can
at least partly be justified by the fact that two of the four metal
components of the enzyme are copper atoms. Since at this Symposium
we will hear at least two lectures exclusively devoted to cytochrome
oxidase, I shall only briefly summarize those aspects where interactions between electron carriers and conformational effects have
become obvious. While there is ample evidence - from criteria such
as sensitivity to proteolytic enzymes (37), electron microscopy (38),
from the appearance of a ferric high spin EPR signal on reduction
(39) and from changes of oxidation-reduction midpoint potentials
during reduction (40) - that the conformation of the reduced enzyme must differ from that of the oxidized form, it became clear
in the recent past that the oxidized enzyme itself can occur in an
unexpected variety of states (41-47). As we will probably hear, one
of the present problems is, which of these states are naturally
occurring and which of these again is (or are) involved in active
turnover of the enzyme. With an enzyme operating at the speed at
which cytochrome oxidase reacts, this is no trivial matter. Most
of the studies on cytochrome oxidase understandably, but maybe unfortunately, were carried out on, or involving the oxidized, resting form of the enzyme, as it is obtained on purification. This is
the form that shows EPR signals for two of its metal components
only, which led us to suggest years ago that the EPR silent components interact with each other so as to form a diamagnetic coupled
species (39). Subsequently, this has been made very probable by
measurement of magnetic susceptibility over a broad range of temperatures (49, 50). We also know now that this resting form of the
enzyme is a relatively unreactive species with respect to ligands
as well as toward reduction by its substrate (51, 52).

Since two of the lectures at this Symposium will deal with
early reaction products of the reduced enzyme with oxygen, I will
discuss here only forms of the oxidase which can exist following
either complete reaction with oxygen in a previous turnover, i.e.

---

* T.A. Kent, J.-L. Dreyer, M.C. Kennedy, E. Münck and H. Beinert,
unpublished

after formation of two molecules of water, or after anaerobic reoxidation by chemical oxidants. To such species also belongs the
classical "oxygenated" form discovered years ago by Okunuki and
his collaborators (53), which is now known not to involve oxygen
and which in fact includes a number of different forms of the oxidized enzyme. Some of these have been described by Tiesjema et al.
(42) and by Orii and King (41). The oxygen "pulsed" oxidase of
Antonini et al. (45), which is obtained in a way similar to the
"oxygenated" form, namely by reoxidation of reduced oxidase, most
likely then also contains different components and one must ask the
question, whether all of these contribute to the high reactivity
with ferrocytochrome $\underline{c}$ that has been observed with the "pulsed"
enzyme. Unfortunately, precise criteria are missing by which the
various components of these oxidized forms of the enzyme can be
characterized. Mainly spectrophotometry has been used for this purpose, but this largely relies on minor band shifts or intensity
differences. We have observed a specific and unusual EPR signal for
one of these components, namely at g = 5; 1.78; 1.69 (47). This
form is seen immediately (<5 msec) on reoxidation of reduced oxidase by oxygen and has a half life of 2 min. It can also be detected
several minutes after anaerobic reoxidation of reduced oxidase by
porphyrexide *. Reoxidation by oxygen does, therefore, not seem to
be an absolute requirement for its formation. Since this form is
rapidly and maximally produced with oxygen and its EPR signal can be
seen in mitochondria after aeration (47) we have given this species
specific attention (54). The fact that an EPR signal is present in
addition to those observed for the $\underline{a}^{3+}$ and Cu_A^{2+} components suggests that $\underline{a}_3^{3+}$ and Cu_B^{2+} are not as strongly coupled as in the
resting form of the enzyme. The peculiar temperature dependence of
the signal, namely disappearance of the signal at low temperature
at non-saturating microwave power, further indicates that the signal
originates from an excited state. We have recorded EPR spectra at
three frequencies, namely 15, 9 and 3 GHz and found a very strong
frequency dependence of the resonances (54) **. None at all were
found at 3 GHz. Plots of this information, essentially according to
Aasa (55), are compatible with those expected from a system with
spin 2. The ground state would then have S = 1. No signal for the
ground state was ever detected. A possibility considered by us **
is that , as suggested by Carter et al. (56), the $\underline{a}_3$ component
occurs with S = $3/2$, which may couple with Cu_B^{2+} to a S = 2 or S = 1
state. In the resting form $\underline{a}_3$ would then have S = $5/2$ and be coupled
to Cu_B^{2+} to a system spin of 2 (56).

Yet another species often observed in oxidized cytochrome oxidase, which is not identical with the resting form is that showing
an EPR signal at very low field, often called the g = 12 signal,
although according to our studies the signal lies at yet lower fields.

---

* R.W. Shaw and F.A. Armstrong, unpublished.
** W.R. Dunham, R.H. Sands, R.W. Shaw and H. Beinert, unpublished.

By an original and unusual approach, using largely NO treatment of the oxidase, Chan and his collaborators (43) have identified (and proposed active site structures for three) species of the oxidized enzyme, including most of the species mentioned above, while Nicholls (44, 57), on the basis of spetrophotometry of photoreduced species in the presence and absence of CO has identified species by these means, which may or may not be identical to or include forms mentioned above.

We have described and partly characterized an oxidized species obtained by anaerobic reoxidation of reduced cytochrome oxidase (46, 58). This species is very reactive towards ligands and reductants. It shows EPR signals for three of the four metal components in their oxidized state, namely $\underline{a}^{3+}$, a_3^{3+} and Cu_A and has a lifetime of a few seconds at room temperature. The a_3^{3+} signal disappears in that time. We are now inclined to interpret these data to mean that Cu_B is not readily oxidized by the oxidants we used, so that initially only three electrons are removed from the enzyme by the oxidant, and that Cu_B then becomes oxidized within seconds and spin coupling between a_3^{3+} and Cu_B^{2+} is reestablished with disappearance of the a_3^{3+} EPR signal.Within minutes during anaerobic oxidation the species with EPR signal at g = 5.0; 1,78; 1.69 arises, but only to about one fifth the intensity observed on oxidation with oxygen. Nevertheless, such a finding in an anaerobic system indicates that oxygen per se is not involved in producing the g = 5; 1,78;1.69 species (59); but aerobic oxidation yields the highest concentration of it.

As with succinate dehydrogenase and aconitase we have here with cytochrome oxidase another example of reductive activation. Neither with succinate dehydrogenase nor with cytochrome oxidase is the "resting" form totally inactive, as it is with aconitase. Another significant difference is that in succinate dehyrogenase and cytochrome oxidase the electron acceptor which has to be reduced for activity to become optimal is itself participating in electron transfer in the reaction catalyzed by the enzyme, whereas in aconitase that electron acceptor is an additional component which, to our knowledge, does not turn over during catalysis but only during activation-inactivation cycles. How far the resting forms are artefacts of human intervention remains to be determined.

Concluding Remarks

Obviously there are a number of principles according to which one might group electron transfer proteins together or compare structural and functional aspects. I have tried to give here a crossection according to interactions of electron accepting sites with each other or with their associated protein moieties, as seen with the eyes of an EPR practitioner. One lesson seems to be implicit in most of what I presented, namely that the enzymes, as we prepare them, often are not in the form in which they carry out the reactions which we intend to study. It will be a task for the future and indeed a difficult one in many instances, to apply our present-day physical tools to the active states of these enzymes.

ACKNOWLEDGEMENTS

The experimental work done in our own and the collaborating laboratories, which forms the basis for this chapter, was carried out jointly with the authors of references 4 - 6, 14, 17 - 19, 24, 26, 27, 47, 50, 54, and those mentioned in the footnotes. I thank those colleagues who permitted me to cite unpublished work. Support of our own work by the Institute of General Medical Sciences, National Institutes of Health, by a Research Grant (GM-12394) and a Research Career Award (5 - K o6 - GM - 18492) is gratefully acknowledged.

REFERENCES

1. S. De Vries, S. P. J. Albracht, and F. J. Leeuwerik, The multiplicity and stoichiometry of the prosthetic groups in QH_2: cytochrome _c_ oxidoreductase as studied by EPR, Biochim. Biophys. Acta 546: 316 (1979).

2. H. G. Heidrich, S. P. J. Albracht and D. Bäckström, Two iron-sulfur centers in mitochondrial membranes from beef heart as prepared by free-flow electrophoresis, FEBS Letters 95: 314 (1978).

3. J. S. Rieske, R. E. Hansen, and W. S. Zaugg, Studies on the electron transfer system. LVII. Properties of a new oxidation-reduction component of the respiratory chain as studied by electron paramagnetic resonance spectroscopy, J. Biol. Chem. 239: 3017 (1964).

4. F. J. Ruzicka and H. Beinert, The soluble "high potential" type iron-sulfur protein from mitochondria is aconitase, J. Biol. Chem. 253: 2514 (1978).

5. T. A. Kent, J. L. Dreyer, M. H. Emptage, I. Moura, J. J. G. Moura, B. H. Huynh, A. V. Xavier, J. LeGall, H. Beinert, W. H. Orme-Johnson, and E. Münck, Evidence for novel three iron center in two ferredoxins, aconitase and glutamate synthase, in: "Interaction between iron and protein in oxygen and electron transport," C. Ho, ed., Pergamon Press, London (1981).

6. H. Beinert, B. A. C. Ackrell, E. B. Kearney, and T. P. Singer, EPR studies on the mechanism of action of succinate dehydrogenase in activated preparations, Biochem. Biophys. Res. Commun. 58: 564 (1974).

7. H. Beinert, B. A. C. Ackrell, E. B. Kearney, and T. P. Singer, Iron-sulfur components of succinate dehydrogenase: stoichio-

metry and kinetic behaviour in activated preparations, Eur.
J. Biochem.54: 185 (1975).

8. K. Dus, H. Deklerk, K. Sletten, and R. G. Bartsch, Chemical
characterization of high potential iron proteins from
Chromatium and Rhodopseudomonas gelatinosa, Biochim. Biophys.
Acta 140: 291 (1967).

9. H. Beinert and S. P. J. Albracht, Acta Rev. Bioen. in press
(1982).

10. T. Ohnishi, T. E. King, J. C. Salerno, H. Blum, J. R. Bowyer,
and T. Maida, Thermodynamic and electron paramagnetic resonance characterization of flavin in succinate dehydrogenase,
J. Biol. Chem. 256: 5577 (1981).

11. C. Paech, J. G. Reynolds, T. P. Singer, and R. H. Holm, Structural identification of iron-sulfur clusters of the respiratory chain-linked NADH dehydrogenase, J. Biol. Chem. 256:
3167 (1981).

12. S. P. J. Albracht, The prosthetic groups in succinate dehydrogenase: number and stoichiometry, Biochim. Biophys. Acta
612: 11 (1980).

13. T. Ohnishi, J. C. Salerno, D. B. Winter, J. Lim, C. A. Yu, L.
Yu, and T. E. King, Thermodynamic groups in the succinate-
ubiquinone reductase segment of the respiratory chain, J.
Biol. Chem. 251: 2094 (1976).

14. H. Beinert, B. A. C. Ackrell, A. D. Vinogradov, E. B. Kearney,
and T. P. Singer, Interrelations of reconstitution activity,
reactions with electron acceptors, and iron-sulfur centers
in succinate dehydrogenase, Arch. Biochem. Biophys. 182: 95
(1977).

15. J. C. Salerno, J. Lim, T. E. King, H. Blum,and T. Ohnishi, The
spatial relationships and structure of the binuclear iron-
sulfur clusters in succinate dehydrogenase, J. Biol. Chem.
254: 4828 (1979).

16. J. C. Salerno, T. Ohnishi, J. Lim, W. R. Widger, and T. E.
King, Spin coupling between electron carriers in the dehy-
drogenase segments of the respiratory chain, Biochem.Biophys.
Res. Commun. 75: 618 (1977).

17. D. J. Steenkamp, T. P. Singer and H. Beinert, Participation of
the iron-sulfur cluster and the covalently bound coenzyme
of trimethylamine dehydrogenase in catalysis, Biochem. J.
169: 361 (1978).

18. D. J. Steenkamp, H. Beinert, W. C. McIntire and T. P. Singer,
in: "Mechanisms of Oxidizing Enzymes", T. P. Singer and R.
N. Ondarza ed., Elsevier, North Holland, Amsterdam (1978).

19. H. Beinert, R. W. Shaw, D. J. Steenkamp, T. P. Singer, R.
Stevenson, W. R. Dunham and R. H. Sands, in: "Flavins and
Flavoproteins", V. Massey and C. H. Williams ed., Elsevier,
North Holland, Amsterdam (1981).

20. B. L. Trumpower, New concepts on the role of ubiquinone in the
mitochondrial respiratory chain, J. Bioenerg. Biomembr. 13:

1 (1981).

21. D. Bäckström, B. Norling, A. Ehrenberg, and L. Ernster, Electron spin resonance measurement on ubiquinone-depleted and ubiquinone-replenished submitochondrial particles, Biochim. Biophys. Acta 197: 108 (1970).

22. A. Kröger, The interaction of the radicals of ubiquinone in mitochondrial electron transport, FEBS Letters, 65: 278 (1976).

23. C. A. Yu, S. Nagaoka, L. Yu, and T. E. King, Evidence for the existence of a ubiquinone protein and its radical in the cytochromes b and c_1 region in the mitochondrial electron transport chain, Biochem. Biophys. Res. Commun. 82: 1070 (1978).

24. F. J. Ruzicka, H. Beinert, K. L. Schleper, W. R. Dunham, and R. H. Sands, Interaction of ubisemiquinone with a paramagnetic component in heart tissue, Proc. Natl. Acad. Sci. U.S.A. 72: 2886 (1975).

25. W. J. Engledew, J. C. Salerno, and T. Ohnishi, Studies on electron paramagnetic resonance spectra manifested by a respiratory chain hydrogen carrier, Arch. Biochem. Biophys. 177: 176 (1976).

26. B. A. C. Ackrell, E. B. Kearney, C. J. Coles, T. P. Singer, H. Beinert, Y. P. Wan, and K. Folkers, Kinetics of the reoxidation of succinate dehydrogenase, Arch. Biochem. Biophys. 182: 107 (1977).

27. N. R. Orme-Johnson, R. E. Hansen, and H. Beinert, EPR studies of the cytochrome b-c_1 segment of the mitochondrial electron transfer, Biochem. Biophys. Res. Commun. 45: 871 (1971).

28. K. Matsuura, N. K. Packman, P. Mueller and L. P. Dutton, The recognition and redox properties of a component, possibly a quinone, which determines electron transfer rate in ubiquinone. Cytochrome c oxidoreductase of mitochondria, FEBS Letters 131: 17 (1981).

29. B. A. C. Ackrell, E. B. Kearney, and D. Edmondson, Mechanism of the reductive activation of succinate dehydrogenase, J. Biol. Chem. 250: 7114 (1975).

30. T. Ohnishi, W. J. Ingledew and S. Shiraishi, Existence of two distinct hipip-type iron-sulfur centers in mitochondria, Biochem. Biophys. Res. Commun. 63: 894 (1975).

31. F. J. Ruzicka and H. Beinert, A mitochondrial iron protein with properties of a high-potential iron-sulfur protein, Biochem. Biophys. Res. Commun. 58: 556 (1974).

32. T. Suzuki, S. Akiyama, S. Fujimoto, M. Ishikawa, Y. Nakao, and H. Fukuda, The aconitase of yeast IV. Studies on iron and sulfur in yeast aconitase, J. Biochem. 80: 799 (1976).

33. C. Kennedy, R. Rauner, and O. Gawron, On pig heart aconitase, Biochem. Biophys. Res. Commun. 47: 740 (1972).

34. J. J. Villafranca and A. S. Mildvan, The mechanism of aconitase action I. Preparation, physical properties of the enzyme, and activation by iron (II), J. Biol. Chem. 246: 772 (1971).

35. O. Gawron, M. C. Kennedy, and R. A. Rauner, Properties of pig

aconitase, Biochem. J. 143: 717 (1974).

36. H. Beinert, F. J. Ruzicka, and J. L. Dreyer, in: "Membrane
     Bioenergetics", C. P. Lee, G. Schatz and L. Ernster ed.,
     Addison-Wesley, Reading, Mass. (1979).

37. T. Yamamoto and K. Okunuki, Redox state of cytochrome a and
     its sensitivity to proteinases, J. Biochem. 67: 505 (1970).

38. T. Wakabayashi, A. E. Senior, O. Hatase, H. Hayashi and D. E.
     Green, Conformational changes in membranous preparations
     of cytochrome oxidase, J. Bioenerg. 3: 339 (1972).

39. B. F. van Gelder and H. Beinert, Studies of the heme components
     of cytochrome c oxidase by EPR spectroscopy, Biochim. Bio-
     phys. Acta 189: 1 (1969).

40. P. Nicholls and L. C. Petersen, Haem-haem interactions in
     cytochrome aa_3 during the anaerobic-aerobic transition,
     Biochim. Biophys. Acta 357: 462 (1974).

41. Y. Orii and T. E. King, On the nature of the three intermediate
     species formed after reaction of reduced cytochrome oxidase
     with oxygen, J. Biol. Chem. 251: 7487 (1976).

42. R. H. Tiesjema, A. O. Muijsers, and B. F. van Gelder, Bioche-
     mical and biophysical studies on cytochrome aa_3 IV. Some
     properties of oxygenated cytochrome aa_3, Biochim. Biophys.
     Acta 256: 32 (1972).

43. G. W. Brudvig, T. H. Stevens, R. H. Morse, and S. I. Chan,
     Conformations of oxidized cytochrome c oxidase, Biochemistry
     20: 3912 (1981).

44. P. Nicholls and G. A. Chanady, Reactivity of photoreduced cyto-
     chrome aa_3 complexes with molecular oxygen, Biochem. J. 194:
     713 (1981).

45. E. Antonini, M. Brunori, A. Colosimo, C. Greenwood, and M.T.
     Wilson, Oxygen "pulsed" cytochrome c oxidase: functional
     properties and catalytic relevance, Proc. Natl. Acad. Sci.
     USA 74: 3128 (1977).

46. H. Beinert and R. W. Shaw, On the identity of the high spin
     heme components of cytochrome c oxidase, Biochim. Biophys.
     Acta 462: 121 (1977).

47. R. W. Shaw, R. E. Hansen, and H. Beinert, A novel electron
     paramagnetic resonance signal of "oxygenated" cytochrome c
     oxidase, J. Biol. Chem. 253: 6637 (1978).

48. R. W. Shaw, R. E. Hansen and H. Beinert, The oxygen reactions
     of reduced cytochrome c oxidase. Position of a form with
     unusual EPR signal in the sequence of early intermediates,
     Biochim. Biophys. Acta 548: 386 (1979).

49. M. F. Tweedle, L. J. Wilson, L. Garcia-Iniguez, G. T. Babcock,
     and G. Palmer, Electron state of heme in cytochrome oxidase
     III. The magnetic susceptibility of beef heart cytochrome
     oxidase and some of its derivatives from 7-200 K. Direct
     evidence for an antiferromagnetically coupled Fe(III)/Cu
     (II) pair, J. Biol. Chem. 253: 8065 (1978).

50. T. H. Moss, E. Shapiro, T.E. King, H. Beinert, and C. R. Hart-
     zell, The magnetic susceptibility of cytochrome oxidase in

STRUCTURE AND FUNCTION OF AMINE OXIDASES

Bruno Mondovì and Alessandro Finazzi Agrò

Institutes of Biological Chemistry and Applied

Biochemistry, University of Rome, 00185 Rome, Italy

INTRODUCTION

Amine oxidases are enzymes widely distributed among living organisms. They catalyze the oxidative deamination of many biologically important amines with the formation of the corresponding aldehyde, hydrogen peroxide and ammonia according to the equation

$$RCH_2NH_2 + O_2 + H_2O \longrightarrow RCHO + H_2O_2 + NH_3$$

Two main types of amine oxidases have been described on the basis of their ability to oxidize only primary amines or both primary and secondary amino groups. Zeller[1] first proposed to call them diamine oxidases and monoamine oxidases respectively. However this proposal has not been fully adopted by the scientific community. As a result a diffuse misunderstanding arose about the subject.°

However classification on the basis of substrate specificity[1,3] or of sensitivity to particular inhibitors[1,3,4] appear to be unsatisfactory since both criteria have been found less specific than previously believed.

A more useful parameter of distinction can be the nature of the prosthetic group involved in the catalytic mechanism, that is a flavin moiety or a copper ion. Thus amine oxidases can be clas-

---

° Anarchy is reigning supreme in the nomenclature of amine oxidases, sadly stated Zeller in a recent review.[2]

sified as FAD-AOs (flavin-containing amine oxidases) and Cu-AOs
(copper-containing amine oxidases).

In addition to the reaction stoichiometry, the two types of
amine oxidases show various common features. For instance inhibitors like hydrazine derivatives are effective toward all amine
oxidases. In particular phenylhydrazine binds irreversibly to both
FAD-AO[5] and Cu-AO.

To the FAD-AO group belong mitochondrial amine oxidase (E.C.
1.4.3.4), pyridoxamine oxidase (E.C.1.4.3.5)dicarbamide-protein
ethanolamine oxidase (E.C.1.4.3.8) and tyramine oxidase (E.C.1.4.
3.9). To the copper type amine oxidases belong tissue diamine
oxidase (E.C.1.4.3.6), serum benzylamine oxidase (E.C.1.4.3.6) and
lysyl oxidase (1.13.12.aa).

Both classes of enzymes are present in higher organisms. FAD-
AOs are also present in bacteria (for instance Micrococcus rubens[6])
while a Cu AO has been isolated from a fungus (Aspergillus niger[7]).

This review will be devoted mainly to Cu-AOs. However a short
account of FAD-AO will also be given in view of the many common
features shown by these two types of enzymes.

FLAVIN-CONTAINING AMINE OXIDASES

These enzymes contain either covalently (mitochondrial type)
or non-covalently bound FAD (bacterial type). The former type only,
which is very important for the biochemistry of the nervous system,
will be further dealt with in this review.

Known under the trivial name of monoamine oxidases (MAO), they
play an essential role in the metabolism of central nervous system
neurotransmitters like norepinephrine, dopamine and serotonin. A
role of MAO in some psychiatric syndromes has been clearly established.

From a molecular point of view FAD-AO is a membrane-bound
enzyme located in the outer membrane of mitochondria. This enzyme
has been isolated from various tissues and seems to be present in
different molecular forms[8] called MAO-A, MAO-B and MAO-C. While the
first two subforms are widely acknowledged, the presence of a type
C MAO is questioned. All these forms are immunologically identical,
but quite distinct from Cu-AO.[8] The difference among them is the
specificity toward substrates and inhibitors. In fact MAO-A oxidizes
better 5-hydroxytryptamine, noradrenaline and adrenaline and is
inhibited by clorgyline, while MAO-B is active mainly on tyramine,
dopamine, benzylamine, phenylethylamine and is inactivated selectively by deprenyl. However many substrates and inhibitors are common
to both types of FAD-AOs.

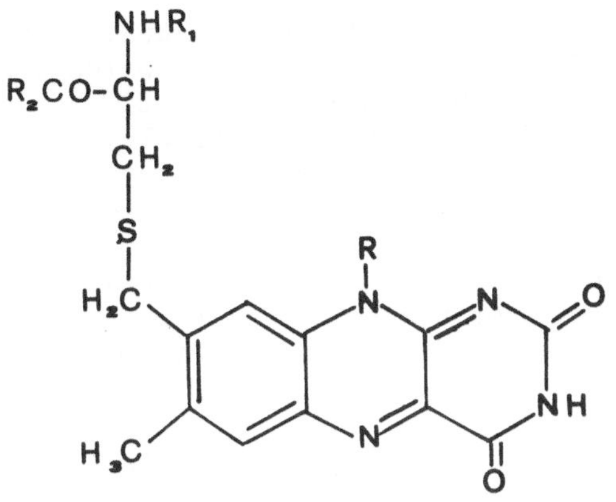

Fig. 1. Flavin moiety covalently bound to bovine liver MAO.[10]

MAO-B was purified from various sources like beef, pig and rat liver, bovine kidney, porcine and human brain, blood platelets°, etc.

The isolated MAO-B is a lipo-glycoprotein°° with M_r = 420,000 that may polymerize up to molecular weights greater than 1,000,000.

Hiramatsu and Yasunobu (ref. 8) obtained from bovine liver an enzyme preparation with M_r = 100,000 containing 1 mole of covalently bound FAD per mole. Kearney et al.[9] showed that FAD is bound to a cysteinyl residue of the protein through the position 8 of the iso-alloxazine ring. Walker et al.[10] were able to isolate a flavin peptide from bovine liver having the sequence Ser-Gly-Cys (FAD)-Tyr (Fig. 1). Both MAO-A and MAO-B have the same flavin peptide.[11] Under anaerobic conditions substrates can reduce the flavin to $FADH_2$.

COPPER AMINE OXIDASES

Physiological Role

Copper amine oxidases have been investigated mainly in connection with the control of histamine level which may be responsible for many pathological situations, like allergic diseases, peptic ulcer and anaphylactic reactions. Hence the name "histaminase" still

---

° Interestingly platelets contain only the type B enzyme.

°° A different content of lipids has been often indicated as causing the difference between A-type and B-type enzymes.

largely used by pharmacologists and physicians. It should however
be pointed out that histaminase and diamine oxidase are the same
enzyme.[12] The activity of histaminase shows significant variations
in some physiopathological conditions. The blood of pregnant women
has a very high histaminase activity, originating from placenta.[13]
Variations of diamine oxidase, serum amine oxidase and polyamine
oxidase were also found in association with the growth of normal
and transformed cells.[14] These variations appear to be correlated
with the sharp increase of biogenic amines in proliferating tissues.[15] It seems therefore conceivable that amine oxidases are
involved in the regulation of cell proliferation.[16] It has been in
fact demonstrated that the aldehydes produced by amineoxidases from
their substrate are powerful cytostatic agents.[17] Some amine oxidases appear to be appropriately located in order to prevent the
entrance of amines in the blood stream. This is the case of pla-
cental,[13] intestinal and renal diamine oxidases. The latter is
localized in the basal portion of first order convoluted tubules
of kidney.[18]

Lysyl oxidase is devoted to the formation of cross linkages
in collagen and elastine. The first step in this formation is the
oxidative deamination of ε-amino groups of lysyl residues in
collagen or elastin precursor proteins.

An increased or a decreased lysyl oxidase activity may have
pathological implications since the extent of cross-linking in
connective tissue is fundamental for their mechanical properties.

Molecular Properties

Cu-AOs have been purified from various sources (fungi, plant
and animal tissues). All these enzymes have molecular weights in
excess of 100,000 and contain at least two Cu-atoms. The enzyme
isolated from pig kidney has a minimum molecular weight of 90,000
calculated on the basis of copper content, but the actual molecular
weight measured by hydrodynamic methods does not fit any multiple
of this figure and depends on the experimental conditions of the
measure.[8,19-21] Bovine and porcine plasma amine oxidases are composed of two equivalent subunits,[22-24] which do not appear to be
linked by disulfide bridges.[25]

All these proteins do not show immunological similarity, despite
striking similarity in amino acid composition. In fact antibodies
raised against each enzyme do not react with any other amine oxidase.[26,27] As reported above FAD-AOs are also immunologically
distinct from copper-containing enzymes. Interestingly enough the
enzyme-antibody complexes are still active. All these observations
seem to indicate that the active sites of these enzymes are not
immunogenic.[27]

The bovine plasma enzyme shows three active forms on disc-gel electrophoresis[8,25] but the amount of each component seems to be different from a preparation to another.

Animal amine oxidases show little amount of α-helix as determined by optical rotatory dispersion.

Prosthetic Groups

Copper amine oxidases contain at least two copper ions per molecule. Electron spin resonance spectra show that copper is 100% in the cupric form and is liganded to the protein in a tetragonal symmetry.[19] The anaerobic addition of substrates induces subtle changes in the esr spectrum of pig kidney amine oxidase, which are typical of the substrate added. However the use of ^{15}N-putrescine instead of ^{14}N-putrescine, demonstrated that substrate is not directly coordinated to copper.

Similar conclusions were reached on the basis of proton relaxation experiments.[28,29]

Pig kidney and beef plasma amine oxidases show superhyperfine lines in the $g_\perp$ region of the esr spectrum, probably arising from nitrogenous ligands.[30,31]

Esr spectra at 35 GHz indicate that the two Cu^{2+} in pig plasma Cu-AO are not identical.[32] Equal proportions of components with axial and rhombic symmetry have been described and it was suggested that oxygen and nitrogen are ligands to copper. Proton relaxation of water indicated the presence of two water molecules coordinated to copper, one in axial and one in equatorial position.[32] The binding of strong ligands like N_3^- and CN^- makes spectroscopically equivalent the two copper atoms. In our laboratory[33] it was shown that the binding of phenylhydrazine to beef plasma amine oxidase induces a change in the esr parameters of only one of the two coppers. These findings suggest that the difference between the two copper atoms has functional significance.

Copper is not steadily reduced by excess substrate in amine oxidases,[8,19] but a transient oxido-reduction cycle has been suggested.[30] Copper in beef plasma AO is reduced by nearly stoichiometric dithionite and rapidly reoxidized by oxygen suggesting a catalytic role of the metal in oxygen activation.[33] In fact other copper proteins devoided of oxidase activity like plastocyanin, azurin or superoxide dismutase are not reoxidized by oxygen.

The most controversial issue concerning the amine oxidases is the presence and the possible nature of a second prosthetic group.

The presence of pyridoxal phosphate proposed by several authors[34-37] was challenged by others[38,39] on the basis of chemical determinations and mechanicistic considerations. More recently in our
laboratory the problem of the chromophore absorbing at 500 nm was
studied.[40] It was clearly demonstrated that this absorption is not
due to copper and that various carbonyl reagents give adducts with
the enzyme with properties superimposable to those observed with
authentic PLP-enzymes, like transaminase. The absence of fluorescence
and other spectroscopic considerations led to the proposal of a
direct binding of pyridoxal phosphate (if present) to copper (see
below). Phenylhydrazine gives an irreversible adduct with pig kidney
diamine oxidase that is enzymatically inactive. This "suicide"
reaction has been studied in detail with bovine plasma AO.[33] It
requires the presence of cupric copper. In fact the reduction by
dithionite or the removal of copper obtained by dialysis against CN^-
after reduction prevents the appearance of the absorption at 440 nm
linked to the formation of the DAO-phenylhydrazine adduct while the
enzymic activity, once reoxidized or reinserted the copper, is recovered. Copper itself does not directly participate to this chromophore as its removal after the formation of the yellow DAO-phenylhydrazine adduct does not abolish the absorption.[33]

The apoprotein obtained by removing copper with CN^- after
reduction in the absence of phenylhydrazine, is practically colorless.
In any case the epr spectrum of copper is modified after the reaction
with phenylhydrazine by the appearance of a new species with a smal-
ler $A_{||}$ (134 gauss) and a greater $g_{||}$. This change accounts for
only half of the copper signal and indicates a change of symmetry
and/or ligands upon binding of phenylhydrazine. It is important to
recall that also the addition of substrate in the absence of air
somewhat affects the epr spectrum of amine oxidases.[19,33]

The presence of a second cofactor which binds very near or
even to copper is indicated by many different lines of evidence. If
this cofactor were pyridoxal phosphate, a possible arrangement of
the active site accounting for the spectroscopical features could
be that reported in fig. 2.

Mechanism of action of copper amine oxidases

No matter what is the second prosthetic group of amine oxidases,
a few points on the catalytic mechanism can be done. Amine oxidases
appear to follow a ping-pong bi-ter mechanism[41-43] on the basis of
steady state and pre-steady state kinetics. However conflicting
reports have appeared concerning the order of products' release.[44-46]

In fact some authors report that only aldehyde is released in
the absence of air, i.e. before the binding of the second substrate
(oxygen), while others find that, at least with bovine serum amine
oxidase, ammonia is also released in the anaerobic part of the

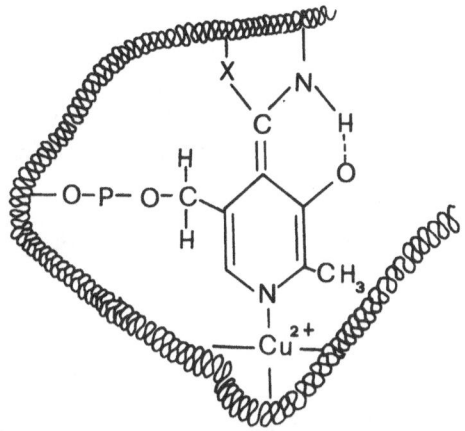

Fig. 2.   Hypothetical structure of amine oxidase active site.

Scheme 1

$$E\text{-}Pyr\text{-}CHO + RCH_2NH_2 \longrightarrow E\text{-}Pyr\text{-}CH_2\text{-}NH_2 + R\text{-}CHO$$

$$E\text{-}Pyr\text{-}CH_2NH_2 + O_2 \longrightarrow E\text{-}Pyr\text{-}CHO + H_2O_2 + NH_3$$

reaction. It must be underlined that the determination of stoichiometric ammonia production is a difficult task and that different
experimental approaches could give different results. These discrepancies led to the proposal of two different mechanisms. The first
(Scheme I) corresponds to a classical transaminase mechanism where
the pyridoxamine-like intermediate is reoxidized by oxygen libe-
rating ammonia and hydrogen peroxide. The second (Scheme II) implies
the formation of an imine as intermediate that hydrolyses in the
absence of air to give aldehyde, ammonia and a reduced enzyme form.

A sulfhydryl group, involved in the redox cycle of scheme II,
was indeed found under anaerobic conditions in the presence of
substrate in bovine plasma amine oxidase.[47] Similar results have
been more recently obtained by Floris et al.[48] with Euphorbia diamine

Scheme   2

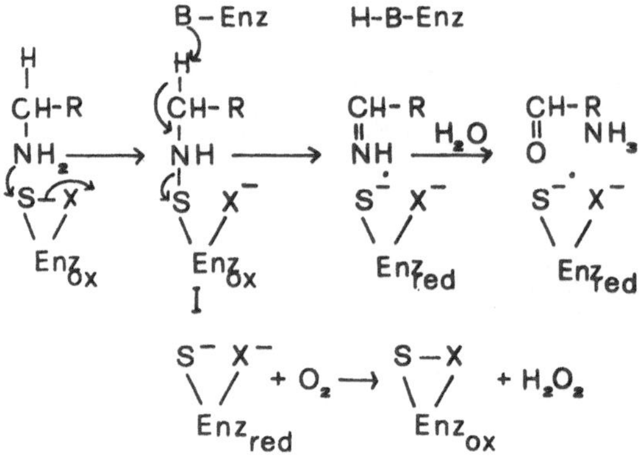

oxidase. Suva and Abeles[49] did not find evidence for the incorporation of non exchangeable protons from either tritiated substrate
or water in the enzyme that appears to favour an imine mechanism.
However such negative evidence has been often questioned.

Steady-state[43,50] and pre-steady state[51-55] kinetics data fit
a transaminase-type mechanism. Ammonia was found to be a competitive inhibitor of the amine substrate[43] and was shown to be bound to
the enzyme throughout the catalytic cycle.[50] On the other hand
aldehyde production was found to occur in the same time scale with
the bleaching of enzyme absorption. Phenylhydrazine is a competitive
inhibitor of the bleaching supporting the formation of a Schiff-
base between the enzyme and the substrate as the first catalytic
event. No ammonia (or hydrogen peroxide) was found in the absence
of air.[45]

Azide and cyanide, well known copper reagents, bind to amine
oxidases preventing the reoxidation of the reduced form of the
enzyme.[54,56] These findings indicate that at least one of the two
copper ions is involved in the catalytic mechanism. It could either
serve as the oxygen binding site or as a Lewis acid catalyst in the
reduction steps leading to hydrolytic release of aldehyde. Azide
and cyanide can displace one of the two water molecules coordinated
to each copper in pig plasma amine oxidase and behave as competitive
inhibitors toward oxygen.[32] Their presence allows the detection of
a radical intermediate[57] previously found only at sub-zero temperatures.[58]

Most recently Knowles et al.[59] put forward an elegant mechanism
for pig plasma amine oxidase where a hydroxyl coordinated to Cu^{2+}
is postulated to act as a nucleophile facilitating the transfer of

Scheme 3

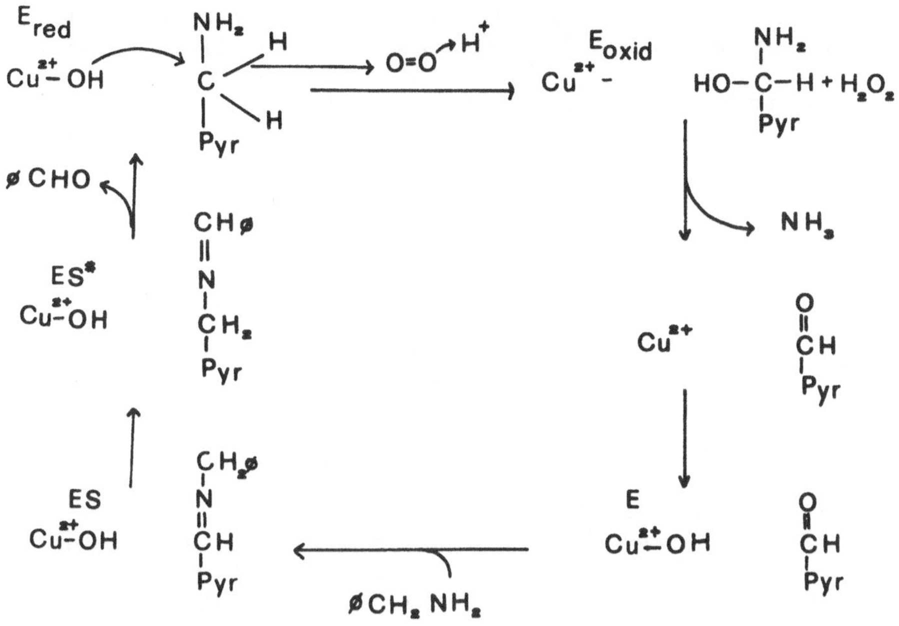

a hydride ion to oxygen, bound at the second copper ion (Scheme 3).

However a full understanding of the catalytic mechanism of these enzymes must await a better knowledge of their structural properties (subunit composition, number of active sites, etc.) and a definite answer about the presence and the chemical nature of the second prosthetic group.

REFERENCES

1. E. A. Zeller, Oxidation of Amines in The Enzymes Vol. II, part I, J. B. Summer and K. Myrbäck eds., Acad. Press New York and London, 1st ed. (1951).
2. T. P. Singer, R. W. Von Korff and D. L. Murphy eds., Monoamine oxidase: structure, function and altered functions. Acad. Press, New York (1979).
3. B. G. Malmström, L. E. Andreasson and B. Reinhammar, Copper containing oxidases and superoxide dismutase in The Enzymes, Vol. 12, 3rd ed., Boyer P.D. ed., Academic Press, New York and London, (1975).
4. H. Blaschko, Amine oxidase, in:"The Enzymes,"Vol. 8, P. D.

Boyer, H. Lardy, K. Myrbäck, eds. Acad. Press, New York
and London (1963).

5. T. P. Singer, Active-site-directed, irreversible inhibitors of
    monoamine oxidase in: "Monoamine Oxidase: Structure, Function
    and altered Functions", T.P. Singer, R. W. Van Korff and D.
    L. Murphy eds., Acad. Press New York (1974).

6. O. Adachi, H. Yamada and K. Gata, Purification and properties
    of putrescine oxidase of Micrococcus rubens, Agr. Biol. Chem.
    30: 1202 (1966).

7. H. Suzuki, Y. Ogura and H. Yamada, Stoichiometry of the reaction
    by amine oxidase from Aspergillus niger, J. Biochem. 69:
    1065 (1971).

8. K. T. Yasunobu, H. Ishizaki and N. Minamiura, The molecular
    mechanistic and immunological properties of amine oxidases,
    Molecular and Cell. Biochem. 13: 3 (1976).

9. F. B. Kearney, J. I. Sakachi, W. H. Walker, R. Seng, L. Kenney,
    W. Zeszotek and T. P. Singer, Eur. J. Biochem. 24: 32 (1971).

10. W. H. Walker, F. B. Kearney, R. L. Seng and T. P. Singer, The
    covalently bound flavin of hepatic monoamine oxidase I, Eur.
    J. Biochem. 24: 328 (1971).

11. B. Gomes, I. Igane, H. G. Kloepfer and K. T. Yasunobu, Amine
    oxidase. XIV. Isolation and characterization of the multiple
    beef liver amine oxidase components, Arch. Biochem. Biophys.
    132: 16 (1969).

12. B. Mondovì, A. Scioscia-Santoro, G. Rotilio, M. T. Costa and
    A. Finazzi-Agrò, In vivo anti-histaminic activity of hista-
    minase, Agents and Actions 5: 460 (1975).

13. W. G. Bardsley, M. J. Crabbe and I. V. Scott, The amine oxidases
    of human placenta and pregnancy plasma, Biochem. J. 139: 169
    (1974).

14. G. Quash, T. Keolonangkhot, L. Gazzolo, H. Ripoll and S. Saez,
    Diamine oxidase and polyamine oxidase activities in normal
    and transformed cells, Biochem. J. 177: 275 (1979).

15. J. Jänne, H. Pösö, A. Raina, Polyamines in rapid growth and
    cancer, Biochim. Biophys. Acta 473: 241 (1978).

16. B. Mondovì, P. Guerrieri, M. T. Costa and S. Sabatini, Amine
    oxidase inhibitors and biogenic amines metabolism, in:
    "Advances in polyamine research", Vol. 3, C. M. Calderera,
    V. Zappia and U. Bachrach eds., Raven Press, New York (1981).

17. U. Bachrach and S.Persky, Interaction of oxidized polyamines
    with DNA. V.Inhibition of nucleic acid synthesis, Biochim.
    Biophys. Acta 179: 484 (1969).

18. M. P. Argento-Cerù, A. Oratore, B. Mondovì and A. Finazzi-Agrò,
    Localization of diamine oxidase in pig kidney: immunofluore-
    scence method, Cell. Mol.Biol. 27: 359 (1981).

19. B. Mondovì, G. Rotilio, M. T. Costa, A. Finazzi-Agrò, E. Chian-
    cone, R. E. Hansen and H. J. Beinert, Diamine oxidase from
    pig kidney: improved purification and properties, J. Biol.
    Chem.242: 1160 (1967).

20. J. M. Pionetti, Analytical-band centrifugation of the active

form of pig kidney diamine oxidase, Biochem. Biophys. Res. Comm. 58: 495 (1974).

21. M. D. Kluetz and P. G. Schmidt, Diamine oxidase: molecular weight and subunit analysis, Biochem. Biophys. Res. Comm. 76: 40 (1977).

22. F. M. Achee, C. Chervenka, R.A. Smith and K. T. Yasunobu, Amine oxidase XII. The association and dissociation and number of subunits of beef plasma amine oxidase, Biochemistry 7: 4379 (1968).

23. F. Buffoni and H. Blaschko, Benzylamine oxidase and histaminase: purification and crystallization of an enzyme from pig plasma, Proc. Roy. Soc. 161: 153 (1964).

24. R. Barker, N. Boden, G. Cayley, S. C. Charlton, K. Henson, M. C. Holmes, I. D. Kelly and P. F. Knowles, Properties of cupric ions in benzylamine oxidase from pig plasma as studied by magnetic resonance and kinetic methods, Biochem. J 177: 289 (1979).

25. P. Turini, S. Sabatini, O. Befani, F. Chimenti, C. Casanova, P. L. Riccio and B. Mondovì, A new method for purification of bovine plasma amine oxidase. To be published.

26. A. Oratore, P. Guerrieri, A. Ballini, B. Mondovì and A. Finazzi-Agrò, Preparation and properties of antibody against pig kidney diamine oxidase, FEBS Letters 104: 154 (1979).

27. A. Oratore, G. Banchelli, F. Buffoni, S. Sabatini, B. Mondovì and A. Finazzi-Agrò, Reaction between mammalian amine oxidases and their antibodies, Biochem. Biophys. Res. Comm. 98: 1002 (1981).

28. M. D. Kluetz, P. G. Schmidt, Proton relaxation study of the pig kidney diamine oxidase active centre, Biochemistry 16: 5191 (1977).

29. M. D. Kluetz, P. G. Schmidt, Proton magnetic relaxation dispersion in solutions of the cuproprotein diamine oxidase, Biophys. J. 29: 283 (1980).

30. B. Mondovì, G. Rotilio, A. Finazzi-Agrò, M. P. Vallogini, B. G. Malmström and E. Antonini, Copper reduction by substrate in diamine oxidase, FEBS Letters, 2: 182 (1969).

31. S. Suzuki, T. Sakurei, A. Nakapora, O. Oda, T. Manabe and T. Okuyana, Spectroscopic aspects of copper binding site in bovine serum amine oxidase, FEBS Letters 116: 17 (1980).

32. R. Barker, N. Boden, G. Cayley, S. C. Charlton, R. Henson, M. C. Holmes, I. D. Kelly and P. F. Knowles, Properties of cupric ions in benzylamine oxidase from pig plasma as studied by magnetic resonance and kinetic methods, Biochem. J. 177: 289 (1979).

33. A. Finazzi-Agrò, B. Mondovì, G. Rotilio and S. Sabatini, Cu-amine oxidase. Interdependence of the copper and the phenylhydrazine-reacting group. To be published.

34. H. Blaschko and F. Buffoni, Pyridoxalphosphate as costituent of the histaminase (benzylamine oxidase) of pig plasma, Proc. Roy. Soc. B, 163: 45-60 (1965).

35. H. Yamada and K. T. Yasunobu, Monoamine oxidase. IV. Nature of
     the second prosthetic group of plasma monoamine oxidase,
     J. Biol. Chem. 238: 2669 (1963).
36. E. V. Goryachenkova and E. A. Ershova in: "Chemical and biolo-
     gical aspects of pyridoxal catalysis, Snell E.E., Fasella P.
     M., Braunstein, A. and Rossi Fanelli, A. eds., Pergamon Press
     Oxford (1963).
37. B. Mondovì, M. T. Costa, A. Finazzi-Agrò and G. Rotilio, Pyri-
     doxal phosphate as a prosthetic group of pig kidney diamine
     oxidase, Arch. Biochem. Biophys. 119: 373 (1967).
38. M. Inamasu, K. T. Yasunobu and W. A. Konig, Cofactor investi-
     gation of bovine plasma amine oxidase. NaBH$_4$ reduction of
     enzyme-substrate mixture, J. Biol. Chem. 249: 5265 (1974).
39. R. Neumann, R. Hevey and R. M. Abeles, The action of plasma
     amine oxidase on β-haloamines. Evidence for proton abstrac-
     tion in the oxidative reaction, J. Biol. Chem.250: 6362
     (1975).
40. A. Finazzi-Agrò, P. Guerrieri, M. T. Costa, and B. Mondovì, On
     the nature of chromophore in pig kidney diamine oxidase,
     Eur. J. Biochem. 74: 435 (1977).
41. A. Finazzi-Agrò, G. Rotilio, M. T. Costa and B. Mondovì, Evi-
     dence for a ping-pong mechanism in the diamine oxidase
     reaction, FEBS Letters 14: 31 (1969).
42. S. Oi, H. Inamasu and K. T. Yasunobu, Mechanistic studies of
     beef plasma amine oxidase, Biochemistry 9: 3378 (1970).
43. W. C. Bardsley, M. J. C. Crabbe and J. Schindher, Kinetics of
     diamine oxidase reaction, Biochem. J. 131: 459 (1973).
44. K. A. Berg and R. H. Abeles, Mechanism of action of plasma amine
     oxidase. Products released under anaerobic conditions,
     Biochemistry 19: 3186 (1980).
45. A. Lindström and G. Petterson, The order of substrate addition
     and product release during the catalytic action of pig-plasma
     benzylamine oxidase, Eur. J. Biochem. 84: 479 (1978).
46. C. E. Taylor, R. S. Taylor, C. Rasmussen and P. F. Knowles, A
     catalytic mechanism for the enzyme benzylamine oxidase from
     pig plasma, Biochem. J. 130: 713 (1972).
47. H. Zeidan, K. Watanabe, L. A. Piette and K. T. Yasunobu,
     Electron spin resonance studies of bovine plasma amine
     oxidase, J. Biol. Chem. 255: 7621 (1980).
48. G. Floris, A. Giartosio and A. Rinaldi, Essential sulfhydryl
     groups in diamine oxidase from Euphorbia characias latex,
     to be published.
49. R. H. Suva and R. H. Abeles, Studies on the mechanism of action
     of plasma amine oxidase, Biochemistry 17: 3538 (1978).
50. I. D. Kelly, P. F. Knowles, K. D. S. Yadov, W. G. Bardsley,
     P. Lepp and R. D. Wright, Steady state kinetic studies on
     benzylamine oxidase from pig plasma, Eur. J. Biochem. 114:
     133 (1981).
51. B. Mondovì, G. Rotilio, A. Finazzi-Agrò and E. Antonini, Amine
     oxidases: a new class of copper oxidases, in: "Magnetic

resonance in biological research", C. Franconi ed., Gordon
and Breach Sci. Publ. (1971).

52. A. Lindström, B. Olsson, and G. Petterson, Transient kinetics
of benzaldehyde formation during the catalytic action of
pig plasma benzylamine oxidase, Europ. J. Biochem. 42: 377
(1973).

53. A. Lindström, B. Olsson, G. Petterson and J. Szymanska, Kinetics
of the interaction between pig-plasma benzylamine oxidase
and various monoamines, Eur. J. Biochem. 47: 99 (1974).

54. B. Olsson, J. Olsson and G. Petterson, The kinetics of reoxidation of reduced benzylamine oxidase, Eur. J. Biochem. 74: 329
(1977).

55. A. Lindström, B. Olsson and G. Petterson, Kinetics of interaction between benzylamine oxidase and hydrazine derivatives, Eur. J. Biochem. 42: 177 (1974).

56. A. Lindström, B. Olsson and G. Petterson, Effect of azide on
some spectral and kinetic properties of pig-plasma benzylamine oxidase, Eur. J. Biochem. 48: 237 (1974).

57. G. Rotilio, A. Rinaldi, G. Floris and A. Finazzi-Agrò, A free
radical intermediate in the reduction of diamine oxidase
from Euphorbia, to be published.

58. M. D. Kluetz, K. Adamson and J. E. Glynn, Cryoenzymology and
spectrophotometry of pea seedling diamine oxidase, Biochemistry 19: 1617 (1980).

59. K. D. S. Yadav and P. F. Knowles, A catalytic mechanism for
benzylamine oxidase from pig-plasma. Stopped-flow kinetic
studies, Eur. J. Biochem. 114: 139 (1981).

THE MECHANISM OF ACTION OF Cu,Zn SUPEROXIDE DISMUTASE

Giuseppe Rotilio[+], Lilia Calabrese[++], Adelio Rigo[°] and E.M. Fielden[°°]

+ Institute of Biological Chemistry, University of Rome, Rome, Italy; ++ Laboratory of Applied Biochemistry, Institute of Organic Chemistry, University of Messina, Messina, Italy; ° Laboratory of Biophysics, Institute of General Pathology, University of Padua, Padua, Italy; °° Division of Physics and Biophysics, Institute of Cancer Research, Sutton, Surrey, U.K.

STRUCTURE OF THE ACTIVE SITE OF THE ENZYME

Superoxide dismutases are nearly ubiquitous metallo-enzymes of strict specificity. The only known reaction catalyzed by them is:

$$O_2^- + O_2^- + 2H^+ \longrightarrow O_2 + H_2O_2;$$

that is dismutation of the anionic form of the perhydroxyl radical, usually referred to as superoxide. The catalytic metal is either copper, manganese or iron. The copper enzyme is by far the most catalytically efficient and contains equimolar amount of zinc (Cu,Zn superoxide dismutase). It is a dimeric enzyme made up of two identical subunits of M_r 16,000 each one containing 1 copper and 1 zinc ion. Superoxide dismutases are rather conservative proteins, as for structure, spectroscopic properties and catalytic activity, throughout all organisms from yeast to mammals. The present article will be dealing with the mechanism of Cu,Zn superoxide dismutase, with reference to data obtained with the bovine enzyme, the best known one. No discussion of the possible biological relevance of the catalytic ability to dismute O_2^- will be made in this context.

The case of Cu,Zn superoxide dismutase is unique among metalloproteins as far as the contribution of spectroscopic studies of the metal-binding sites to the knowledge of their structural

properties is concerned. In fact, extremely detailed information
was obtained from spectroscopy, especially magnetic resonance,[1] much
before the X-ray analysis of the protein at the proper resolution
level[2] and the complete primary structure[3] were available. The active
site region can thus be described in detail, and the principal features are the following (Scheme 1):

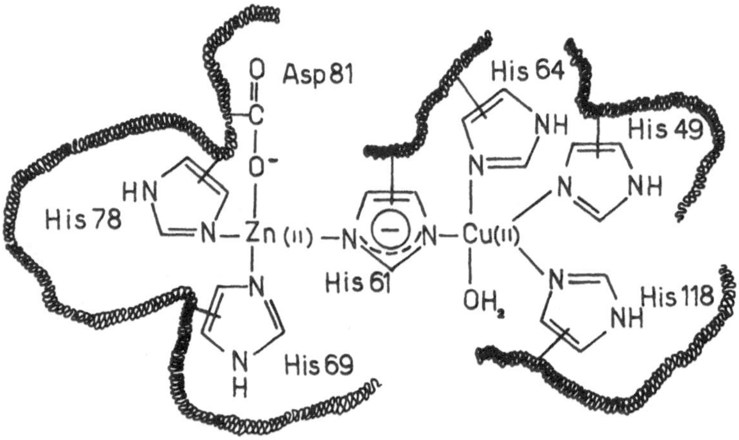

The copper and zinc on each subunit are very near to each other,
about 6 Å apart, and the imidazole ring of his 61 appears to be a
ligand for both metals. The copper appears to be located in a distorted square planar arrangement of his 44, 46, 61 and 118, with an
extra coordinating position providing access of the solvent through
a cavity formed by two "loops" of the protein backbone extending out
of the compact "core" of the protein structure. Arg 141 is one of
the side groups lining this cavity. This critically located positively charged residue is now supposed to be essential for the approach
of O_2^- to the copper in the catalytic mechanism (page 160). Zinc is
screened by the copper from the solvent and appears to be in an
approximately tetrahedral arrangement with his 61, his 69, his 78
and asp 81 as ligands. Metal substitution has been extensively and
successfully used for a better characterization of the metal-binding
sites. The zinc site can accept a number of metal ions without altering the enzyme activity: cobalt substitution for the zinc[4] has
proved the most useful for mechanistic studies (see page 162). The
catalytically active copper site is strictly specific; only in the
case of cobalt substitution a superoxide dismutase activity has
been observed, although it is orders of magnitude lower than in the
native enzyme.[5]

MECHANISM OF ACTION: GENERAL FEATURES

The mechanism of action of Cu,Zn superoxide dismutase brings together within a single protein site mechanistic elements that, independently, operate in each type of uncatalyzed or non-enzymatically catalyzed dismutation of superoxide. This fact results in the highest reaction rates known for these processes. Pulse radiolysis work[6-9] has shown that the enzyme carries out a very simple cycle of its copper between Cu(II) and Cu(I):

$$E-Cu(II) + O_2^- \longrightarrow E-Cu(I) + O_2$$

$$E-Cu(I) + O_2^- \xrightarrow{+2H^+} E-Cu(II) + H_2O_2$$

The rates of the two half reactions are identical and are the same as that of the overall reaction ($k = 2 \times 10^9 M^{-1}s^{-1}$) which is limited by diffusion.[9] pH affects the reaction rate only very slightly.[6,7]

Let us compare the reaction scheme shown above and its kinetic parameters to the relevant features of non-enzymic reactions of the same type. First of all the non-enzymic reactions are very much dependent on pH. In particular the uncatalyzed reaction (otherwise referred to as "spontaneous" dismutation or dismutation in pure water) depends on the concentration of the protonated form of O_2^- with pK = 4.8:

$$HO_2 + HO_2 \longrightarrow H_2O_2 + O_2 \qquad k = 8.6 \times 10^5 M^{-1}s^{-1}$$

$$HO_2 + O_2^- \xrightarrow{+H^+} H_2O_2 + O_2 \qquad k = 1.0 \times 10^8 M^{-1}s^{-1}$$

$$O_2^- + O_2^- \xrightarrow{+2H^+} H_2O_2 + O_2 \qquad k < 0.4\ M^{-1}s^{-1}$$

This pH dependence indicates that charge neutralization of the two negatively charged reacting molecules is an essential factor to allow efficient electron exchange between them. Since the enzymic reaction is extremely efficient ($k \simeq 10^9 M^{-1}s^{-1}$) at pH values where superoxide is completely unprotonated, one obvious function of the active site metal is to take the place of a proton in eliminating the charge repulsion of the two electron exchanging O_2^- molecules. Moreover, the copper being essential to the catalysis by the enzyme, this metal ion acts as an electron carrier in the enzyme catalysis, analogously to its general function in copper proteins.[10] The ping-pong reaction with O_2^- shown above is due to the substantial kinetic inertness of the superoxide dismutase copper to other redox reagents as compared to O_2^-, in particular to H_2O_2 and O_2^-.[11] In fact, low molecular weight complexes of either copper, iron or manganese

are able to catalyze dismutation of O_2^- but only the copper incorporated in natural Cu,Zn superoxide dismutase can do it at very
high efficiency in a wide range of pH; moreover it reacts in redox
side reactions with rates that are kinetically irrelevant.
This latter point is particularly important in order to define a
metalloprotein as a "true" superoxide dismutase. Other metalloproteins react with superoxide at very high rates, as, for instance,
oxidized cytochrome c, laccase or ceruloplasmin, but they, once
reduced by O_2^-, can be reoxidised by H_2O_2 or O_2 much faster than by
O_2^-. The strict specificity of Cu,Zn superoxide dismutase for O_2^-
is associated with the peculiar structural features of its copper
(see page 162) and may be relevant to the physiological role of
the enzyme. For example a Fenton type of reaction (Cu(I) + $H_2O_2 \rightarrow$
Cu(II) + OH· + OH$^-$), which is almost the rule for inorganic complexes of redox metals and gives rise to the indiscriminate oxidant
OH·, does not occur during the catalytic cycle of the superoxide
dismutase copper with O_2^-. On the contrary,when the superoxide
dismutase copper is reduced noncatalytically by the presence of
reducing agents other than O_2^- it can be reoxidized by H_2O_2 and
produce OH· radicals. Since H_2O_2 itself reduces the copper ion,[12]
incubation with excess peroxide slowly leads to inactivation of
the enzyme[13] through a Fenton mechanism. When, however, the enzyme
is under a flux of O_2^- the relatively low concentrations of H_2O_2
produced in the catalytic process never displace the substrate from
reaction with the enzyme.[9]

MECHANISM OF ACTION: FINE DETAILS

    Although the general feature of the enzymic reaction are well
known, the detailed mechanism remains to be firmly defined. In particular, debated points are the following: a) the steady-state
valence of the copper sites and the related problem of the possible
interaction of copper sites on different subunits; b) the actual
formation of an inner sphere O_2^--copper complex and the related
problem of the nature of the enzyme inhibition by singly-charged
anions; c) the role in catalysis of the protein moiety outside the
copper center; d) the role of the zinc in assisting the catalysis
by the nearby copper, and the related problem whether the geometry
of the native copper site is absolutely specific in determining
the features of the catalytic act.

Do the enzyme subunits interact between each other?

    The value of the ratio [E-Cu(II)]/[E-Cu(I)] at steady-state
should be 1, in view of the equal rates[9] of reduction of ECu(II) by
O_2^- and reoxidation of ECu(I) by another O_2^-. However, it was found[9]
that in going from the fully oxidized enzyme to the steady-state
(that is in conditions where $[O_2^-] \gg$ [enzyme]) only approx. 25% of
the 680 nm optical band, typical of the enzyme-bound Cu(II), was
bleached. Similarly, if the H_2O_2-reduced protein was exposed to

excess O_2^- only approx. 25% of the 680 nm band was recovered. Thus there is a "gap" of 50% in accounting for the amount of copper actually functioning. Further work,[14] and very recent experiments with ^{19}F NMR,[15] which selectively "sees" only ECu(II), showed that this "gap" varies with the enzyme batch and can even approach zero in very fresh preparations, in particular those not subjected to lyophilization. However it was clear that there is no relationship between the enzyme activity of the protein and the amount of copper apparently "working". On the contrary, the copper of protein samples having [E-Cu(II)] / [E-Cu(I)] ≃ 1, seems to be "less active" than that of protein samples with higher values of this ratio; in these the constant value of enzyme activity, irrespective of the steady-state value of Cu(II),[15] would be explained by an anticooperative model. Other results,[16-18] suggest that superoxide dismutase subunits actually interact during reconstitution of the holoprotein from the copper-free protein and increasing amount of copper up to the stoichiometric copper/protein ratio. In fact these results showed that stability constants of copper binding, kinetics of copper binding and enzyme activity of one site changed, whether or not the other site was occupied by the metal. This finding was later confirmed[19] (see page160) by results obtained with the isolated dimeric molecule containing copper bound on one subunit only.

All this evidence prompted the search for isolated, and possibly active subunits, an attempt that proved particularly difficult, as most Cu,Zn superoxide dismutases are dissociated into subunits only by denaturing procedures, as, for instance, the combined exposure to SDS, mercaptoethanol, and heat. A new approach was opened by the report[20] that wheat germ contains two isoenzymes, one of which is dissociated into subunits by incubation in SDS alone. This process was studied by Rigo et al.[21] who observed reversible inactivation by 4% SDS and described the kinetics of this inactivation and of reactivation following removal of SDS. It was found that the monomeric protein exists in two states. One state is 500 times less active than the native enzyme, displays an axially altered EPR spectrum, and is likely to correspond to unfolded subunits. The other state, half as active as the native enzyme and apparently representing folded monomers, is observed as a transient state in the reactivation process. These results provided the first experimental evidence for activity in the isolated subunits. However, the instability of the active monomers under the conditions used and the possible interference of protein-bound SDS did not allow to reach a definite conclusion on the question whether subunits are fully active. Urea (8.0 M) was considered as a possible candidate for obtaining active subunits, as it was shown to produce hybrid dimers detectable by gel electrophoresis, upon incubation in urea of two superoxide dismutases from different sources and subsequent removal of urea.[22] However it was later demonstrated, by sedimentation equilibrium analysis,[23] that under these conditions the enzyme

is completely dimeric. Urea just increases the small fraction of
monomeric molecules to an extent which is not observable by transport
methods but allows the detection of hybrids by gel electrophoresis.
Through this urea effect hybrids were produced consisting of an
active subunit and of another subunit inactivated by diazonium
reagents and H_2O_2.[24] The specific activity of the hybrid was approximately half that of the native enzyme. A similar result was obtained by Cocco et al.[19] who produced, by an entirely different procedure, including graded copper removal by diethyldithiocarbamate[25],
an enzyme containing active copper on one subunit and practically
inactive cobalt on the other subunit. Therefore, in these cases,
no mutually inhibitory interaction could be detected. However this
type of dimer with the inactive subunit bound to Cu(II) or Co(II)
may behave differently from that having only one subunit bound to
copper and a specific activity (per copper) significantly higher
than in the native enzyme.[17-19] In fact, it is established, mainly
from work on cobalt derivatives[26,27] that the conformation of the
active site is identical whether copper is reduced or completely
absent, but is different from the conformation with the copper in
the oxidized state (see also page 162). In conclusion, one subunit
may "feel" the conformational influence from the other subunit only
in certain conditions.

    Stable and active subunits can be obtained by extensive succi-
nylation of bovine superoxide dismutase.[28] Succinylated subunits
display the same spectroscopic properties and 10% of the catalytic
activity of the native enzyme. This low catalytic activity is very
likely to be imputable to a charge effect on the efficiency of the
O_2^--enzyme interaction (see page 161). In fact the chemical modification of Arg 141, a positively charged residue positioned nearly
the active site in all known Cu,Zn superoxide dismutases, leads
to an enzyme just 10% as active as the native enzyme.[29] However,
in this case, arginine modification caused extensive changes in the
visible absorbance at 680 nm typical of the copper at the active
site. From the results with the succinylated protein it is defini-
tively established that the integrity of the active site copper is
not the only factor determining the catalytic efficiency of superoxide dismutase but that the charge of the protein plays a role in
the enzyme-substrate interaction (see page 161).

Does the enzyme copper work by an inner-sphere mechanism involving the exchange of the coordinated water?

    Both general considerations on the nature of the copper site
and direct experimental evidence favor an inner-sphere mechanism for
the electron transfer between O_2^- and the superoxide dismutase copper.[30] Firstly, the copper site of superoxide dismutase (Scheme 1)
is characterized by the access of its coordination sphere to the
solvent; secondly, NMR work on the water relaxation rate[30] has de-

mostrated easy exchange of the coordinated water with anionic singly
charged molecules and has allowed to estimate the exchange time
between 10^6 and 10^9 s^{-1}. This value is compatible, at least in the
shorter limit, with the involvement of water-exchange in the catalytic process. As a matter of fact the rates of dismutation by all
known metal types of superoxide dismutases, including the artificial
protein where copper is substituted by cobalt,[5] closely parallel
the rate constants for substitution of their inner-sphere water.[31]

Inner sphere complexes of the superoxide dismutase copper with
singly charged anions are spectroscopically well characterized,[32]
and their association constants are linearly correlated with their
inhibitory power.[33] For all anions, the process appears to affect
the coordinated water molecule.[30] The decrease of enzyme activity,[34]
the rhombic-axial transition of the EPR spectrum,[35] and the change
of water relaxation rate,[36] which reversibly occur between pH 10
and 12, can be explained as well by deprotonation of the copper-bound
water.

The existence of an enzyme-O_2^- complex would be supported by
saturation kinetics of the enzyme. Such kinetics were indeed
observed in polarographic experiments[37] with 1 mM O_2^- which allowed
to estimate $K_M \simeq 3.6 \times 10^{-4}$ M and $V_{max} \simeq 1.0 \times 10^6 s^{-1}$. With the
same technique it was shown that all the inhibitory anions display
a competitive type of inhibition.[33] These data, if confirmed by
other methods, could demonstrate the presence of a monomolecular step
in the mechanism, very likely to involve an inner sphere copper-O_2^-
complex. In relation to this point it should be mentioned that the
inhibitory effect of ionic strength on Cu,Zn dismutase[38] was interpreted in terms of weakened electrostatic interactions between
superoxide and the enzyme.

Do amino acid side chains participate in catalysis?

Strictly related to the last point of the preceeding section
is the question whether charged aminoacid side chains may play a
role in assisting the approach of O_2^- to the copper. This matter has
already been treated to some extent in relation to the activity of
succinylated superoxide dismutase (see page 160). The results
discussed there actually show that neutralization of positive charges
on amino acid side chain, in particular those near the copper, like
Arg 141 in the bovine enzyme, inhibits the catalytic activity.
This effect favors the hypothesis that electrostatic guidance of O_2^-
to the active site copper is important in the enzyme catalysis and
is controlled by some charges of the protein. More questionable is
a role of amino acid side chains in facilitating proton conduction
in the second catalytic step. In fact, absence of any kinetic isotope effect on catalysis[39] upon deuteration of the enzyme and the
pH independence of the activity[6] between pH 5 and 10 indicate that
proton transfer is not rate limiting in the action of superoxide

dismutase. This point will be touched upon again in the next section.

Does the zinc play a role in assisting copper catalysis?

The geometry of the copper site of superoxide dismutase is unique as evidenced by the parameters of its EPR spectrum.[40] This peculiar geometry has been repeatedly associated with the thermo-dynamic and kinetic parameters of the reactions catalyzed. In particular the redox potential of the copper is altered in chemical situations associated with axial conversion of its EPR spectrum, which is typically rhombic in the native enzyme. An example is the low-pH transition, where zinc is released from the imidazolate nucleus bridging it to the copper.[41] Under these conditions, the copper is not reduced by ferrocyanide, as it is at pH > 5, but binds the ligand in a cupric complex with unique EPR spectrum.[42] This and other facts have focused the attention of several investigators on the possible involvment of the zinc in the specific nature of the geometry and catalytic activity of the superoxide dismutase copper.

An interesting finding has been made in pulse radiolysis studies of the protein where zinc is substituted by Co(II) without affecting the functional and catalytic properties of the enzyme.[43] It was found[44] that, upon reduction of copper by O_2^-, the cobalt, which does not react directly with O_2^- in this type of protein, changes its optical spectrum to that typical of cobalt-superoxide dis-mutases, either free of copper or with copper in the cuprous state.[26] The original spectrum is recovered on reoxidation of the copper. These events occur within the same time range of the valence change of the copper during the catalytic cycle of the native enzyme with O_2^-. It was concluded that the imidazolate bridge between copper and zinc (Scheme 1) cyclically releases and rebinds the copper during catalysis in parallel with its reduction and reoxidation by O_2^-. Coincidentally, the zinc-bound imidazolate protonates and deprotonates. The following scheme (Scheme 2) accounts for this mechanism, and includes the exchange of water by the two incoming O_2^- molecules as well.

In relation to this mechanism two important points are to be mentioned: the source of the proton to be provided to the emerging peroxide dianion ($O_2^=$) product in the second step of catalysis (first pK of H_2O_2 = pH 11.6); the change of coordination number of copper during catalysis.

Protonation of the peroxide dianion is essential because charge repulsion makes it very unstable (see page 157). The imidazolate bridge could be the source of protons in the process of H_2O_2 formation. The high rate and the substantial pH-independence of this process require a special mechanism. Reduction of the copper leads to binding of a proton with pK > 9,[46] which is compatible with the pK of the second nitrogen of a zinc-bound imidazole.[47] The expe-

Scheme 2

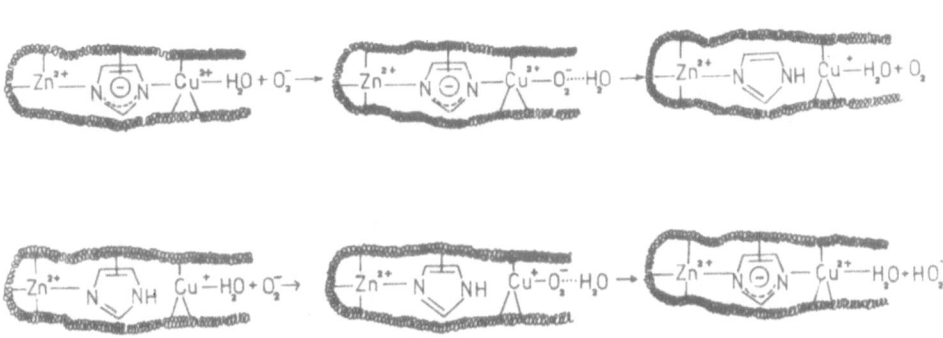

riment[44] discussed above demonstrates the kinetic competence of this proton to be involved in the process of product formation with maximal rates between pH 5 and 10. The reversible decrease of activity[34] between pH 10 and 12 can be correlated with the titration of this imidazole nitrogen in the reduced enzyme as well as with the deprotonation of the copper-bound water (see page 161). It can be concluded that the imidazolate bridge can be the ideal proton source for the reoxidation step, being so close to the copper as to allow this reaction not to be limited by proton transfer.

According to the mechanism shown in Scheme 2, the copper changes its coordination number from five-coordinate in the oxidized state to four-coordinate in the reduced protein. From the X-ray structure[2] (Scheme 1), the remaining four ligands of the reduced copper should be in an approximately tetrahedral geometry. Thus in superoxide dismutase the redox-active metal finds a structural situation such as to have optimal coordination numbers for both its valence states, without appreciable rearrangement of the position of surrounding nuclei.[47] This unique structure can explain why superoxide dismutase undergoes a redox cycle at rates even higher than those of the "blue" copper centers, where the tetrahedral coordination of the cupric ion is considered to be a major factor for rapid valence change.[47]

In conclusion, sufficiently rapid water exchange for entrance of O_2^- in the copper coordination sphere and structural facilitations of rapid reduction and reoxidation of the copper, with a nearby proton source for the forming product, may be the essential points of Cu, Zn superoxide dismutase catalysis.

It is still unclear whether the coordination of the zinc, or
of any other metal replacing it, to the imidazole nitrogen opposite
to the copper, is essential in providing this condition. The zinc-free
protein has a more axial EPR spectrum in either frozen or liquid
solution and, accordingly, is not reduced by either H_2O_2 or ferrocyanide.[48] However, it seems to be fully active below pH 7 . Above
this pH value migration of copper to the vacant zinc site occurs,
resulting in disproportionation of the zinc-free molecules into
half apoprotein and half protein with both sites occupied by copper.[49]
The latter protein is as active as the native enzyme and thus this
process halves the original enzyme activity. The mechanism outlined
in Scheme 2 can well operate even in the absence of the second coordinating metal, being the coordination sphere of the copper not
altered, beside the plausible closer position of his 61, which is
reflected by the more axial line shape of the EPR spectrum. However
it is clear that the full biological potentiality of the enzyme can
be displayed only with the dimetal cluster typical of the native
enzyme. In fact, without zinc, the protein is fully active only
below pH 7 and is much less stable than with zinc or cobalt in the
nearby metal site. Even the protein with both sites occupied by
copper is not as stable as other dimetal derivatives: in particular,
copper is released from the zinc site when the adjacent ion is absent
or reduced.

ACKNOWLEDGEMENTS

We would like to express thanks to all our colleagues who have
contributed their valuable collaboration to our work in this field.
In particular, we should mention Dina Cocco, Laura Morpurgo, Paolo
Viglino, Peter B. Roberts, Michael E. Mc Adam and the decisive role
of Robert C. Bray in the pioneer studies of the mechanism of action.

REFERENCES

1.   G. Rotilio, Magnetic resonance of biological copper centers
      with special emphasis on superoxide dismutases, in "VIth
      International Conference on Magnetic Resonance in Biological
      systems", Kandersteg (1974).
2.   J. S. Richardson, K. A. Thomas, B. H. Rubin, and D. C. Richard-
      son, Crystal structure of bovine Cu, Zn superoxide dismutase
      at 3 Å resolution: chain tracing and metal ligands, Proc.
      Natl. Acad.Sci. U.S.A.72: 1349 (1975).
3.   H. M. Steinman, V. R. Naik, J. L. Abernathy and R. L. Hill,
      Bovine erythrocyte superoxide dismutase. Complete amino acid
      sequence. J. Biol. Chem. 249: 7326 (1974).
4.   G. Rotilio and L. Calabrese, The cobalt derivatives of bovine
      superoxide dismutase in: "Superoxide and superoxide dismu-
      tase", A. M. Michelson, J. M. McCord and I. Fridovich eds.,
      Academic Press, London (1977).
5.   P. O'Neill, E. M. Fielden, D. Cocco, G. Rotilio and L. Calabre-

se, Evidence for catalytic dismutation of superoxide by Co(II) derivatives of bovine superoxide dismutase in aqueous solution as studied by pulse radiolysis, Biochem. J., in press.

6. G. Rotilio, R. C. Bray, and E. M. Fielden, A pulse radiolysis study of superoxide dismutase, Biochim. Biophys. Acta 268: 605 (1972).

7. D. Klug, J. Rabani and I. Fridovich, A direct demonstration of the catalytic action of superoxide dismutase through the use of pulse radiolysis, J. Biol. Chem. 247: 4839 (1972).

8. D. E. Klug-Roth, I. Fridovich, and J. Rabani, Pulse radiolytic investigations of superoxide catalyzed disproportionation, mechanism for bovine superoxide dismutase, J.Amer. Chem. Soc. 95: 2786 (1973).

9. E. M. Fielden, P. B. Roberts, R. C. Bray, D. J. Lowe, G. N. Mautner, G. Rotilio, and L. Calabrese, The mechanism of action of superoxide dismutase from pulse radiolysis and electron paramagnetic resonance, Biochem. J. 139: 49 (1974).

10. G. Rotilio, Copper proteins: problems and prospects, in: "Metalloproteins", U. Weser ed., Thieme Verlag, Stuttgart (1979).

11. A. Rigo and G. Rotilio, Kinetic factors affecting the steady state concentration of superoxide in biological systems and their relation to the production of hydroxyl radicals by metal ions and to the role of superoxide dismutase, in: "Chemical and Biochemical aspects of Superoxide and Superoxide dismutase", J. V. Bannister and H. A. O. Hill eds., Elsevier, North Holland, New York (1980).

12. G. Rotilio, L. Morpurgo, L. Calabrese, and B. Mondovì, On the mechanism of superoxide dismutase. Reaction of the bovine enzyme with hydrogen peroxide and ferrocyanide, Biochim. Biophys. Acta 302: 229 (1973).

13. R. C. Bray, S. A. Cockle, E. M. Fielden, P. B. Roberts, G. Rotilio and L. Calabrese, Reduction and inactivation of superoxide dismutase by hydrogen peroxide, Biochem. J. 139: 43 (1974).

14. S. A. Cockle and R. C. Bray, Do all the copper atoms in bovine superoxide dismutase function in catalysis? in: "Superoxide and superoxide dismutase", A. M. Michelson, J.N. McCord and I. Fridovich eds., Academic Press, London (1977).

15. P. Viglino, A. Rigo, E. Argese, L. Calabrese, D. Cocco and G. Rotilio, ^{19}F relaxation as a probe of the oxidation state of Cu, Zn superoxide dismutase. Studies of the enzyme in steady-state turnover, Biochem. Biophys. Res. Commun. 100: 125 (1981).

16. A. Rigo, P. Viglino, L. Calabrese, D. Cocco and G. Rotilio, The binding of copper ions to copper-free bovine superoxide dismutase: copper distribution in protein samples recombined with less than stoichiometric copper ion/protein ratios, Biochem. J. 161: 27 (1977).

17. A. Rigo, M. Terenzi, P. Viglino, L. Calabrese, D. Cocco and G.

Rotilio, The binding of copper ions to copper-free bovine superoxide dismutase. Properties of the protein recombined with increasing amounts of copper ions, Biochem. J. 161: 31 (1977).

18. A. Rigo, P. Viglino, M. Bonori, D. Cocco, L. Calabrese, and G. Rotilio, The binding of copper ions to copper-free bovine superoxide dismutase. Kinetic aspects, Biochem. J., 169: 277 (1978).

19. D. Cocco, L. Calabrese, A. Rigo, F. Marmocchi and G. Rotilio, Separation of selectively metal free and metal-substituted derivatives by reactions of Cu, Zn superoxide dismutase with diethyldithiocarbamate, Biochem. J. 199: 675(1981).

20. C. O. Beauchamp and I. Fridovich, Isozymes of superoxide dismutase from wheat germ, Biochim. Biophys Acta 317: 50 (1973).

21. A. Rigo, F. Marmocchi, D. Cocco, P. Viglino and G. Rotilio, On the quaternary structure of copper-zinc superoxide dismutase. Reversible dissociation into protomers of the isozyme I from wheat germ, Biochemistry 17: 534 (1978).

22. F. Marmocchi, G. Venardi, F. Bossa, A. Rigo and G. Rotilio, Dissociation of Cu-Zn superoxide dismutase into monomers by urea. Evidence from gel filtration and molecular hybridization, FEBS Lett. 94: 109 (1978).

23. A. Malinowski and I. Fridovich, Subunit association and side-chain reactivities of bovine erythrocyte superoxide dismutase in denaturing solvents, Biochemistry 18: 5055 (1979).

24. D. P. Malinowski and I. Fridovich, Bovine erythrocyte superoxide dismutase. Diazo coupling, subunit interactions and electrophoretic variants, Biochemistry 18: 237 (1979).

25. D. Cocco, L. Calabrese, A. Rigo, E. Argese, and G. Rotilio, Re-examination of the reaction of diethyldithiocarbamate with the copper of superoxide dismutase, J. Biol. Chem. 256: 8983 (1981).

26. L. Calabrese, D. Cocco, L. Morpurgo, B. Mondovì, and G. Rotilio, Cobalt superoxide dismutase. Reactivity of the cobalt cromophore in the copper containing and the copper free protein, Eur. J. Biochem. 64: 465 (1976).

27. A. Desideri, F. Comin, L. Morpurgo, D. Cocco, L. Calabrese, B. Mondovì, W. Maret and G. Rotilio, X-ray absorption edge spectroscopy of Co(II)-binding sites of copper- and zinc-containing proteins, Biochim. Biophys. Acta 670: 312 (1981).

28. F. Marmocchi, I. Mavelli, A. Rigo, R. Stevanato, F. Bossa and G. Rotilio, Succinylated Cu, Zn superoxide dismutase. A novel approach to the problem of active subunits, Biochemistry, in press.

29. D. P. Malinowski and I. Fridovich, Chemical modification of arginine at the active site of bovine erythrocyte superoxide dismutase, Biochemistry 18: 5909 (1979).

30. G. Rotilio, A. Rigo and L. Calabrese, Recent developments on the active-site structure and mechanism of bovine copper-

and zinc-containing superoxide dismutase, in: "Frontiers in Physicochemical Biology" B. Pullman, ed., Academic Press, New York (1978).

31. F. A. Cotton and G. Wilkinson, "Advanced Inorganic Chemistry", Interscience, New York (1972).

32. G. Rotilio, L. Morpurgo, G. Giovagnoli, L. Calabrese, and B. Mondovì, Studies of the metal sites of copper proteins. Symmetry of copper in bovine superoxide dismutase and its functional significance, Biochemistry 11: 2187 (1972).

33. A. Rigo, R. Stevanato, P. Viglino and G. Rotilio, Competitive inhibition of Cu, Zn superoxide dismutase by monovalent anion, Biochem. Biophys. Res. Commun. 79: 776 (1977).

34. A. Rigo, P. Viglino and G. Rotilio, Polarographic determination of superoxide dismutase, Anal. Biochem. 68: 1 (1975).

35. G. Rotilio, A. Finazzi-Agrò, L. Calabrese, F. Bossa, P.Guerrieri, and B. Mondovì, Studies of the metal sites of copper proteins. Ligands of copper in hemocuprein, Biochemistry 10: 616 (1971).

36. M. Terenzi, A. Rigo, C. Franconi, B. Mondovì, L. Calabrese, and G. Rotilio, pH dependence of the nuclear magnetic relaxation rate of solvent water protons in solution of bovine superoxide dismutase, Biochim. Biophys. Acta 351: 230 (1974).

37. A. Rigo, P. Viglino, and G. Rotilio, Kinetic study of O_2^- dismutation by bovine superoxide dismutase. Evidence for saturation of the catalytic sites by O_2^-, Biochem. Biophys. Res. Commun. 63: 1013 (1975).

38. A. Rigo, P. Viglino, G. Rotilio, and R. Tomat, Effect of ionic strength on the activity of bovine superoxide dismutase, FEBS Letters 50: 86 (1975).

39. E. K. Hodgson and I. Fridovich, The interaction of bovine erythrocyte superoxide dismutase with hydrogen peroxide inactivation of the enzyme, Biochemistry 14: 5294 (1975).

40. A. Desideri, L. Morpurgo, G. Rotilio, and B. Mondovì, Stereochemistry of anion complexes of Type 2 Cu(II) in Rhus vernicifera laccase. Analogy with superoxide dismutase and Cu(II) carbonic anhydrase, FEBS Letters 98: 339 (1979).

41. L. Calabrese, D. Cocco, L. Morpurgo, B. Mondovì, and G. Rotilio, Reversible uncoupling of the copper and cobalt spin systems in cobalt bovine superoxide dismutase at low pH, FEBS Letters 59: 29 (1975).

42. L. Morpurgo, I. Mavelli, L. Calabrese, A. Finazzi Agrò, and G. Rotilio, A ferrocyanide charge-transfer complex of bovine superoxide dismutase. Relevance of the zinc imidazolate bond to the redox properties of the enzyme, Biochem. Biophys. Res. Commun. 70: 607 (1976).

43. L. Calabrese, G. Rotilio, and B. Mondovì, Cobalt erythrocuprein: preparation and properties, Biochim. Biophys. Acta 263: 827 (1972).

44. M. E. McAdam, E. M. Fielden, F. Lavelle, L. Calabrese, D. Cocco

and G. Rotilio, The involvement of the bridging imidazolate in the catalytic mechanism of action of bovine superoxide dismutase, Biochem. J. 167: 271 (1977).

45. J. A. Fee, and P. E. DiCorleto, Observations on the oxidation-reduction properties of bovine erythrocyte superoxide dismutase, Biochemistry 12: 4893 (1973).

46. R. T. Sundberg and R. B. Martin, Interactions of histidine and other imidazole derivatives with transition metal ions in chemical and biological systems, Chem. Rev. 74: 471 (1974).

47. G. Rotilio, L. Morpurgo, L. Calabrese, A. Finazzi Agrò, and B. Mondovì, Metal-ligand interactions in Cu-proteins, in: "Metal Ligand Interactions in Organic Chemistry and Biochemistry" part 1, B. Pullman and N. Goldblum eds., D. Reidel Publishing Co., Dordrecht-Holland (1977).

48. L. Calabrese, D. Cocco and G. Rotilio, manuscript in preparation.

49. J. V. Valentine, M. W. Pantoliano, P. J. McDonnell, A. R. Burges and S. T. Lippard, pH-dependent migration of copper(II) to the vacant zinc-binding site of zinc-free of bovine erythrocyte superoxide dismutase, Proc. Natl. Acad. Sci. U.S.A. 76: 4245 (1979).

BIOENERGETICS, MEMBRANE STRUCTURE, AND MULTIENZYME COMPLEXES

PROTON AND ELECTRIC CHARGE TRANSLOCATION IN MITOCHONDRIAL

ENERGY TRANSDUCTION

Albert L. Lehninger

Department of Physiological Chemistry
Johns Hopkins University School of Medicine
Baltimore, MD

Despite the fact that oxidative phosphorylation was first discovered over 40 years ago, we do not yet know precisely how ATP is generated from ADP and phosphate during mitochondrial respiration. This may seem rather paradoxical, since the synthesis of ATP coupled to mitochondrial electron transport is one of the most dynamic processes in cells and necessary for most of their activities. For example, a 70 kg adult with a caloric intake of 3000 kcal per day must generate almost 400 moles of ATP per day, if we therefore assume the standard free energy is +7.5kcal/mol. This amounts to about 190 kg of ATP, or 2.5 times his body weight. Since the body contains altogether only about 50 gm of ATP, the terminal phosphate group of body ATP must turn over several thousand times per day.

Although much progress has been made in the study of oxidative phosphorylation, the problem also appears to be growing more complex. A little over a dozen years ago there appeared to be only 7 or 8 electron carriers in the respiratory chain. Today, however, at least 17 redox centers in the respiratory chain are known, half of them iron-sulfur centers; all are kinetically compatible with the rate of oxygen reduction. Why so many redox centers should be required in the electron transport process is still a mystery. Similarly, we have today much new evidence regarding the structure of the F_1ATPase of the inner mitochondrial membrane, certainly one of the most complex enzymes, with a molecular weight of 380,000 and a total of 10 subunits with multiple binding sites for adenine nucleotides. The ever-increasing complexity has given us very little mechanistic information as to how these membrane-bound systems actually synthesize ATP from ADP and phosphate at the expense of redox energy.

THE INTERMEDIATE HIGH-ENERGY STATE

At the center of the problem is the nature of the modality or
vehicle by which the free energy of electron transport is coupled to
the synthesis of ATP. Today we have two different concepts as to how
this energy transfer is brought about. The chemiosmotic hypothesis
postulates that the required high-energy intermediate or state is a
negative-inside acid-outside electrochemical proton gradient generated
across the inner membrane by electron transport, which is used as
the driving force for the synthesis of ATP. The other concept is
the conformational coupling hypothesis, which in its purest form
postulates that the coupling intermediate is an energy-rich conformational state of the membrane or some membrane component that is
generated by electron transport and utilized as energy source for
synthesis of ATP. The chemiosmotic hypothesis has been very
explicitly developed and has been very successful in the interpretation of experimental data and in the design of new experiments. It
has very solidly established the proton as one of the two elementary
particles of respiratory energy transduction by membrane systems, the
other being of course the electron. The conformational coupling
hypothesis, in contrast, is not very highly developed, has been
stated in many variants, and has not found wide use in either
interpretation or design of experiments. Nevertheless, conformational
coupling can by no means be excluded from consideration. However,
to be valid as a possible mechanism it must account for the
characteristic proton movements and membrane potentials that are
known to be developed by respiring mitochondria.

THE PROTON TRANSLOCATION STOICHIOMETRY OF ELECTRON TRANSPORT

Our approach to the identity of the intermediate high-energy
state is through further analysis of the stoichiometry and mechanism
of H^+ and electric charge movements in the inner membrane that are
driven by electron transport. Our starting point was our discovery
in the early 1960's that close to 2 Ca^{2+} ions, i.e. 4 positive
electric charges, are transported into the mitochondrial matrix from
the medium per electron-pair passing through each of the 3 energy-
conserving segments of the respiratory chain.[1-3] This was the first
demonstration of a definite stoichiometric relationship of a membrane
transport process coupled to electron transport in mitochondria and
has been confirmed many times since. It was also found in the 1960's
that 3 or more K^+ ions are transported into respiring mitochondria
per electron-pair per site in the presence of the K^+ ionophore valinomycin.[4,5] For some years these observations on the stoichiometry
of Ca^{2+} and K^+ uptake by respiring mitochondria have been in conflict
with the conclusion of Mitchell and Moyle in 1965-1967 that 2 H^+ are
translocated outward per pair of electrons per site,[6,7] since much
evidence has shown that respiration-coupled uptake of Ca^{2+} or K^+
proceeds by electroneutral exchange with ejected H^+. Thermodynamic

evidence also strongly suggested that the H^+/site ratio of 2 is too low,[8,9] as did many observations that the influx of P_i and ADP and efflux of ATP alone requires 1 H^+ per external ATP formed,[10] leaving only 1 H^+ per $2e^-$ per site to carry out ATP synthesis. We therefore set out to re-examine the basic proton stoichiometry of mitochondrial electron transport in order to resolve these and other long-standing discrepancies regarding the proton stoichiometry.

We were readily able to repeat the O_2 pulse measurements of Mitchell and Moyle, from which they had concluded that the H^+/site ratio is 2. However, we soon found that their method grossly under-estimated the H^+/site ejection ratio, because a large fraction of the H^+ ejected was almost as quickly reabsorbed by the respiring mitochondria and thus escaped measurement.[11] The interfering H^+ re-uptake was caused by obligatory inward co-transport of H^+ and medium $H_2PO_4^-$ via the phosphate transport system of the inner membrane, in response to the alkalinity developed in the mitochondrial matrix during electron flow from substrate to oxygen.[11] This very fast interfering H^+ back-decay could, however, be inhibited by N-ethylmaleimide, which is a rather specific inhibitor of the H^+/$H_2PO_4^-$ co-transport system when used in low concentrations.[11,12] When the interfering H^+ re-uptake was abolished in this way or by other procedures, the measured H^+/site ratio in oxygen pulse experiments rose from 2 to at least 3.[12] However, the oxygen pulse method requires a number of assumptions and a large upward correction of data. We therefore developed a new method, which utilized measurements of both O_2 consumption and H^+ ejection under aerobic conditions, the reaction being initiated by adding the substrate. The ratio of the initial rates of H^+ ejection and O_2 consumption yields the H^+/O ratio. With this procedure, which eliminates the assumptions and corrections needed in the oxygen pulse method, we found that the average H^+/site ratio during NADH and succinate oxidation is close to 4.0.[12]

These early observations then served as a basis for further refinement of H^+/site stoichiometric measurements. We set ourselves the goal of determining the H^+/site ratio for all 3 sites, not only individually but also in combinations. For example, sites 2 + 3 were examined with succinate as electron donor, oxygen as electron acceptor, and rotenone to prevent electron flow through site 1. Sites 1 + 2 were examined with NAD-linked electron donors in the presence of cyanide to block electron flow through site 3, using either ferricyanide or ferricytochrome c as electron acceptor (Figure 1). Site 2 was isolated by use of rotenone and cyanide, with succinate as electron donor and ferricyanide as electron acceptor. In particular, we wished to examine the H^+ stoichiometry of site 3, the cytochrome oxidase reaction, which Mitchell and Moyle have long insisted cannot translocate protons.[13] For such experiments we employed several different techniques of measurement, different combinations of electron donors and acceptors, and

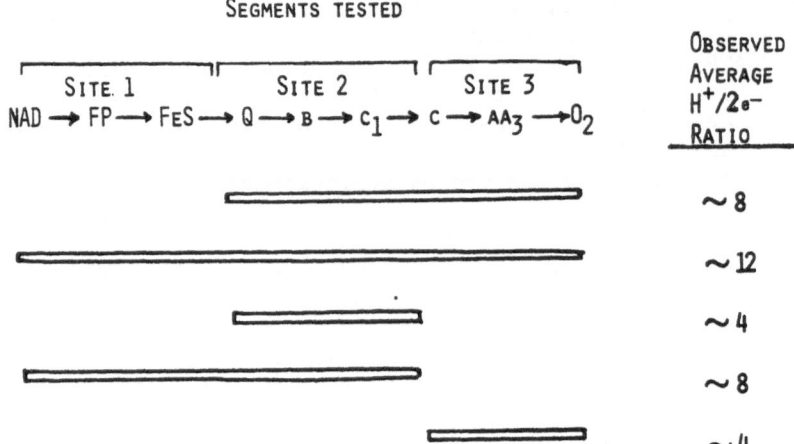

Figure 1.   Diagram showing the segments of the respiratory chain
for which the $H^+/2e^-$ ratio was determined.

mitochondria from different tissues (reviews [14,15]).   Moreover, these
data provided cross-checks for the internal consistency of H^+/site
values, as is indicated by Figure 1.   Our "first generation"
experiments, largely completed in 1979, led to the conclusion that
the maximum H^+/site ratio is at least 3 and very likely close to 4
for each of the 3 energy-transducing segments of the chain, in
contrast to the older value 2.   In addition, we have been able to
show that the H^+/site stoichiometry is the same in mitochondria from
rat liver,[12,14,15] heart,[16] and brain and from Ehrlich ascites tumor
mitochondria.[17]   The latter findings are quite important since all
earlier work had been carried out on rat liver mitochondria, which
are known to promote a number of metabolic reactions other than
electron transport and oxidative phosphorylation, some of which might
result in H^+-absorbing or H^+-releasing side reactions.

"SECOND GENERATION" MEASUREMENTS OF THE H^+/SITE RATIO

Over the past 2 years we have repeated many of our experiments
using better methods and more precise measurements, in response to
criticisms that our use of Clark oxygen electrodes resulted in underestimation of oxygen consumption, due to their sluggish response, and
thus overestimation of the H^+/site ratios.[18,19]   However, this criticism was unfounded, since we had already published data on P/O and
Ca^{2+}/O ratios obtained with our Clark-type electrodes that were in

full agreement with well-established literature values. Moreover,
we had also assured ourselves that our oxygen electrodes, on which
we employed stretched ultra-thin membranes, were readily capable of
accurate recording of oxygen uptake rates in the range required, by
calibrating them against enzymatic reactions in which the H^+/O ratio
was obligatorily stoichiometric, for example, the oxidation of an
aldehyde to an acid or the oxidation of NADH to NAD^+ by submitochondrial particles in the presence of a protonophore.[15] Moreover, we
had also reported $H^+/$site ratios well above 3 for the cytochrome
oxidase reaction using dual-wavelength spectrophotometry, which has
no instrumental dead time, to monitor the rate of ferrocyanide
oxidation.[20] Nevertheless, we embarked on a program to develop an
oxygen electrode much faster in response than the Clark electrode.
Such an electrode is very badly needed in current studies not only
of mitochondrial respiration but also oxygen evolution in photosynthesis research. We have developed a fast-responding oxygen electrode
in which the platinum tip is covered by microscopic layers of sintered
glass rather than Teflon membranes.[21] Such electrodes give 90%
response times as short as 10 milliseconds, compared to several
seconds for commercially available Clark electrodes with standard
Teflon membranes.

The new oxygen electrodes have made possible careful examination
of the kinetics of the early stage of oxygen consumption and H^+
ejection under the special conditions in which H^+/O ratios are
measured, i.e. with K^+ (+ valinomycin) moving inward to provide
charge compensation for the ejected H^+. It was found that the rates
of both oxygen reduction and H^+ ejection during succinate are not
linear, but decline in an exponential manner with time. From the
rate law followed by the two processes it was possible to determine
the true initial rates of O_2 reduction and H^+ ejection, i.e. at zero
rate of H^+ back-decay. Again we found that the average $H^+/$site ratio
is close to 4.0 for succinate oxidation.[21]

Of special importance is the H^+/O ratio for site 3, the cytochrome oxidase reaction, which Mitchell,[13] as well as the group of
Papa,[18] claim does not translocate H^+. Other laboratories have found
cytochrome oxidase to translocate H^+, but with a H^+/O ratio of
2,[19,22-25] in contrast to our finding of close to 4.[20] We have found,
using ferrocytochrome c as substrate (instead of non-biological
electron donors such as ferrocyanide or TMPD) and mitoplasts instead
of mitochondria (to remove the permeability barrier of the outer
membrane to ferrocytochrome c) and either dual-wavelength spectrophotometry of ferrocytochrome c oxidation[26] or measurement of O_2
consumption with the new fast oxygen electrode,[21] that the H^+/O ratio
from initial rates obtained from precise kinetic measurements again
was in substantial excess of 3 and often close to 4.0. Very recently
other work in our laboratory utilizing a new K^+ uptake method for
determination of the charge translocation ratio (see Table 1) has
provided independent verification.[27] Thus the maximal H^+/O ratio for

Table 1. Probable Maximum Stoichiometric Ratios for
          Electron Transport and ATP Synthesis

	H^+/site	Charges/site	ATP/site
Site 1	4	4	~1.0
Site 2	4	2	0.5
Site 3	4	6	1.5

the cytochrome oxidase reaction under optimal conditions is close
to 4.0, a value which suggests that the cytochrome oxidase reaction
can be in near-equilibrium with the phosphorylation potential of
close to 16 kcal per mol ATP. This suggestion has been verified
independently[28] and is supported by the demonstration of a partial
reversal of cytochrome oxidase.[29]

We have therefore concluded on the basis of our very comprehen-
sive measurements by different methods that the average H^+/site
ratio for the 3 sites is close to 4 under optimal conditions. This
value has been independently arrived at by the groups of Azzone[30,31]
and Lemasters.[28] This does not necessarily mean that mitochondrial
electron transport always delivers 4 H^+ per site, since it is
possible that the H^+ stoichiometry may be regulated, particularly
in the cytochrome oxidase reaction, where variable stoichiometry
between 2 and 4 seems quite likely, dependent upon conditions. The
three energy-conserving sites of the respiratory chain are not
equally effective in generating ATP. This is because the net
stoichiometry of electric charge translocation is more basic than
the H^+ stoichiometry in the energy economy of electron transport.
Although the average number of electric charges translocated is 4
per site, site 2 translocates only 2, but site 3 translocates up to
6 electric charges per $2e^-$ (Table 1). Thus the predicted or
determined[30,31] $ATP/2e^-$ synthesis ratio is close to 1.0 for site 1,
close to 0.5 for site 2, and close to 1.5 for site 3 under ideal or
optimal conditions. However, actual measurements of the P/O ratio
in isolated mitochondria are usually less than the theoretical
value 3.[32] As has been shown by non-equilibrium thermodynamic
analysis,[33] the observed P/O ratio is always less than 3.0 depending upon the rate of ATP synthesis and the load against which the
ATP is made, i.e. the phosphorylation potential. Moreover, the
P/O ratio is also depressed by H^+ back-flow, electrical leaks, and
"slippage" of the H^+ pumps.[34]

IMPLICATIONS OF 4 H^+ PER SITE

The average H^+/site ratio approaching 4 is not merely a trivial
matter of changing some numbers. On the contrary, it necessitates a
radical change in a fundamental axiom of the chemiosmotic hypothesis,
namely, that H^+ ejection is driven by ligand conduction, i.e. by
"loops" of electron-carrier molecules in the membrane.[35] The
chemiosmotic hypothesis postulates that 3 pairs of H^+ are translocated outward through the membrane by 3 ligand-conducting loops,
each consisting of a hydrogen-transferring carrier, for example,
$FMNH_2$ and ubiquinol (QH_2), which carry 2 reducing equivalents in the
form of 2 hydrogen atoms from the inner face of the membrane to be
discharged as 2 H^+ at the outer face (Figure 2). The 2 corresponding reducing equivalents are then viewed as passing back to the inner
face of the membrane as electrons by pure electron carriers, such as
iron-sulfur centers or cytochromes. It is clear, however, that such
a mechanism, as is true for all protolytic redox reactions, whether
scalar or vectorial, can deliver no more than 1 H^+ per electron
carried or 2 H^+ per electron-pair. The observation that 4 H^+ rather
than 2 H^+ are translocated per $2e^-$ per site calls for a radically
different molecular process than that proposed in the chemiosmotic
hypothesis.

The most obvious alternative is a conformational H^+ pump which
functions through a thermodynamic linkage between the redox state of
an integral electron-carrying protein and the protonation and
deprotonation of specific ionizing groups on the outer and inner
faces of the protein, as in the linkage between the Bohr protons and
the oxygenation state of hemoglobin, studied in such detail here in
this Institute. Such a redox-driven conformational proton pump must,
however, have a vectorial property; it must accept H^+ at ionizing
groups on the inner face of the membrane on reduction of a specific
redox center and must discharge H^+ from ionizing groups on the outer
face on oxidation of the redox center (Figure 2). Such a conformational pump can in principle transport any number of H^+ per $2e^-$; the
maximum and minimum number of H^+ ions pumped per $2e^-$ would obviously
be determined by the quantitative relationships between the mid-point
potential of its redox center, the magnitude of the electrochemical
H^+ gradient developed across the membrane by electron transport,
the pK' values of the ionizing groups on both faces of the carrier
protein.

But there are other possible mechanisms[36] to account for the
observed $H^+/2e^-$ ratio of 4 per site (Figure 2). One could be a
combination of an H^+-conducting loop, capable of translocating
2 $H^+/2e^-$, plus a conformational H^+ pump also giving 2 H^+/site, to
yield a total of 4 $H^+/2e^-$ per site. It is also possible that each
of the 3 energy-conserving sites might consist of 2 successive H^+-
conducting loops (miniloops ?), each yielding 2 H^+ per $2e^-$. In this
case the full respiratory chain would then consist of six loops,

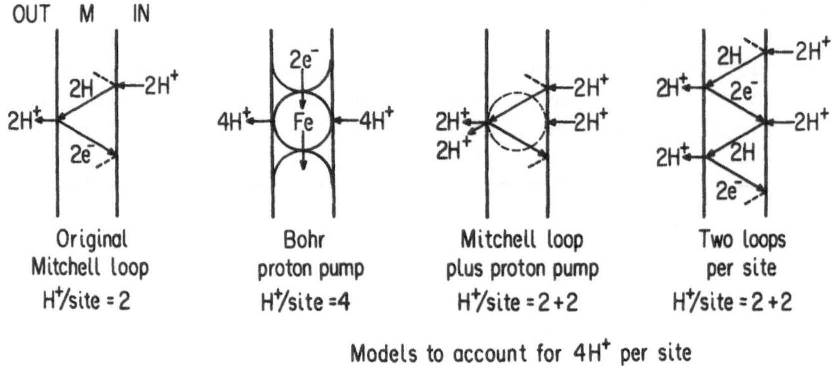

Figure 2. Models to account for the observed average H^+/site ejection ratio of 4.

each translocating 2 H^+. The known occurrence of at least 16 redox centers in the respiratory chain would allow the minimum number of carriers needed, but it would also require a specific alternating sidedness of the electron carriers, as well as appropriately spaced mid-point potentials. We have recently reported some support for the view that an energy-conserving site may consist of 2 H^+-transporting reactions, each delivering 2 H^+. We have found that site 2 of the respiratory chain, which translocates 4 H^+ per $2e^-$, ejects them in 2 separable steps, 2 H^+ via oxidation of ubiquinol (QH_2) plus 2 H^+ in a subsequent step, presumably oxidation of some reduced species in the cytochrome b – cytochrome c_1 segment of the chain.[37]

As pointed out above, the outward translocation of a maximum of 12 H^+ or inward translocation of 12 negative charges per atom of oxygen reduced, if they occur against a membrane potential of about 155–175 mV, negative-inside, could be taken to mean that the fully coupled respiratory chain under ideal conditions is thermodynamically reversible. Moreover, partial reversal of cytochrome oxidase with the formation of spectroscopically detectable oxygenated forms has been recently reported by Wikström[29] and independently observed in our laboratory by A. Alexandre (unpublished data). These observations therefore raise interesting questions regarding the biological regulation of electron transport and oxidative phosphorylation. Until now much evidence has clearly shown that electron flow through sites 1 and 2 is in near-equilibrium with ATP synthesis and thus thermodynamically regulated, whereas electron flow through cytochrome oxidase has been thought to be irreversible and kinetically controlled. If the cytochrome oxidase reaction has a variable H^+/0 stoichiometry, as suggested above, it is possible that both thermodynamic and kinetic regulation of ATP-producing electron transport take place.

H^+ REQUIREMENTS FOR ATP SYNTHESIS

If the electron transport chain translocates a total of 12 H^+ rather than 6 H^+ across the inner membrane, then it is necessary to reconsider the number of H^+ equivalents required to drive the synthesis of 1 molecule of ATP from ADP and P_i during oxidative phosphorylation. Since the maximum P/O ratio for electron transport from NADH to oxygen is 3, as we have recently verified by kinetic experiments and non-equilibrium thermodynamic formalism,[38] then the synthesis of each molecule of medium or cytosolic ATP must require uptake of more than 2 H^+. It is now well established that 1 H^+ per site is required to translocate one molecule each of P_i^- and ADP^{3-} from the cytosol into the matrix in exchange for a molecule of ATP coming out.[10] If the overall H^+/ATP ratio is 2, as was proposed in the chemiosmotic hypothesis, only 1 H^+ would be available per site to carry out the work of ATP synthesis. We have made a direct determination of the H^+/ATP ratio by measuring the initial rate of vectorial H^+ ejection during hydrolysis of added ATP by intact rat liver mitochondria under conditions appropriate for such measurements, with K^+ as charge-compensating cation in the presence of valinomycin. Close to 3 H^+ were ejected and 3 K^+ were taken up for each ATP hydrolyzed by respiration-inhibited mitochondria. Since no transport work is done when ATP undergoes hydrolysis by intact mitochondria under these conditions (presence of valinomycin) the H^+/ATP ratio for synthesis of external ATP, for which 1 H^+ is required for the transport process alone, must be 3 + 1 = 4. This stoichiometry for ATP synthesis has been confirmed by Azzone and his colleagues,[30,31] by Sorgato and Ferguson,[39] and also by Lemasters.[28] This value is also in full agreement with data from several laboratories on the H^+/ATP ratio for the analogous CF_0CF_1-ATPase of chloroplasts. We may therefore conclude that the stoichiometric relationships originally postulated in the chemiosmotic hypothesis are incorrect and must be replaced by an H^+/site ejection ratio of 4 for the 3 sites and an uptake of 4 H^+ to generate each molecule of external ATP from ADP and P_i, three H^+ to make the ATP from ADP and P_i and 1 H^+ to transport it outward.

Little can be said about the mechanism of ATP synthesis, despite much debate over different mechanistic ideologies. However, some interesting progress is being made. F_1ATPase can bind at least 4 molecules of ADP or ATP, two very strongly and the rest less so. Secondly, it has been proposed that the energy needed by the ATPase is not primarily for the synthesis of the new bond of ATP from ADP and P_i, but for release of the newly-formed ATP from the binding site on F_1.[40] Repke[41] and Moudrianakis[42] some years ago proposed that the bound nucleotides participate in an "alternating site" reaction mechanism, which has since been further elaborated by Boyer,[43] in which the binding of ADP and P_i by one site in the respiration-activated enzyme is accompanied by the release of ATP by the other. A third advance has come from the increasing evidence that the subunit structure of F_1 involves 3 α and 3 β subunits in addition to single γ, δ, and ϵ subunits. Since 3 H^+ are involved in the synthesis of ATP, cooperative interactions of the 3 H^+ with the 3 α- or 3 β-subunits appears to be an attractive idea.

Recent research at Johns Hopkins by Pedersen and his colleagues[44] has led to isolation of highly purified F_0F_1ATPase,[44] which when examined by electron microscopy, looks exactly like pictures drawn in textbooks. But more important, Pedersen and Amzel have obtained rat liver F_1 in crystalline form, sufficiently large for single-crystal x-ray analysis.[45,46] This will be a major effort, since the F_1ATPase is among the largest oligomeric proteins for which x-ray analysis has ever been undertaken. A model of the results to 9 Å resolution shows 2-fold symmetry, but higher resolution will be required to confirm the $\alpha_3\beta_3\gamma\delta\epsilon$ subunit structure currently favored.[47]

THE CENTRAL ISSUE AGAIN

But now I must return to the central issue of this discussion. Is a proton gradient across the membrane really the means of conveying redox energy from the electron transport carriers to the ATP synthetase? This may seem to be an unnecessary question to ask, just after I have described the measurement of H^+ translocation during electron transport and ATPase activity. But here I must point out that these H^+ movements are seen and measurable only because of the presence of valinomycin, to allow K^+ to enter the mitochondria to replace the ejected H^+ on an electroneutral basis. If valinomycin is absent, no detectable H^+ would be ejected, nor would K^+ enter mitochondria at a significant rate. Indeed, observation of respiration-coupled H^+ translocation across the mitochondrial membrane requires conditions that do not allow oxidative phosphorylation to take place. We have here an example of Heisenberg's uncertainty principle – to measure H^+ movements we must perturb the system so seriously that it no longer carries out its normal function. It is well established that H^+ <u>can</u> be ejected into the external bulk phase during electron transport and that an H^+ gradient <u>can</u> cause ATP synthesis. However, the question is: Does the system really work this way? Actually, the proton gradient in energized respiring mitochondria is very small, only about 0.3 pH unit. In respiring mitochondria the negative-inside membrane potential $\Delta\Psi$ constitutes 80-85% of the total electrochemical H^+ gradient and appears to be the more directly involved energy term.

The question has been raised whether protons actually cycle between the two bulk phases separated by the inner membrane, as was originally proposed in the chemiosmotic hypothesis. An alternative is that protons may, at least in part, be transferred from the electron transport carriers to the ATPase within or on the surface of the membrane itself, as first suggested by Williams many years ago.[48] Recent observations that open, non-vesicular fragments of the inner membrane of skeletal muscle mitochondria can be energized and cause reverse electron flow support this view.[49] Other recent work has shown that open, non-vesicular sheets of the purple membrane of H. halobium can be energized by illumination.[50] Moreover, intact illuminated H. halobium cells, can under some conditions transfer energy by a pathway not involving the external bulk phase.[51] Increasingly the possibility is being considered that at least some of the H^+ may pass from the respiratory chain to ATP synthetase on or in the membrane, rather than via the external bulk phase.

Yet another question emerges. Is H^+ translocation and/or movement the primary means of recovering energy delivered by electron transport, or is it secondary to some more fundamental process? The total electrochemical H^+ gradient $\Delta\tilde{\mu}_{H^+}$ across the membrane is the sum of two separately measurable terms, the negative-inside membrane potential ($\Delta\Psi$) and the acid-outside pH gradient ($-Z\Delta pH$):

$$\Delta\tilde{\mu}_{H^+} \;=\; \Delta\Psi - Z\Delta pH$$

We may now raise the question as to which is developed first and thus is primary, ΔpH or $\Delta\Psi$, as indicated by the following diagrams, which suggest that either $\Delta\Psi$ or ΔpH may be generated in a side reaction:

One possible criterion is the relative rates at which $\Delta\Psi$ and ΔpH are generated in the transient situation in which electron flow is turned on. This type of experiment can be done in photosynthetic systems. Witt[52] has established the rates of ΔpH and $\Delta\Psi$ formation following a light flash, using optical methods applied to chloroplasts. He has found that $\Delta\Psi$ is generated in about 20 nanoseconds whereas ΔpH generation requires about 20 ms, orders of magnitude slower than $\Delta\Psi$. More recently, Melandri and Melandri[53] have done similar experiments on chromatophores from R. spheroides. On comparing the rates of formation of ATP, $\Delta\Psi$, and ΔpH in the transient from dark to light they found that $\Delta\Psi$ is generated at a rate great enough to provide energy for ATP, which is formed at a surprisingly high rate, much faster than the rate and extent of ΔpH formation. $\Delta\Psi$ is formed at a

high enough rate to provide the energy for ATP synthesis. Although
the rate of diffusion of H^+ within the thylakoids of chloroplasts,
which normally develop $\Delta\tilde{\mu}_{H}+$ almost entirely as ΔpH and very little
as $\Delta\Psi$, is certainly high enough to allow a bulk phase H^+ gradient to
be the normal means of energy transfer, nevertheless, $\Delta\Psi$ is generated first and may be responsible for at least some of the energy
transfer, with ΔpH formed as a possible side reaction.

The two components of $\Delta\tilde{\mu}_{H}+$, namely, $\Delta\Psi$ and ΔpH, differ considerably in their physical characteristics. The membrane potential $\Delta\Psi$,
which is about 150–175 mV in state 4 mitochondria, is a high-
intensity energy state, since it represents an electrical field of
many thousands of volts per cm, but it has a very low capacity. On
the other hand, the pH gradient across the energized inner membrane
is of very low intensity, i.e. a H^+ gradient of no more than about
4:1 across the membrane, but it has rather high capacity because of
the buffering power on both sides of the membrane. Perhaps $\Delta\Psi$ and
ΔpH perform different functions: $\Delta\Psi$ may be used as a high-intensity
source for ATP synthesis and ΔpH as a low intensity form suitable for
energy storage, smoothing of fluctuations, and for driving H^+-coupled
metabolite transport systems. Increasing attention is being focussed
on the possible role of surface potentials and intra-membrane potential differences, instead of transmembrane potentials, in energy-
transduction during ATP synthesis.

From these considerations it is clear that we have much to learn
about oxidative phosphorylation. The chemiosmotic hypothesis, with
its recognition of the proton as one of the elementary particles of
energy transduction has brought us a very long way toward recognizing
the overall pattern of energy transduction in mitochondria. However
the molecular "black box" in which oxidative phosphorylation occurs
is still very much a mystery, which may require much more penetrating
understanding of the molecular structure of energy-transducing
membranes, as well as the molecular physics of electrical and conformational interactions within ordered lipid-protein clusters in
membranes. We all recognize that ATP can be generated by soluble
enzyme systems, as in glycolysis. But the real mystery is why a
membrane system, not merely an enzyme complex, was selected by
evolution to carry out this vital energy transforming activity.

REFERENCES

1. C. S. Rossi and A. L. Lehninger, Stoichiometric relationships
        between accumulation of ions by mitochondria and the energy-
        coupling sites in the respiratory chain, Biochem. Z. 338:
        698 (1963).
2. C. S. Rossi and A. L. Lehninger, Stoichiometry of respiratory
        stimulation, accumulation of Ca^{++} and phosphate, and oxidative
        phosphorylation in rat liver mitochondria, J. Biol. Chem. 239:
        3971 (1964).

3. B. Reynafarje and A. L. Lehninger, the K^+/site and H^+/site stoichiometry of mitochondrial electron transport, J. Biol. Chem. 253:6331 (1978).

4. R. Cockrell, E. J. Harris, and B. C. Pressman, Energetics of potassium transport in mitochondria induced by valinomycin, Biochem. 5:2326 (1966).

5. E. Rossi and G. F. Azzone, Ion transport in liver mitochondria, Eur. J. Biochem. 7:418 (1969).

6. P. Mitchell and J. Moyle, Stoichiometry of proton translocation through the respiratory chain and adenosine triphosphatase systems of rat liver mitochondria, Nature 208:147 (1965).

7. P. Mitchell and J. Moyle, Acid-base titration across the membrane system of rat-liver mitochondria, Biochem. J. 104:588 (1967).

8. H. Rottenberg, The measurement of transmembrane electrochemical proton gradients, J. Bioenerg. 7:61 (1975).

9. D. G. Nicholls, The influence of respiration and ATP hydrolysis on the proton-electrochemical gradient across the inner membrane of rat liver mitochondria as determined by ion distribution, Eur. J. Biochem. 50:305 (1974).

10. A. C. Schoolwerth and K. F. LaNoue, Metabolite transport in mitochondria, Ann. Rev. Biochem. 48:871 (1979).

11. M. D. Brand, B. Reynafarje, and A. L. Lehninger, Re-evaluation of the H^+/site ratio of mitochondrial electron transport with the oxygen pulse technique, J. Biol. Chem. 251:5670 (1976).

12. B. Reynafarje, M. D. Brand, and A. L. Lehninger, Evaluation of the H^+/site ratio of mitochondrial electron transport from rate measurements, J. Biol. Chem. 251:7442 (1976).

13. J. Moyle and P. Mitchell, Cytochrome c oxidase is not a proton pump, FEBS Letts. 88:268 (1978).

14. A. L. Lehninger, B. Reynafarje, and A. Alexandre, Stoichiometry of proton movements coupled to mitochondrial electron transport and ATP hydrolysis, in: "Structure and Function of Energy-Transducing Membranes," K. van Dam and B. V. van Gelder, eds., Elsevier, Amsterdam (1977), p. 95.

15. A. L. Lehninger, B. Reynafarje, A. Alexandre, and A. Villalobo, Proton stoichiometry and mechanisms in mitochondrial energy transduction, in: "Membrane Bioenergetics," C-P. Lee, G. Schatz, and L. Ernster, eds., Addison-Wesley, Reading, MA (1979), p. 393.

16. A. Vercesi, B. Reynafarje, and A. L. Lehninger, Stoichiometry of H^+ ejection and Ca^{2+} uptake coupled to electron transport in rat heart mitochondria, J. Biol. Chem. 253:6379 (1978).

17. A. Villalobo and A. L. Lehninger, The proton stoichiometry of electron transport in Ehrlich ascites tumor mitochondria, J. Biol. Chem. 254:4352 (1979).

18. S. Papa, F. Guerrieri, M. Lorusso, G. Izzo, D. Boffoli, F. Capuano, N. Capitanio, and N. Altamura, The H^+/e$^-$ stoichiometry of respiration-linked proton translocation in the cytochrome system of mitochondria, Biochem. J. 192:203 (1980).

19. M. Wikström and K. Krab, Respiration-linked H^+ translocation in mitochondria: Stoichiometry and mechanism, Curr. Topics in Bioenerg. 10:51 (1980).

20. A. Alexandre, B. Reynafarje, and A. L. Lehninger, Stoichiometry of vectorial H^+ movements coupled to electron transport and to ATP synthesis in mitochondria, Proc. Natl. Acad. Sci. 75: 5296 (1978).

21. A. L. Lehninger, B. Reynafarje, P. Davies, A. Alexandre, A. Villalobo, and A. Beavis, The stoichiometry of H^+ ejection coupled to mitochondrial electron flow, measured with a fast-responding oxygen electrode, in: "Mitochondria and Microsomes," C. P. Lee, G. Schatz, and G. Dallner, eds., Addison-Wesley, Reading, MA (1979), p. 459.

22. M. D. Brand, W. G. Harper, D. G. Nicholls, and W. J. Ingledew, Unequal charge separation by different coupling spans of the mitochondrial electron transport chain, FEBS Letts. 95:125 (1978).

23. E. Sigel and E. Carafoli, The proton pump of cytochrome c oxidase and its stoichiometry, Eur. J. Biochem. 89:119 (1978).

24. E. Sigel and E. Carafoli, The charge stoichiometry of cytochrome c oxidase in the reconstituted system, J. Biol. Chem. 254: 10572 (1979).

25. R. P. Casey, J. B. Chappell, and A. Azzi, Limited-turnover studies on proton translocation in reconstituted cytochrome c oxidase-containing vesicles, Biochem. J. 182:149 (1979).

26. A. Alexandre, The stoichiometry of H^+ ejection coupled to oxidation of ferrocytochrome c by rat liver mitoplasts, Fed. Proc. 39:1706 (1980).

27. A. Beavis, The stoichiometry of K^+ accumulation coupled to mitochondrial respiration - a solution to the controversy, Fed. Proc. 40:1563 (1981).

28. J. J. Lemasters, Near thermodynamic equilibrium of oxidative phosphorylation by inverted inner membrane vesicles of rat liver mitochondria, FEBS Letts. 110:96 (1980).

29. M. Wikström, Energy-dependent reversal of the cytochrome oxidase reaction, Proc. Natl. Acad. Sci. 78:4051 (1981).

30. T. Pozzan, V., Miconi, F. Di Virgilio, and G. F. Azzone, H^+/site, charge/site, and ATP/site ratios at coupling sites I and II in mitochondrial e^- transport, J. Biol. Chem. 254:10200 (1979).

31. G. F. Azzone, T. Pozzan, and F. Di Virgilio, H^+/site, charge/ site, and ATP/site ratios at coupling site III in mitochondrial electron transport, J. Biol. Chem. 254:10206 (1979).

32. P. C. Hinkle and M. L. Yu, The phosphorus/oxygen ratio of mitochondrial oxidative phosphorylation, J. Biol. Chem. 254: 2450 (1979).

33. J. W. Stücki, The optimal efficiency and the economic degrees of coupling of oxidative phosphorylation, Eur. J. Biochem. 109:269 (1980).

34. D. Pietrobon, G. F. Azzone, and D. Walz, Effect of funiculosin and antimycin A on the redox-driven H^+-pumps in mitochondria:

On the nature of "leaks," Eur. J. Biochem. 117:389 (1981).

35. P. Mitchell, Compartmentation and communication in living systems. Ligand conduction: A general catalytic principle in chemical, osmotic, and chemiosmotic reaction systems, Eur. J. Biochem. 95:1 (1979).

36. A. L. Lehninger, Proton transport and charge separation across the mitochondrial membrane coupled to electron flow, in: "Protons and Ions Involved in Fast Dynamic Phenomena," P. Laszlo, ed., Elsevier, Amsterdam (1978), p. 435.

37. A. Alexandre, F. Galiazzo, and A. L. Lehninger, On the location of the H^+-extruding steps in site 2 of the mitochondrial electron transport chain, J. Biol. Chem. 255:10721 (1980).

38. A. Beavis, Re-evaluation of the P/O ratios of rat liver mitochondria, Fed. Proc. 39:2056 (1980).

39. M. C. Sorgato and S. H. Ferguson, unpublished observations.

40. P. D. Boyer, B. Chance, L. Ernster, P. Mitchell, E. Racker, and E. C. Slater, Oxidative phosphorylation and photophosphorylation, Ann. Rev. Biochem. 46:955 (1977).

41. K. R. H. Repke and R. Schön, Flip-flop model of energy interconversion by ATP synthetase, Acta Biol. Med. Ger. 33:K27 (1974).

42. R. Adolfsen and E. Moudrianakis, Molecular polymorphism and mechanism of activation and deactivation of the hydrolytic function of the coupling factor of oxidative phosphorylation, Biochem. 15:4163 (1976).

43. P. D. Boyer, The coupling of proton translocation to ATP formation, in: "Mitochondria and Microsomes," C. P. Lee, G. Schatz, and G. Dallner, eds., Addison-Wesley, Reading, MA (1981), p. 407.

44. J. W. Soper, G. L. Decker, and P. L. Pedersen, Mitochondrial ATPase complex. A dispersed, cytochrome-deficient, oligomycin-sensitive preparation from rat liver containing molecules with a tripartite structural arrangement, J. Biol. Chem. 254:11170 (1979).

45. M. L. Amzel and P. L. Pedersen, Adenosine triphosphatase from rat liver mitochondria. Crystallization and x-ray diffraction studies of the F_1-component of the enzyme, J. Biol. Chem. 253:2067 (1978).

46. M. L. Amzel, Structure of F_1ATPases, J. Bioenerg. & Biomem. 13: 109 (1981).

47. M. L. Amzel, M. McKinney, P. Narayanan, and P. L. Pedersen, Structure of the mitochondrial F_1-ATPase at 9 Å resolution, Nature (submitted).

48. R. J. P. Williams, Possible functions of chains of catalysts, J. Theor. Biol. 1:1 (1961).

49. B. T. Storey, D. M. Scott, and C-P. Lee, Energy-linked quinacrine fluorescence changes in submitochondrial particles from skeletal muscle mitochondria. Evidence for intramembrane H^+ transfer as a primary reaction of energy coupling, J. Biol. Chem. 255:5224 (1980).

50. D. B. Kell, On the functional proton current pathway of electron transport phosphorylation, Biochim. Biophys. Acta 549:55 (1979).

51. H. Michel and D. Oesterhelt, Electrochemical proton gradient across the cell membrane of Halobacterium halobium: Effect of N,N'-dicyclohexylcarbodiimide, relation to intracellular adenosine triphosphate, adenosine diphosphate, and phosphate concentration, and influence of the potassium gradient, Biochem. 19:4607 (1980).

52. H. T. Witt, Energy conversion in the functional membrane of photosynthesis. Analysis by light pulse and electric field, Biochim. Biophys. Acta 505:355 (1979).

53. B. A. Melandri and A. Baccarini-Melandri, Bioenergetics of the early events of bacterial photophosphorylation, in: "Cation Flux Across Biomembranes," Y. Mukohata and L. Packer, eds., Academic Press, New York (1979), p. 219.

THE MOLECULAR SLIPPING IN THE REDOX-DRIVEN H$^+$ PUMPS

Giovanni F. Azzone, Daniela Pietrobon and Dieter Walz°

C.N.R. Center for the Study of Physiology of Mitochondria
and Institute of General Pathology, University of Padova
Italy. °Biozentrum, University of Basel, Switzerland

In mammalian mitochondria, as well as in bacteria, respiration
is coupled to H$^+$ ion extrusion due to operation of the redox-driven
H$^+$ pumps. This leads[1,2] to formation of a H$^+$ electrochemical gradient,
$\Delta\tilde{\mu}_H$. Two rates of respiration are distinguished, that of fully
uncoupled mitochondria and that of coupled mitochondria in the stationary state, denoted as state 4 or static head. In uncoupled mitochondria the respiratory rate J_e is limited purely by the kinetics
of e$^-$ transfer in the respiratory chain. On the other hand in coupled
mitochondria the respiratory rate in static head, J_e^{sh}, is limited
partly by the kinetics of e$^-$ transfer and partly by the thermodynamics of the H$^+$ pump (energetic control). The nature of the energetic
control requires some clarification.

Isolated mitochondria incubated aerobically with a substrate
reach spontaneously a stationary state of minimal entropy production
and therefore of low respiration[3]. A similar state is reached when
ADP is fully phosphorylated to ATP. In this state the rate of e$^-$
transfer J_e^{sh} is minimal, the H$^+$ electrochemical gradient $\Delta\tilde{\mu}_H^{sh}$
is maximal, and the net extrusion of H$^+$ ions is negligible. In this
stationary state, denoted static head, the dimension of $\Delta\tilde{\mu}_H^{sh}$

187

originates from the balance of H^+ ion pumping and passive influx of H^+ ions along the $\Delta\tilde{\mu}_H$ gradient.

According to the chemiosmotic hypothesis the rate of e^- transfer in static head J_e^{sh} is essentially determined by the passive influx of H^+ ions along the $\Delta\tilde{\mu}_H$ gradient.[1,2] The influx occurs through "leaks" which have no physical or functional connection to the redox driven H^+ pump. Then the passive influx $J_H inf$, may be taken as proportional to $\Delta\tilde{\mu}_H$, with a proportionality factor being the leak conductance, L_H^1:

$$J_H inf = L_H^1 \, \Delta\tilde{\mu}_H \tag{1}$$

This rate of passive H^+ influx is balanced in static head by an equal rate of H^+ ion efflux, $J_h eff$. However since the H^+ efflux is due to the redox driven H^+ pump, it may be taken as proportional to the e^- flow, J_e, with a proportionality factor[4] being the H^+/e^- stoichiometry, n:

$$J_H eff = n \, J_e \tag{2}$$

At static head $J_H inf = J_H eff$, hence

$$L_H^1 \Delta\tilde{\mu}_H^{sh} = n \, J_e^{sh} \tag{3}$$

Thus according to equation (3) addition of a respiratory inhibitor, which causes an inhibition of J_e^{sh} should result, if L_H^1 and n remain constant, in a proportional depression of $\Delta\tilde{\mu}_H$. Furthermore addition of a respiratory inhibitor should cause a depression of the ratio between uncoupled J_e and coupled J_e^{sh} e^- flow, also denoted as respiratory control ratio. This is due to the fact that in the uncoupled e^- flow J_e, the control is purely kinetic while in the coupled e^- flow J_e^{sh}, the control is both kinetic and energetic.

Since the inhibitors depress $\Delta\tilde{\mu}_H^{sh}$ and then the energetic control one expects a more marked inhibition of J_e than of J_e^{sh}. This would result in a depression of the respiratory control ratio.

In 1974 Nicholls[5] reported that addition of increasing malonate concentrations to intact liver mitochondria resulted first in a depression of J_e^{sh} with no change of $\Delta\tilde{\mu}_H^{sh}$ and then a proportional depression of J_e^{sh} and $\Delta\tilde{\mu}_H^{sh}$. To explain an inhibition of J_e^{sh} without parallel depression of $\Delta\tilde{\mu}_H^{sh}$ Nicholls proposed the hypothesis that the H^+ conductance, L_H^1, is not constant but rather a function of $\Delta\tilde{\mu}_H$

$$L_H^1 = f(\Delta\tilde{\mu}_H) \tag{4}$$

with f strongly increasing above a critical value of $\Delta\tilde{\mu}_H$, say when $\Delta\tilde{\mu}_H$ is close to static head. Nicholls's view has been extensively adopted[6,7] to explain the relative independence of $\Delta\tilde{\mu}_H$ to inhibition of the respiration. Below we will show some new findings[8] which cannot be easily explained with this concept.

Fig. 1 shows the effect of increasing funiculosin concentrations, an inhibitor of respiration in the b-c$_1$ complex[9], on three parameters namely the electron flow in static head, J_e^{sh}, the $\Delta\tilde{\mu}_H^{sh}$ in static head, and the ratio of the electron flow in the presence and absence of uncoupler J_e^u/J_e^{sh}. It may be noted that: i) at about 60 pmol funiculosin x mg prot^{-1} the inhibition of respiration J_e^{sh} was more than 90% while the inhibition of $\Delta\tilde{\mu}_H^{sh}$ was only about 8%; ii) the electron flow ratio J_e^u/J_e^{sh}, first increased up to a funiculosin concentration of 25 pmol x mg prot^{-1} where J_e^{sh} was about 50% inhibited, and then steeply declined to approach the value of 1 at inhibitor concentration depressing J_e^{sh} more than 90%. Experiments like those shown in Fig. 1 have been made by using other inhibitors such as antimycin or malonate, or other e$^-$ acceptor, instead of oxygen, such as ferricyanide. The relationship between $\Delta\tilde{\mu}_H$ and H^+ conductance L_H^1 can be assessed by combining equation (4) with equation (3) and obtaining:

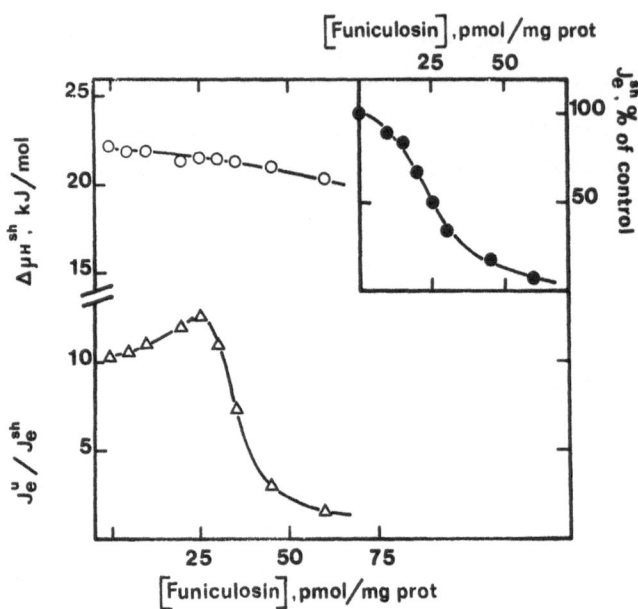

Fig. 1. Effect of funiculosin on $\Delta\tilde{\mu}_H^{sh}$ (o) and electron flow, J_e^{sh} (●) for mitochondria at static head, as well as on the electron flow ratio J_e^i/J_e^{sh} (△), 10 mM Tris-Cl, 10 mM Tris-acetate, 25 mM Tris-succinate, 0.1 mM EDTA, 2 μM rotenone and 1 μg oligomycin x mg prot^{-1}, pH 7.0. Temperature 24°C.

$$f(\Delta\tilde{\mu}_H) = n J_e^{sh}/\Delta\tilde{\mu}_H^{sh} \tag{5}$$

Fig. 2 shows a plot of $n J_e^{sh}/\Delta\tilde{\mu}_H^{sh}$ vs $\Delta\tilde{\mu}_H^{sh}$ as obtained from the experiments carried out by using various respiratory inhibitors and various e$^-$ acceptors. It is important to note that $n J_e^{sh}/\Delta\tilde{\mu}_H$ reflects the H$^+$ conductance and then the plot shows the dependence of the H$^+$ conductance vs $\Delta\tilde{\mu}_H^{sh}$. If the conductance were constant the plot would be a straight line parallel to the abscissa. This is however not the case. In accordance with what already observed by Nicholls[5] at a certain value of $\Delta\tilde{\mu}_H^{sh}$ there seems to be an exponential rise of H$^+$ conductance. The critical point however is that either the $\Delta\tilde{\mu}_H^{sh}$ value at which there begins the rise or the level of the rise are not unique but vary depending on the inhibitors or on the e$^-$ acceptor. This indicates that there is not only one valid function correlating $nJ_e^{sh}/\Delta\tilde{\mu}_H^{sh}$ to $\Delta\tilde{\mu}_H^{sh}$ but rather a set of different functions according to the inhibitor or to the e^- acceptor.

In order to cope with these observations we have proposed a new concept in bioenergetics namely that of the "slipping" H^+ pump. In our view the residual respiration in static head H_e^{sh} is not due to H^+ leaks.L_H^1, but rather to molecular slips in the redox driven H^+ pumps. According to this concept the redox-driven H^+ pumps

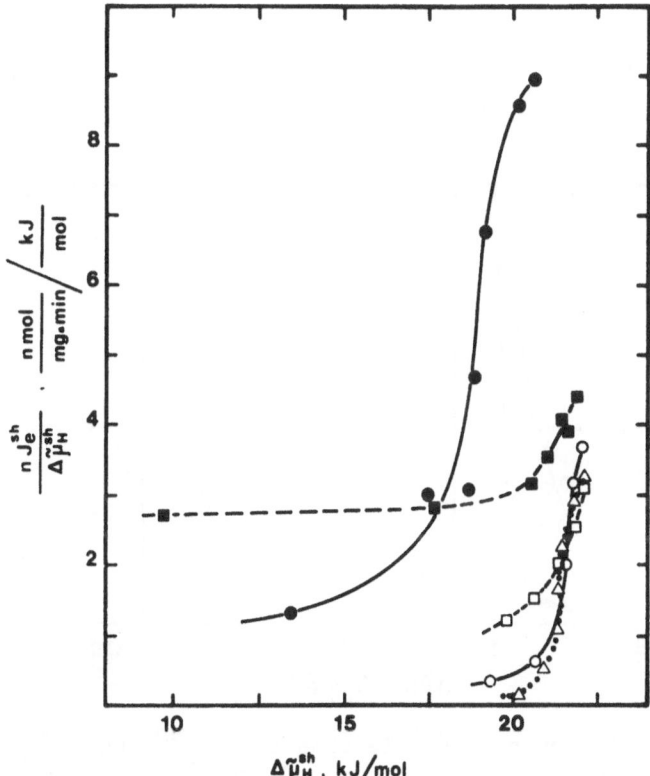

Fig. 2. Effect of the leak conductance L_H^1 on $\Delta\tilde{\mu}_H$. Values of n $J_e^{sh}/\Delta\tilde{\mu}_H^{sh}$ were calculated from various experiments using various types of inhibitors and e^- acceptors, and from the literature; n=2 for ferricyanide and n=4 for oxygen as e^- acceptor, respectively. Inhibitors: funiculo-sin (Δ); antimycin A, with oxygen (o) and ferricyanide (●) as e^- acceptor; malonate, data from Nicholls[5] (■) or Pietrobon et al.[8] (□). J_e^{sh} is in nmol x min⁻¹ and $\Delta\tilde{\mu}_H^{sh}$ as kJ/mol.

are not tightly coupled but rather they display a certain number of
failures. Then during a certain percentage of turnovers of the pumps
there is transfer of e^- without movement of H^+ across the membrane.
The opposite is also true. H^+ ions can move along the $\Delta\tilde{\mu}_H^{sh}$ gradient through the pump without moving e^- backwards in the respiratory chain.

   A further qualitative description of the concept of molecular
slipping in the redox-driven H^+ pump can be obtained by observing
the plot of Fig. 3 where the data have been replotted as J_e^{sh} at
each inhibitor concentration vs. $\Delta\tilde{\mu}_H^{sh}$, both divided by the perti-
nent values without inhibitor for scaling reasons. Suppose there

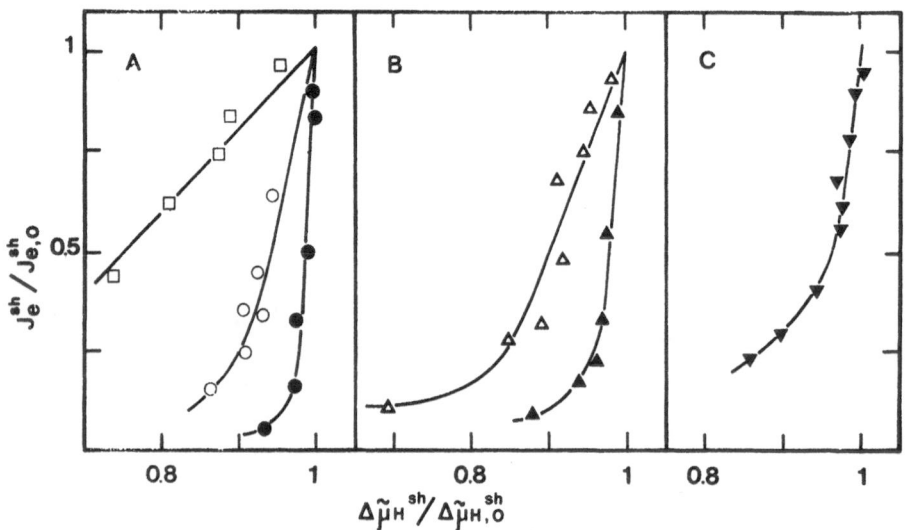

Fig. 3.   Correlation between electron flow J_e^{sh} and $\Delta\tilde{\mu}_H^{sh}$ for
          mitochondria at static head in the presence of different
          inhibitors. For scaling reasons, J_e^{sh} and $\Delta\tilde{\mu}_H^{sh}$ were di-
          vided by the pertinent parameters in the absence of inhibi-
          tors, $J_{e,0}^{sh}$ and $\Delta\tilde{\mu}_{H,0}^{sh}$ respectively. In A inhibition with
          funiculosin; data without uncouplers (●), or with (o) 10
          pmoles FCCP x mg prot^{-1} ($J_{e,0}^{sh}$ = 90 nmolxmin^{-1} x mg prot^{-1};
          $\Delta\tilde{\mu}_{H,0}^{sh}$ = 17.2 kJ/mol) or with (□) 15 pmol FCCPxmg prot^{-1}
          ($J_{e,0}^{sh}$ = 120 nmol x min^{-1} x mg prot^{-1}, $\Delta\tilde{\mu}_{H,0}^{sh}$ = 14.7 kJ/mol).
          In B, inhibition with antimycin A; data with (▲) oxygen, or
          (△) ferricyanide as e^- acceptor. In C, inhibition with
          malonate.

are no H^+ leaks (L_H^1 = 0) and that addition of respiratory inhibitor results in a block of the pumps and of the slip through the pump. In this case it is expected that $\Delta\tilde{\mu}_H^{sh}$ should not change an inhibition of J_e^{sh}, which would be represented by a vertical line at $\Delta\tilde{\mu}_H^{sh}/\Delta\tilde{\mu}_{H,0}^{sh}$ = 1. The data with antimycin and funiculosin nearly represent this situation, thus indicating that the H^+ leaks provide a minor contribution to J_e^{sh}. Limited amounts of uncoupler increase the H^+ conductance and then L_H^1. This leads to a considerable deviation with respect to the vertical line. The deviation also observed with ferricyanide does not need to be explained by larger L_H^i, but can be accounted for in terms of non equilibrium thermodynamics by different phenomenological coefficients. The similar plot observed with malonate as inhibitor is also easily understoood since malonate limits the supply of e^- to the pumps and then limits the contribution of the "redox slip" to J_e^{sh}.

Other data obtained in our laboratory are in accord with this concept and permit the not tightly coupled redox driven H^+ pump to be dealt with by the formalism of non equilibrium thermodynamics. From these data and this treatment the conclusion can be drawn that uncouplers not only increase the H^+ leaks but also uncouple the pumps themselves.

REFERENCES

1. P. Mitchell, Chemiosmotic coupling in oxidative and photosynthetic phosphorylation, Biol. Rev. 41: 445 (1966).
2. P. Mitchell, A commentary on alternative hypotheses of protonic coupling in the membrane systems catalyzing oxidative and photosynthetic phosphorylation, FEBS Letters 78: 1 (1977).
3. D. Walz, Thermodynamics of oxidation-reduction reactions and its application to bioenergetics, Biochim. Biophys. Acta 505: 279 (1979).
4. H. D. Westerhoff and K. Van Dam, Irreversible thermodynamic description of energy transduction in biomembranes, in: "Current Topics in Bioenergetics", D.R. Sanadi ed., vol. 9, Acad. Press, New York, S. Francisco, London (1979).
5. D. G. Nicholls, The influence of respiration and ATP hydrolysis on the proton-electrochemical gradient across the inner

  membrane of rat-liver mitochondria as determined by ion
  distribution, Europ. J. Biochem. 50: 305 (1974).
6. C. Sorgato, S. J. Ferguson, Variable proton conductance of
  submitochondrial particles, Biochemistry 18:5737 (1979).
7. D. B. Kell, P. John, and S. J. Ferguson, On the current-voltage
  relationship of energy-transducing membranes: phosphoryla-
  ting membrane vesicles from Paracoccus dinitrificans,
  Biochem. Soc. Trans. 6: 1292 (1978).
8. D. Pietrobon, G. F. Azzone and D. Walz, Effect of funiculosin
  and antimycin A on the redox-driven H^+-pumps in mitochondria:
  on the nature of "leaks", Europ. J. Biochem. 117: 389 (1981).
9. U. K. Moser and P. Walker, Funiculosin: a new specific inhibitor of the respiratory chain in rat liver mitochondria,
  FEBS Letters 50: 279 (1975).

STRUCTURAL ASPECTS OF THE BINDING OF MELITTIN TO PHOSPHOLIPID

BILAYERS, AS A MODEL FOR PROTEIN-LIPID INTERACTIONS IN MEMBRANES

Roberto Strom,[+] Franca Podo,[++] Carlo Crifò[+] and
Giuseppe Zaccai[^]

(+) Institutes of Biological Chemistry, Universities of
Rome and of L'Aquila, and C. N. R. Center for Molecular
Biology, Rome, Italy; (++) Istituto Superiore di Sanità,
Rome, Italy; (^) Institut Laue-Langevin, Grenoble, France

INTRODUCTION

Membranes can be envisaged[1,2] as a two-dimensional solution of
proteins immersed in an anisotropic lipid-water phase, the characteristics of these proteins depending - as it happens for any other
protein - on their structure, their interaction with the "solvent"
(summarizing in this term the different portions of the phospholipid
bilayer as well as the aqueous region on either side of it), their
rotational and translational mobility in this "solvent", their
ability to undergo conformational changes and/or to exist in a monomeric or multimeric state.

A detailed study of these aspects is often made difficult not
only by solubility problems and by the tight association of the
proteins with other membrane components, but also by the connection between the specific features of these proteins and their
vectorial location in the asymmetric and anisotropic membrane.

Melittin, the most abundant peptide of bee venom, though being
not, strictly speaking, a membrane protein, has the peculiar property of combining a high water solubility to a considerable affinity
for lipid membrane systems. The small molecular weight of this peptide, and the knowledge of its primary sequence (Fig. 1) allow a
fairly detailed investigation of the structural aspects of its
binding to phospholipid bilayers, as a basis for its permeability
-enhancing effect on membranes.

Gly-Ile-Gly-Ala-Val-Leu-Lys-Val-Leu-Thr-Thr-Gly-Leu-
-Pro-Ala-Leu-Ile-Ser-Trp-Ile-Lys-Arg-Lys-Arg-Gln-GlnNH$_2$

Fig. 1:   Aminoacid sequence of melittin from <u>Apis mellifera</u>, ac-
          cording to Habermann and Jentsch.[3]

MELITTIN CONFORMATION IN SOLUTION

        Circular dichroism spectra in the 200-240 nm region indicate
(Fig. 2) that melittin has essentially a random coil conformation
when dissolved in salt-free water at neutral pH. Addition of salts
of most monovalent anions in moderate amounts (e.g. 0.15 M NaCl)
does not appreciably modify the overall CD spectrum, while a clear
shift toward a right-handed helical conformation occurs upon addition of divalent (or multivalent) anions (Fig. 2a) or of phospho-

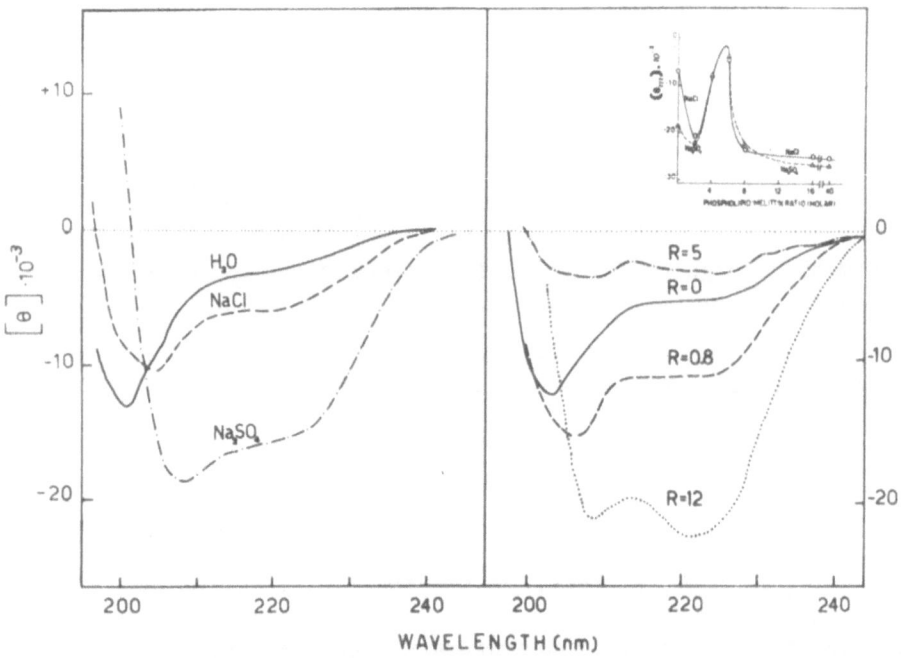

Fig. 2.   Circular dichroism spectra of melittin in aqueous solution
          (a) and upon binding to egg lecithin vesicles (b), R being
          the phospholipid-to-melittin molar ratio. The insert re-
          presents the dependence, on this ratio, of the ellipticity
          at 222 nm.

lipid vesicles (Fig. 2b). In the latter case, a transient flattening
—shown in the insert of Fig. 2b— of the CD spectra[4,5] indicates the
occurrence of optical artifacts, such as described by Urry and
Krivacic,[6] presumably due to co-precipitation of highly hydrophobic
melittin-phospholipid in pairs which re-dissolve in excess phospholipid.

The occurrence of a more complex situation -- i.e. the existence
of more than the two extreme conformations, random coil and helix,
of the peptide -- is indicated by high-resolution NMR studies.[7,8]
^{13}C-NMR spectra (Fig. 3) of melittin dissolved in 0.15 M phosphate
at pH 7 are e.g. characterized (curve c) by a wider spread of sharp
resonances in the 160-180 ppm region(corresponding to signals arising
from the peptide carbonyl groups), as compared to the profiles in
D_2O (curve a) or in 0.15 M NaCl (curve b). By contrast, in the 10-
70 ppm region the profile in 0.15 M NaCl shows greater similarity
to that in phosphate than to the one in D_2O.

Similar conclusions can be reached by analysis of the ^{1}H-NMR
spectra, where the assignment[9,10] of several signals to specific
protons (Fig. 4) clarifies the various conformational changes oc-

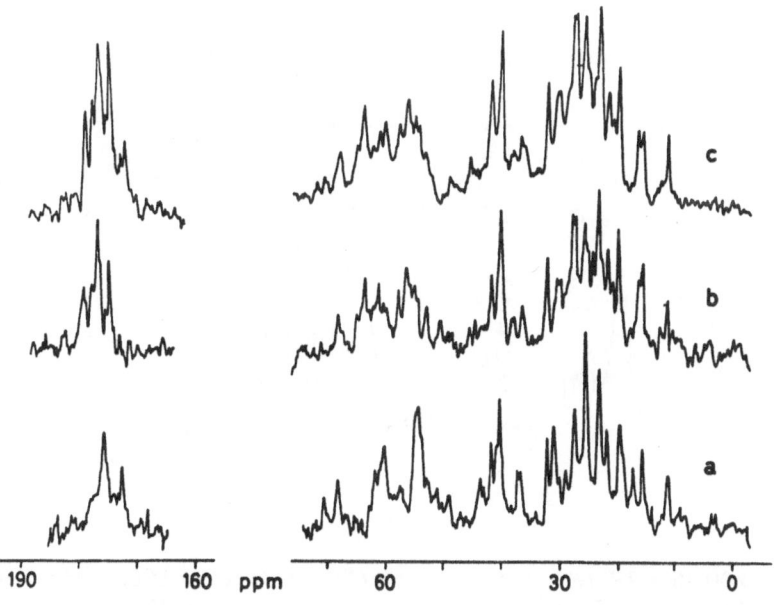

Fig. 3. ^{13}C-NMR spectra of melittin in a) D_2O , pH 6 (melittin
         concentration: 3.5 mM); b) 0.18 M NaCl, pH 6 (melittin
         concentration: 3.5 mM); c) 0.15 M phosphate buffer, pH 7
         (melittin concentration: 21 mM). From Podo et al.[7]

curring in the peptide molecule. In the absence of salts, the occurrence of sharp and well-resolved resonances (lower trace) indicates
that the peptide is in a highly flexible, random conformation --
this conclusion being also supported by photon correlation spectroscopy measurements,[7] which indicate a very low value -- around 5.4
A -- of the equivalent hydrodynamic radius. Melittin assumes instead
a more compact conformation, indicated by a more complex ^{1}H-NMR

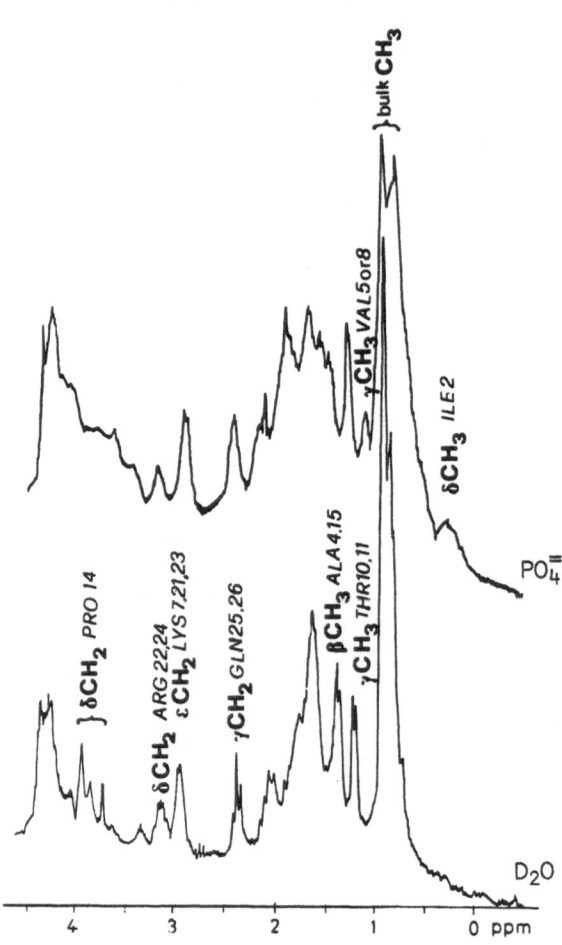

Fig. 4.   200 MHz ^{1}H-NMR spectra at 24°C of 4 mM melittin in D_2O
           (lower profile)and in the presence of 13.75 molar excess
           of phosphate buffer, pH 7.0 (upper profile). Some resonan-
           ces of particular interest are indicated. Slightly modified
           from Podo et al.[7]

profile and by some resonance shifts (upper trace of Fig. 4), in
0.15 M aqueous phosphate buffer; correspondingly, the value of the
equivalent hydrodynamic radius can then be estimated to be around
18 A, as would be expected for a disc-like tetramer.

Correlations can be established between the binding of phosphate
ions (evaluated by ^{31}P-NMR) and the structural variations undergone
by the peptide. It can be shown[7],[8] that some spectral changes, corresponding to an overall ordering of the single peptide protomer or
to changes in the local conformation around Pro 14, occur with a
direct proportionality to the binding of phosphate (Fig. 5a), while
other ^{1}H-NMR signals are modified only upon binding of a larger
number of phosphate ions (Fig. 5b), this phenomenon being presumably related to inter-chain interactions with formation of tetrameric structures.

In dilute (0.15 M) NaCl at neutral pH, the features of the
^{1}H-NMR spectrum resemble somewhat those found at low phosphate concentration, indicating that under these conditions each polypeptide
chain, though non-helical, has a relatively compact and ordered
structure. The corresponding value of the translational diffusion
coefficient is indeed consistent[7] with the assumption of a 44x13x13
prolate ellipsoid formed by a single polypeptide chain, with an
equivalent hydrodynamic radius of approximately 9 Å.

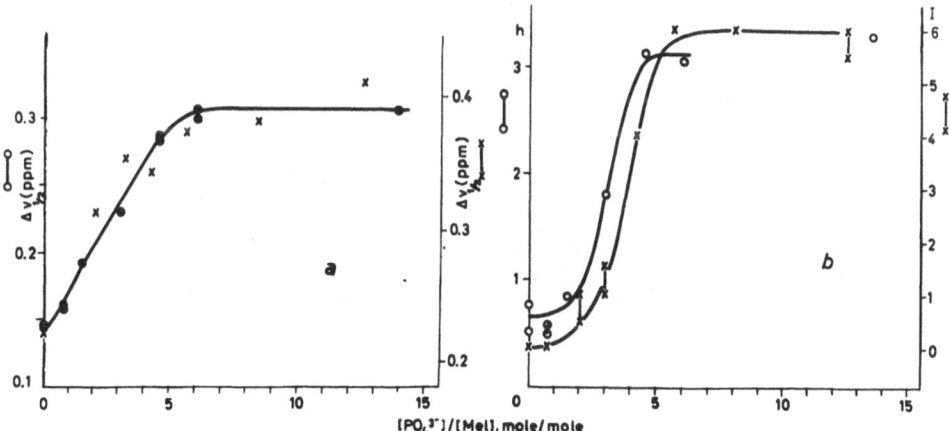

Fig. 5. Variations, as a function of phosphate-to-melittin molar
         ratio, of ^{1}H-NMR signals arising from melittin methyl groups:
         a) width at half-height, $\Delta\nu_{\frac{1}{2}}$, of the main -CH$_3$ band cente-
         red at ~ 0.9 ppm (200 MHz, 24°C: O-O; 100 MHz, 29°C: x-x);
         b) relative height (at 200 MHz: O-O) and intensity (at
         100 MHz: x-x) of the Ile 2 δ CH$_3$ signal appearing at ~0.3
         ppm. From Podo et al.[8]

CONFORMATION OF MELITTIN UPON INTERACTION WITH PHOSPHOLIPIDS

As previously mentioned (see Fig. 2b), melittin assumes a right-handed helical conformation also upon interaction with phospholipids. This effect appears on the whole to occur independently of the fluidity of the alkyl chains, although the extent of the interaction may be different above or below the chains' transition temperature T_m. It can thus be shown (Fig. 6) that, in the presence of egg phosphatidylcholine or of perdeuterated dipalmitoyl phosphatidylcholine vesicles below the T_m, most ^{1}H-NMR spectral features of melittin are similar to those existing in aqueous phosphate, except for a reduction of the ε CH$_2$ signals of Lys 7, 21 and/or 23, for a broadening and/or a reduction of Gln 25, 26 γ CH$_2$ resonances, and for the absence of the upfield shift of the Ile 2 δ CH$_3$ signal.[7]

Due to the overlap with the resonances from phospholipid choline headgroups, nothing can be said about the δ CH$_2$'s of Arg 22, 24. There is also a consistent broadening of the Trp 19 profile (not shown in Fig. 6), which contrasts with the finding[4] of longer

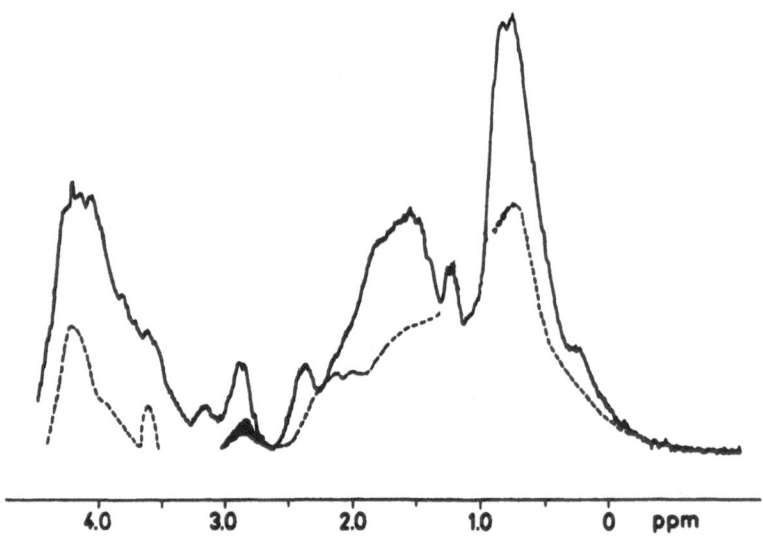

4.0   3.0   2.0   1.0   0   ppm

Fig. 6.   Comparison of the 100 MHz ^{1}H-NMR spectra at 32°C of 4 mM melittin in 0.15 M phosphate buffer, pH 7.2 (full line) and upon addition of a 5-fold excess of perdeuterated dipalmitoylphosphatidylcholine (dashed line), the latter after subtraction of the spectrum of phospholipids alone. (omitting the regions where the spectrum of phospholipids had been altered by the interaction with melittin). Unpublished data by Strom et al.[11]

spin lattice relaxation times of the signal(s) arising from the protons in position 2 and/or 7 of this same Trp 19.

Upon interaction with perdeuterated dipalmitoylphosphatidylcholine vesicles above the T_m, the ^{1}H-NMR profile of melittin is broadened below detection,[7] although the interaction is revealed by the circular dichroism spectra and by the broadening of the ^{31}P-phosphate signal and of the ^{1}H-resonances from the phospholipid headgroups. These results may be tentatively correlated to the conclusions reached by Verma et al.[12] about a tighter interaction of melittin with phospholipids above the T_m.

EFFECTS OF MELITTIN ON PHOSPHOLIPIDS ORGANIZATION

Neutron scattering density profiles

In oriented multilayers of egg phosphatidylcholine or of dimiristoylphosphatidylcholine, addition of melittin even in a relatively large proportion (1/10 of the phospholipids on a molar basis) did not appreciably modify the typical bilayer as revealed by low angle neutron scattering experiments. Fig. 7, where selectively deuterated

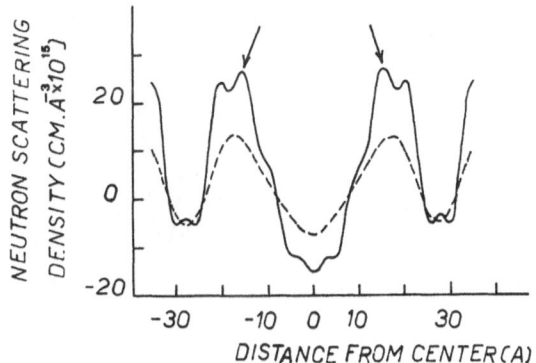

Fig. 7.   Comparison of the neutron scattering density profiles, on an absolute scale, of C6-^{2}H dimiristoylphosphatidylcholine-water multilayers at 15°C and 95% relative humidity with (dashed line) or without (full line) added melittin (phospholipid-to-melittin molar ratio = 1:10, in the melittin-containing-sample). The arrows indicate the position of the selective deuteration in C6 on the phospholipid alkyl chains. Unpublished data by Strom et al.[11]

dimiristoylphosphatidylcholine was used in order to ensure the cor-
rectness of the phase assignments, shows that, when melittin is
present in the sample, the neutron scattering density profile, though
less resolved, is not conspicuously modified.By progressively substituting D_2O to H_2O, it can be shown that exchangeable protons, which
in the absence of melittin exist only in the outer region of the
bilayers -- not beyond a distance of 14 Å from the center -- can
instead be found,when the peptide is present, up to 4-6 Å from the
center of each bilayer.

A comparison of the neutron scattering density profiles on an
absolute scale, show, on the other hand, that melittin is present
throughout the bilayer (Fig. 8).

Modifications of the NMR signals of the phospholipids

Further indications about the effects of melittin on phospho-

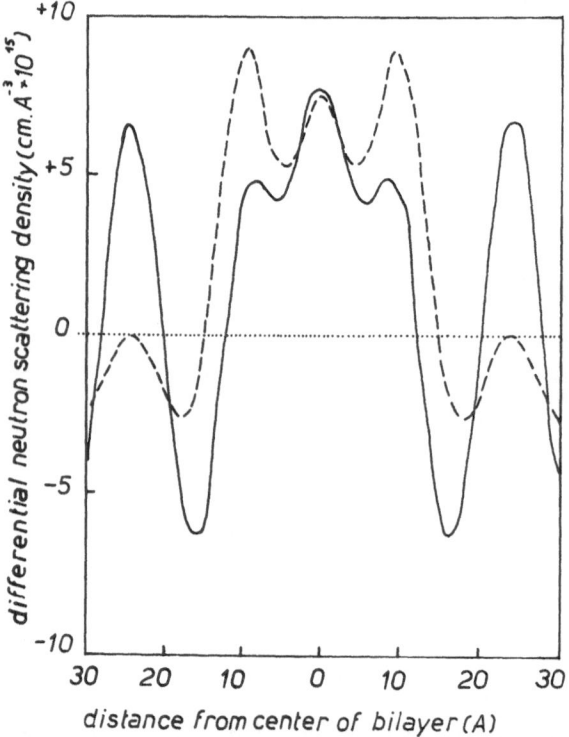

Fig. 8.   Differential neutron scattering density profile, giving the
          position of melittin incorporated in egg phosphatidylcho-
          line multilayers at 20°C and 75% relative humidity, in the
          presence of H_2O (full line) or of a 45% D_2O - 55% H_2O
          mixture (dashed line). From Podo et al.[8]

lipids architecture in the vesicles can be obtained by examining the modifications induced by the peptide on the NMR resonances arising from the phospholipids.

While the resonances from the $>CH_2$ and $-CH_3$ protons of the alkyl chains are not appreciably modified by the addition of melittin, the involvement of the polar headgroups is thus documented by the conspicuous broadening caused by the peptide on the ^{1}H-NMR signals arising from $N^+(CH_3)_3$, $N-CH_2$ and $CO-CH_2-C$ protons, and, even more markedly, by the broadening of the ^{31}P-NMR signals. The orientation of the choline residues of the phospholipids seems however not to be modified, the value of the nuclear Overhauser effect remaining constantly around 40%.

The increase of the permeability of the vesicles to water and to water solutes is evidenced and can be quantitatively evaluated by measuring, according to Hunt et al.[13] the accessibility, to an NMR shift reagent, of the choline headgroups located on the inner and outer surface, respectively, of the vesicles. While in the absence of melittin only the resonances from the outer N-methyl protons are shifted by Pr^{3+} ions, those from the inner compartment remaining unshifted, addition of the peptide causes a progressive fusion of the two signals, due to the penetration of Pr^{3+} in the vesicles. The phenomenon, the kinetics of which can be followed as a function of the phospholipid-to-melittin ratio (Fig. 9), is reversed by addition of EDTA, and restored with more Pr^{3+}, thus documenting that the vesicles still have an inner aqueous compartment, even if its accessibility to Pr^{3+} is considerably increased by the presence of melittin.

DISCUSSION

A most interesting property of melittin resides in its ability to assume, according to the milieu in which it is dissolved, a variety of conformational states. In aqueous solution at neutral pH, the possible conformations range from a highly flexible monomeric random coil to a tetrameric association of right-handed helices -- an additional intermediate well-defined state being that of a compact, though still monomeric and non-helical, structure.[7] Association with phospholipids results in a helical conformation of the peptide, a distinct feature being however the absence of any spectral evidence of inter-chain association (there is notably, in contrast to what occurs in aqueous phosphate, no ring-current upfield shift exerted by Trp 19 on the ^{1}H-NMR resonances of Ile 2 δ CH_3). Ultracentrifugation and quasi-elastic light-scattering analysis by Brown et al.[14] have shown , on the other hand, that mixed dodecyl-phosphocholine-melittin micelles are formed by single peptide chains associated with approximately forty detergent molecules.

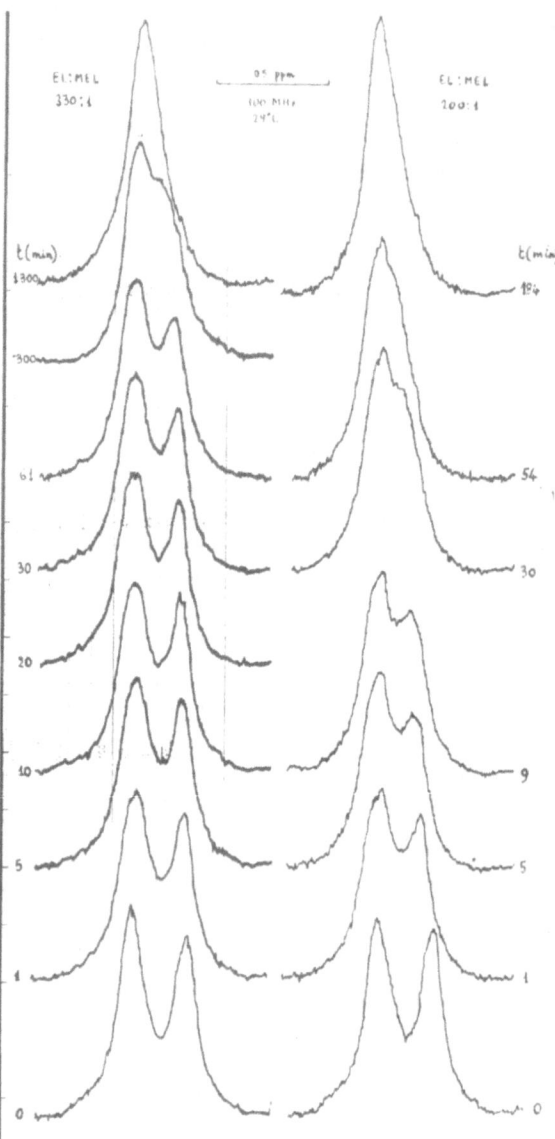

Fig. 9. Kinetics of Pr^{3+} entrance in egg phosphatidylcholine vesicles after addition of melittin. By examining the ^{1}H-NMR
spectra at various times (shown, in minutes, near each
spectrum), the modifications induced by melittin (added
at phospholipid-to-melittin ratios of 330: 1 or 200: 1)
on the accessibility to Pr^{3+} ions of resonances from the
choline headgroups of the inner and the outer compartments
of the vesicles can be followed.

The primary sequence of the N-terminal half of melittin is indeed such that in a α-helical conformation the hydrophilic amino-acids are all confined on one-third of the helix surface, the rest of which is covered by hydrophobic residues. This can easily explain why in aqueous solution the peptide chains -- when structured in α-helices by the presence of divalent anions -- associate in tetramers; this inter-chain association being instead superfluous when the helices are surrounded by a hydrophobic milieu.

The phospholipid-peptide interaction is not solely governed by the ionic attraction between the highly charged cationic C-terminal fragment of melittin and the phosphate group of the phospholipids, though undoubtedly favoured by it -- as shown e.g. by reciprocal effects of Gln and Lys proton resonances on the peptide and on the ^{31}P and choline-^{1}H signals of the phospholipids. In fact, addition of dicetylphosphate to dipalmitoylphosphatidylcholine vesicles can be shown to increase the rate at which the interaction takes place, but not its stoichiometry.[5]

According to Dawson et al.,[15] the "membrane-disrupting" effect of melittin could be due to a wedge-like action of the peptide, which, interposed between the polar heads of the phospholipids, would induce a secondary disordering of the lipid bilayer. The neutron diffraction data show instead that the overall lamellar pattern is preserved also with high melittin-to-phospholipid ratios. The presence of the peptide also in the central region of the bilayer, the existence of exchangeable protons near this region, and the Pr^{3+} permeation experiments, which indicate that melittin-doped vesicles still possess distinct outer and inner layers of phospholipids, substantiate the hypothesis of hydrophilic channels penetrating the bilayers. Similar conclusions have recently been reached also by Tosteson and Tosteson,[16] and are in good agreement with the demonstration, by Lavialle et al.,[17] of melittin-induced alterations of the C-H and C-C vibrational modes of the phospholipid acyl chains.

The presence of a hydrophilic side on the otherwise hydrophobic N-terminal helix of melittin could suggest the formation of such aqueous channels by interaction of two or more peptide units. The 5th power dependence, on melittin concentraiton, of the Pr^{3+} permeation rate appears to substantiate this hypothesis.

Recent experiments by Kempf et al.[18] have shown that, in oxidized cholesterol black lipid membranes, melittin causes an increase of conductance only if a trans-negative voltage difference is imposed; under such conditions - but not if the potential is reversed -- melittin is accessible to the proteolytic action of pronase added to the trans side. The possibility of an electric field-dependent channel formation --somewhat similar to those hypothesized by Urry[19] -- should therefore be considered. Alternatively,

conducting and non conducting "conformations" could be due to a
more or less profound penetration of the peptide in the lipid core.

The limitations of melittin as a model for the study of peptide-
lipid interactions in membranes reside in the dramatic effect it
exerts on membrane permeability. Other peptides, having e.g. a
protected amino-terminal end or lacking the highly charged Lys 7
residue in the otherwise essentially hydrophobic N-terminal stretch,
can indeed resemble melittin in their ability to be both water- and
lipid-soluble, without having however such permeability-increasing
potency.

REFERENCES

1. S. J. Singer and G. L. Nicolson, The fluid mosaic model of the
      structure of cell membranes, Science 175: 720 (1972).
2. G. Vanderkooi, Organization of proteins in membranes with spe-
      cial reference to the cytochrome oxidase system, Biochim.
      Biophys. Acta 344: 307 (1974).
3. E. Habermann and J. Jentsch, Sequenzanalyse des Melittins aus
      den tryptischen und peptischen Spaltstücken, Hoppe-Seyler's
      Z. Physiol. Chem. 348: 37 (1967).
4. R. Strom, C. Crifò, V. Viti, L. Guidoni and F. Podo, Variations
      in circular dichroism and proton-NMR relaxation properties
      of melittin upon interaction with phospholipids, FEBS Letters
      96: 45 (1978).
5. G. Carpinelli, A. Fiori, C. Crifò, F. Podo and R. Strom,
      manuscript in preparation (1982).
6. D. W. Urry and J. Krivacic, Differential scatter of left and
      right circularly polarized light by optically active parti-
      culate systems, Proc. Natl. Acad. Sci. U.S.A. 65: 845 (1970).
7. F. Podo, R. Strom, C. Crifò and M. Zulauf, Dependence of melit-
      tin structure on its interaction with multivalent anions and
      with model membrane systems, Int. J. Peptide Protein Res. 19:
      in press (1982).
8. F. Podo, R. Strom, C. Crifò, C. Berthet, M. Zulauf and G. Zaccai
      The interaction with phospholipids of bee venom melittin: a
      structural study of the peptide and lipid component, Biophys.
      J. 37, in press (1982).
9. J. Lauterwein, L. R. Brown and K. Wüthrich, High-resolution
      ^{1}H-NMR studies of monomeric melittin in aqueous solution,
      Biochim. Biophys. Acta 622: 219 (1980).
10. L. R. Brown, J. Lauterwein and K. Wüthrich, High-resolution
      ^{1}H-NMR studies of self aggregation of melittin in aqueous
      solution, Biochim. Biophys. Acta 622: 231 (1980).
11. R. Strom, F. Podo, C. Crifò, C. Berthet, and G. Zaccai, manu-
      script in preparation (1982).
12. S. P. Verma, D. F. H. Wallach and J. C. P. Smith, The action
      of melittin on phosphatide multilayers as studied by infrared

dichroism and spin labeling, Biochim. Biophys. Acta 345: 129 (1974).

13. G. R. A. Hunt, L. R. H. Tipping and M. R. Belmont, Rate-determining processes in the transport of Pr^{3+} ions by the ionophore A 23187 across phospholipid vesicular membranes. A ^{1}H-NMR and theoretical study, Biophys. Chem. 8: 341 (1978).

14. J. Lauterwein, C. Bösch, L. R. Brown and K. Wüthrich, Physicochemical studies of the protein-lipid interactions in melittin-containing micelles, Biochim. Biophys. Acta 556: 244 (1979).

15. C. R. Dawson, A. F. Drake, J. Helliwell and R. C. Hider, The interaction of bee melittin with lipid bilayer membranes, Biochim. Biophys. Acta 510: 75 (1978).

16. M. T. Tosteson and D. C. Tosteson, The sting. Melittin forms channels in lipid bilayers, Biophys. J. 36: 109 (1981).

17. F. Lavialle, I. W. Levin and C. Mollay, Interaction of melittin with dimiristoyl-phosphatidylcholine liposomes. Evidences for boundary lipid by Raman spectroscopy, Biochim. Biophys. Acta 600: 62 (1980).

18. C. Kempf, R. D. Klausner, J. N. Weinstein, J. Van Renswoude, M. Pincus and R. Blumenthal, Voltage-dependent transbilayer orientation of melittin. Personal communication, to be published (1982).

19. D. W. Urry, A molecular theory of ion-conducting channels: a field-dependent transition between conducting and nonconducting conformations, Proc. Natl. Acad. Sci. U.S.A. 69: 1610 (1972).

ACETYLCHOLINE RECEPTORS FROM ELECTROPLAX MEMBRANES: IN VITRO AND IN SITU PROPERTIES

M. Martinez-Carrion, J.M. Gonzalez-Ros, M. Llanillo and A. Paraschos

Department of Biochemistry, Medical College of Virginia
Virginia Commonwealth University, Richmond, Virginia
23298

INTRODUCTION

Cell surface receptors are capable of recognizing chemical mediators of cellular communication and generate a specific signal for a biological response. At certain neuronal synapses, the vertebrate neuromuscular end plate and the electric tissue of certain elasmobranch, the chemical mediator is acetylcholine and the receptor is a postsynaptic transmembrane protein.[1] Acetylcholine binding induces opening of short lived ($\sim$ 3 ms) membrane channels permeable to monovalent or divalent cations of less than 8 Å in diameter.[1] This event leads to a conductance increase and a depolarization of the membrane. While it is agreed that the acetylcholine receptor (AcChR)* and the ionic channel are coupled in the membrane, the relationship of the receptor binding sites and the ionic channel sites is still unclear. When the neurotransmitter concentration or that of a variety of compounds eliciting similar responses (agonists) is maintained, the number of active receptor channels slowly declines and the phenomenon is known as desensitization. In general, activation of the receptors results from recognition of agonists with low affinity for the resting state of the AcChR. As binding slowly develops desensitization ensues which parallels the appearance of a state of high AcChR affinity for those agonists.[1,2] Voltage clamp

---

*Abbreviations: AcChR, acetylcholine receptor; DPH, 1,6-diphenyl-1,3,5-hexatriene; MBTA, 4-(N-maleimido)benzoyltrimethylammonium; PC, phosphatidylcholine; EP, ethanolamine phosphoglycerides; PySA, pyrenesulfonyl azide; TMA-DPH, 1-(4-trimethylammonium phenyl)-6-phenyl-1,3,5-hexatriene.

and radioactive tracers ion flux studies of the rapid activation
process are consistent with a mechanism in which binding of two or
more agonist molecules to the receptor is required to open the ion
channel.[1,2] Elapid α -neurotoxins appear to share binding domains
with agonists and they have become valuable probes in determining
the degree of occupation of the receptor by agonists and in the
molecular characterization of the receptor.

Because of high concentration of AcChR in the electroplax of
electric fish, this tissue is preferred for molecular studies of
this receptor which is the most extensively characterized neuro-
receptor to date. The electric ray Torpedo californica electroplax
membranes share many analogies with those of muscle receptors and
this tissue has been a favorite for the isolation of solubilized
receptor and the preparation of membrane vesicles in which over 50%
of the protein is represented by the AcChR and in which ligand binding, desensitization and ion flux can be reproduced with responses
similar to those detected through electrophysiologic measurements
in whole cells. In this work, we describe the molecular and functional properties of the AcChR when, after isolation, it is incorporated
into a lipid bilayer in a process of membrane reconstitution in
which the properties of reconstituted membranes are compared to
those of freshly isolated whole electroplax membranes.

MOLECULAR PROPERTIES OF AcChR

AcChR can be solubilized from Torpedo californica and other
electric fish species through the use of detergents. Nonionic
detergents are best and Triton X-100 has been preferred. However,
this detergent binds very tightly to AcChR and can amount to as
much as 50% of the total weight of the receptor preparation and
cannot be easily separated from the protein. More recently, β-D-
octylglucopyranoside has been introduced as an effective solubilizer
of membrane electroplax proteins and removal of this detergent by
dialysis is complete.[3] Solubilized AcChR can be purified through
the use of affinity chromatography methods using agonists or antagonists (compounds which prevent ionic conductance), including
α-neurotoxins, as ligands attached to the solid chromatographic
matrix. SDS polyacrylamide gels reveal four types of subunits
(Fig. 1) of 40, 50, 60, 65,000 daltons for the purified AcChR with
stoichiometry of 2;1;1;1. The molecular weight of the receptor is
260,000 and it may exist as a dimer of about 500,000 daltons in
which the monomers are linked by a disulfide bond between the 65,000
dalton subunits. The Stokes radius of detergent free receptor is
70 Å and has an acidic isoelectric point.[3] The first 60 amino acid
residues in the N-terminal sequence of the anion subunits are highly
homologous, yet, only the 40,000 subunit has been associated, through
covalent affinity binding studies, as that containing the site(s)
for agonist and antagonist. Separation of the subunits is achieved
through the use of SDS detergent, yet subunit interchanging between

two populations of AcChR has been claimed.[4] We have also used 3 M
urea to dissociate the subunits which, after a separation through
gel permeation procedures, can regain the properties of isolated
receptor upon the re-mixing of the components and removal of the
urea. This approach appears promising in the creation of partial
aggregates and hybrid subunit species and/or the correlation of
formation of this complex to agonist and antagonist binding, as
well as membrane reconstitution and generation of selected binding
properties. Intact isolated receptor monomer appears to have two
ligand binding sites for reversible agonists such as carbamylcholine
and for antagonists such as tubocurarine or α-neurotoxin. Only one
of these sites appears easily affinity-labeled after reduction of
a critical disulfide, yet the other available site remains accessible to carbamyl or acetylcholine as can be detected by their
ability to induce a Na^+ ion influx in sample membrane preparations
after the sites had been affinity labeled (with bromoacetylcholine
or MBTA). By using more drastic conditions both sites can be labeled with bromoacetylcholine.[5]

MEMBRANES RICH IN AcChR

Preparation of membrane vesicles rich in AcChR is accomplished
through homogenization of electroplax tissue and centrifugation in
sucrose density gradients. This procedure distinctively separates

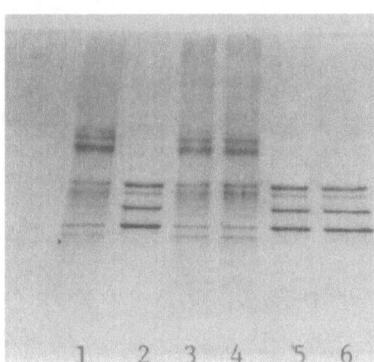

Fig. 1.   Polypeptide patterns on SDS-polyacrylamide exponential
          (5-20% polyacrylamide) gels of crude extracts and purified
          AcChR from β-D-octylglucoside or Triton X-100-solubilized
          preparations. (1) 1% Triton X-100 solubilized extract;
          (2) acetylcholine receptor purified from (1); (3) 30 mM
          β-D-octylglucoside solubilized extract; (4) 30 mM β-D-
          octylglucoside solubilized extract after removal of the
          detergent by extensive dialysis; (5) acetylcholine receptor
          purified from (3); (6) purified acetylcholine receptor
          after removal of β-D-octylglucopyranoside by extensive
          dialysis.

vesicles rich in AcChR from those rich in acetylcholinesterase and
is, itself, an index of the different topographical localization of
these two molecules in the electroplaque. AcChR-rich membranes
(Fig. 1) contain the protein subunit components of isolated receptor
as well as higher molecular weight components and a 43,000 dalton
subunit. Many of the subunits, including the 43,000 subunit, can be
removed by extraction of the membranes at pH 11.[6] The new membranes,
although somewhat more fragile, retain primarily the polypeptide
patterns characteristic of the purified receptor. The 43,000 dalton
subunit has been proposed as the binding site for an ionophore
blocker or local anesthetic binding site.[7] On the other hand, alkali
treated membranes[6] and reconstituted systems of purified AcChR and
lipids all lacking the 43,000 subunit,[8-11] act as functional AcChR
membranes.

AcChR is arranged in the membrane in such a manner that electron density profiles of x-ray scattering patterns and negative
stain electron micrographs of membranes alone[12] or after treatment
with antiAcChR antibodies coupled to colloidal gold[13] indicate it,
forms an organized array of transmembrane proteins protruding 55 A
from the extracellular surface. Three other lines of evidence, low-
angle neutron scattering studies,[14] susceptibility of right-side-
out and inside-out vesicles to external proteases,[15] and binding
of ferritin labeled antiAcChR Fab fragments derived from specific
antibodies against AcChR[16] also point to the transmembrane nature
of the AcChR. Labeling with water soluble affinity labels such as
MBTA[17] and bromoacetylcholine[7,18] as well as photoactive azide affinity labels derived from cholinergic agonists,[19] or partially
hydrophobic agents,[20] always label the 40,000 subunit as the water
exposed extracytoplasmic subunit. Furthermore, iodination by the
lactoperoxidase method indicates that the 50,000 and 60,000 dalton
subunits, in addition to the 40,000 subunit, are easily accessible
from the extracytoplasmic surface.[21] We have investigated which
portions of the AcChR are in contact with lipids in membranes by
designing a fluorescent (and tritium labeled) hydrophobic photoaf-
finity probe, pyrene sulfonyl azide, which would partition quantitatively into the hydrocarbon lipids. Addition of bovine serum
albumin does not remove the membrane incorporated probe and addition
of local anesthetics induces a quenching of the probe fluorescence
characteristic of its lipid environment perturbation.[22] The probe,
PySA can be activated by ultraviolet light with generation of a
nitrene that preferentially reacts with membrane proteins to form
a covalent adduct. After detergent solubilization and separation of
the membrane components the fluorescent probes can be shown to be
attached almost exclusively to the protein with a marked preference
(Fig. 2) for the 50,000 and 60,000 subunits of the AcChR component
as could be expected if these subunits have contacts with the lipids
in the membrane bilayer. The absence of label on the 40,000 subunit
contrasts the results obtained when carrying out the same experiment
with purified AcChR in which detergent is present and lipids are

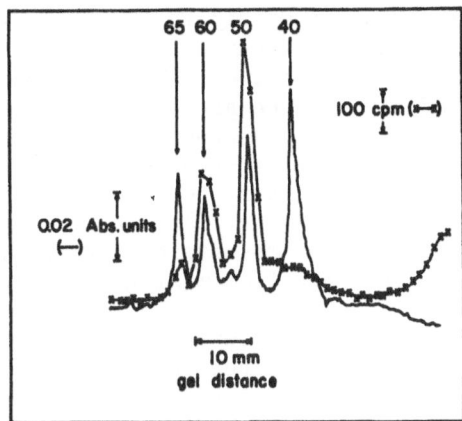

Fig. 2. Distribution of radioactivity and protein in [³H] PySA labeled AcChR purified from membrane fragments enriched in AcChR. After photolabeling, receptor was solubilized from the membrane fragments and isolated by an affinity chromatography procedure. Numbers in the upper part of the figure indicate apparent molecular weights in thousands.

absent.[22] We observe that the fluorescence lifetime of the pyrene probe covalently bound at the protein boundary changes when cholinergic agents bind to the 40,000 dalton subunit. These changes are blocked by preincubation of the membranes with α-neurotoxin which also binds to the 40,000 dalton subunit. Therefore, it appears that polypeptide–polypeptide interactions between the subunits take place upon occupancy of the cholinergic sites at the 40,000 dalton subunit by agonists or neurotoxin. These intersubunit interactions are expressed as environmental changes of the fluorophore-bound regions of the 50,000 and 60,000 subunits accessible to the lipid environment of the membrane bilayer.

AcChR membranes capable of undergoing changes from low affinity to high affinity for agonist can be easily monitored by several procedures: among them the rate of α-bungarotoxin binding between native and carbamylcholine exposed membranes is a good index of this change.[1] Labeling with the PySA probe does not affect the desensitization ability of the membrane. On the other hand, treatment of the membranes with local anesthetics[1,24] and reduction of disulfides with dithiothreitol affect the properties of the membrane.

Other approaches to the study of AcChR structural organization have relied on the use of proteases. Trypsinization of AcChR leads to significant changes in their morphology[23] as well as the production of a 27,000 subunit fragment arising from other subunits which is still capable of binding α-neurotoxins.

AcChR-rich membranes are mostly in the vesicular state and can be loaded with radioactive monovalent cations such as $^{22}Na^+$, $^{86}Rb^+$ and $^{204}Tl^+$. The membranes are leaky to anions in the resting state but the rate of flux can be accelerated by the presence of agonists and this agonist induced process is blocked by the presence of antagonists including α-neurotoxins.[1,2,6,11,25]. Dose concentrations of agonist necessary to induce cation efflux or influx responses can be shown (Fig. 3) to be of the same magnitude as those concentrations needed to obtain electrophysiologic responses in cells. Also, with rapid flow mixing quenching procedures the agonist induced cation flux can be measured on fast time scales and attain initial rates of ion flux comparable to those obtained for physiologically detected ion channel openings.[9,26]

In AcChR membranes, ion permeability increases elicited by agonists are influenced by the presence of α-neurotoxins. Since toxins share binding domains with agonist and antagonists, they are excellent probes to determine the degree of occupation of the sites and how the occupation influences ion permeability response. Furthermore, since toxin binding affinity is high, and when added in substoichiometric amounts can produce random occupation of the sites, the toxin provides an excellent tool for determining the number of ligand binding sites per receptor and the effect of this occupancy on the availability of the remaining sites to agonist. Overall, it is possible to monitor the stoichiometric binding of

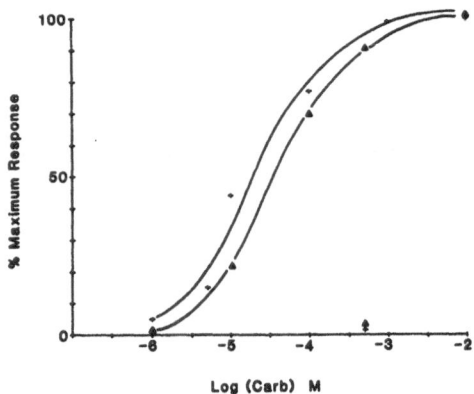

Log (Carb) M

Fig. 3.   Efflux of $^{22}Na^+$ from AcChR-enriched membranes (Δ - Δ) and AcChR reconstituted vesicles (+ - +) at varying carbamylcholine concentrations. The 100% maximum response represents 852 cpm and 3392 cpm $^{22}Na^+$ effluxed at 1x10^{-3}M carbamylcholine per 10 μl native membranes and reconstituted vesicles, respectively. For desensitization, the samples were preincubated with 1x10^{-6}M carbamylcholine for 30 minutes and allowed to efflux at 0.3x10^{-3} M carbamylcholine (Δ , + single points near abscissa).

toxin and agonist to receptor and probe into the number of agonist
molecules needed to activate the ion flux function. Basically, the
problem is to determine whether hybrid occupancy, one site occupied
by toxin, another by agonist, is functional and to what extent. The
experiment is carried out by pre-incubation of AcChR-rich membranes
with progressively increasing concentrations of ^{125}I labeled
α-neurotoxins. After random labeling, there is addition of agonist
sufficient to produce flux from ^{22}Na loaded vesicles, and the ion
flux is measured. The precise measurement of the amount of receptors activated requires that the rate of ion flux be proportional
to the number of AcChR molecules and the duration of the exposure
to agonist. Models which relate occupancy of the toxin sites to
the capacity to inactivate ion permeability changes can be proposed,
and the response given by the different behaviors are different
for each model.[27] In AcChR-membranes, whether freshly isolated or
in reconstituted systems (see below), we obtain responses as represented in Fig. 4. That behavior only fits models in which both
agonist sites must be free to produce a full ion permeability
response while hybrids of 1 site occupied by toxin and the other
by agonist are inactive as are those molecules in which both sites
are occupied by the toxin. Thus, in membrane preparations as in
whole cells,[27] toxin and agonist sites are mutually exclusive but
occupation of either site by the toxin suffices to block the functional capacity of the receptor. These studies are consistent with
independent biochemical studies on detergent solubilized and isolated
receptor which show protection of agonist and antagonist binding by
α-neurotoxins.

LIPID ENVIRONMENT OF THE AcChR RICH MEMBRANES

    Lipids constitute only about one third of the total dry weight
of AcChR-rich membranes and of these lipids, 50% is cholesterol.
The high imbalance of the lipid/protein ratio as compared to other
membranes such as those from cytoplasm, mitochondria or erythrocytes
may be related to AcChR function. It is also known that perturbation
of the lipid environment such as exposure to phospholipase A_2 or
introduction of exogenous unsaturated fatty acids[28] prevents the
large increases of sodium flux induced by the presence of carbamylcholine. The lipid composition of the AcChR-rich membranes is uncha-
racteristic of that found in most membranes and different even from
that found in other membrane fractions prepared from electroplax.
These lipids also show an abundance of EP and PC (Table 1). Presence
of significant amounts of plasmalogens is detected in the ethanolamine phospholipids and a high proportion of long chain polyunsatu-
rated fatty acids appear to be bound preferentially to ethanolamine
phosphoglycerides. Sphingomyelin and cardiolipin are present as well
as abundant levels of cholesterol. This unusual lipid distribution
must play a significant role in AcChR functions and structure stabilization since, as seen in our reconstitution experiments, there are
indications that the nature of the lipid affects the ability of AcChR

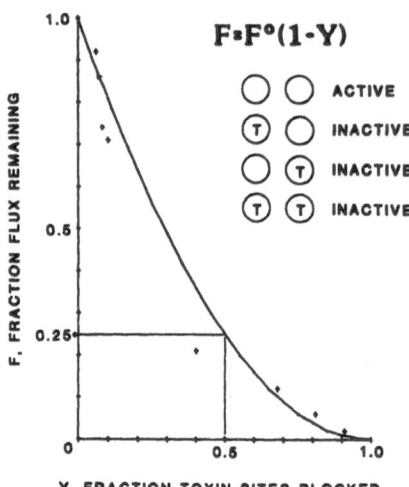

Y, FRACTION TOXIN SITES BLOCKED

Fig. 4. Inhibition of carbamylcholine induced efflux by the progressive occupancy of membrane-bound AcChR by α-Bungarotoxin. A theoretical curve was generated by the equation
F = F° (1-Y) where F° is the fractional efflux in the
absence of toxin and Y is fractional toxin occupancy. This
model predicts that each AcChR contains two binding sites
where activation requires the occupation of both sites by
agonist whereas inhibition requires the occupation of
either site by toxin. Each experimental point was determined from the $^{22}Na^{+}$ efflux from AcChR-rich membranes by
1 mM carbamylcholine.

Table 1. Phospholipid Composition of Electroplax Membrane Fractions
          (molar percentages)

	AcChR-rich membranes	AcChR-poor membranes
Phosphatidyl Choline	38.4	40.1
Ethanolamine Phosphoglyceride	42.2	32.1
Phosphatidyl Ethanolamine	35.8	25.6
Phosphatidyl Serine	10.9	18.3
Sphingomyelin + Lysophosphatidyl		
Choline	2.3	6.4
Cardiolipin	3.3	2.2
Phosphatidic Acid	2.9	0.8

to promote agonist induced Na$^+$ fluxes. Incorporation of two fluorescent probes, DPH and TMA-DPH, into lipid vesicles prepared from
total lipid extracts isolated from AcChR-rich membranes and into
intact membranes themselves, can be used to infer some physical
properties of those lipids. Simultaneous measurements of fluorescence polarization and lifetimes of the singlet excited state of
the probes must be carried out in order to calculate the rotational
relaxation time, $\bar{\varrho}$, of the fluorophore. No distinctive phase changes
occur in the temperature range of 5° to 45°C. The values of the
rotational relaxation times of the probes as a function of temperature is also indicative of a higher degree of restriction of the
polar head groups of the lipid than deeper into lipid bilayer (Fig.
5). Furthermore, the degree of rotational freedom of the probe in
the lipid environment appears to be the influence of the high concentrations of cholesterol in these membranes. Presence of protein
in the lipid membrane bilayer either redistributes the partition
of the probes among the lipids or induces a higher degree of order
among the polar heads and acyl chains of the phospholipids which
in turn restricts the rotation of both probes.

RECONSTITUTION OF AcChR MEMBRANES

To study the properties and functional responses of the various
constituents of the membrane, formation of membranes with a defined
composition is a preferred approach. Most efforts at reconstituting
a functional AcChR into a lipid bilayer when this is the only
protein present, used to fail or exhibited problems of irreprodu-
cibility. We assumed that most of the problems arise from the high
capacity of solubilized AcChR to tightly bind detergent which would
ultimately interfere with its function in the membrane. Others had

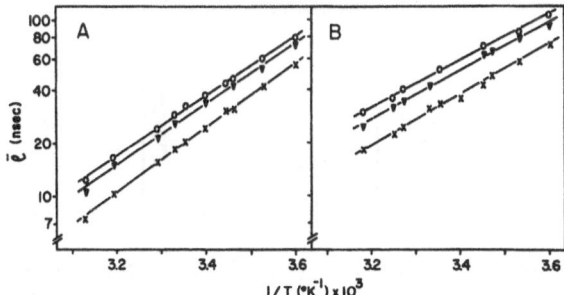

Fig. 5.   Representative temperature dependence of the rotational
relaxation time ($\bar{\varrho}$) of DPH (A) and TMA-DPH (B) in plain
lipid vesicles. Vesicles were produced by the β-D-octylglucoside dialysis method[22] with total lipids from unfrac-
tionated (o), AcChR-rich (▼) and AcChR-poor (x) membranes
from Torpedo electroplax. Fluorescence polarization and
lifetime were measured at the indicated temperatures and
the rotational relaxation time values calculated by the
Perrin equation.

assumed that purification of the receptor itself could irreversibly
damage AcChR and, more recently, that a lipid environment must be
maintained throughout the whole purification procedure.[10,11] We
circumvented the detergent problem by using β–D-octylglucoside as
an inert and easily dialyzable detergent for which AcChR has low
affinity.[8] Membranes extracted with this detergent and subsequent
purification steps carried out in the presence of this detergent
in the absence of exogenous lipids, produce AcChR preparations with
the same electrophoretic patterns (Fig. 1) and subunit stoichiometry as those obtained by other means. Reconstitution of this purified receptor was first attempted using the total lipids extracted
from electroplax tissue to retain the original lipid environment.
The amount of AcChR incorporated into a lipid vesicle is related
to the initial lipid concentration and, as seen in Fig. 6, reaches
completion at 28 mg/ml. The mere presence of lipid restores the
ability of purified receptors to undergo a desensitization-like
process measured as the rate of α–bungarotoxin binding with and
without pre-exposure to carbamylcholine.[9] Reconstituted vesicles

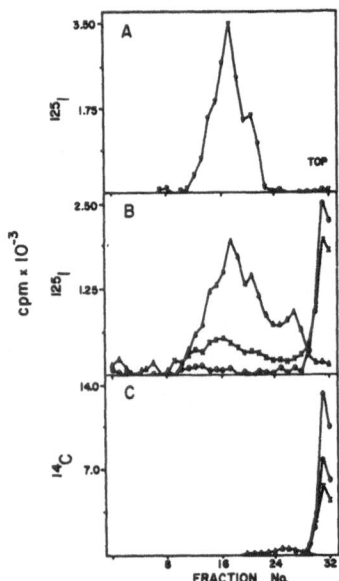

Fig. 6. Sucrose density gradient profiles of purified AcChR protein
         (A), and reconstituted AcChR protein (B) with corresponding
         electroplax lipid (C). Lipids were monitored with the ra-
         dioactive tracer, cholesteryl [1-^{14}C] oleate. ^{125}I-α-
         Bgt binding was measured as ^{125}I bound. The initial lipid
         concentration (mg/ml) and the lipid/protein ratio (w/w),
         respectively, for the different reconstituted samples were
         as follows: (Δ), 3,5/1; (x) 15, 30/1; (o) 28, 60/1. The
         initial lipid concentration for the plain lipid vesicles (o)
         was 15 mg/ml and the sedimentation profile is also depicted
         in panel C.

can also be loaded with $^{22}Na^+$, and the rate of loading as well as
the amount of efflux induced by agonist can be quantitatively measured. The octylglucoside reconstitution procedure is most efficient
in yielding large vesicles with large internal volumes. This allows
for incorporation of high numbers of ions, and the size of the
agonist induced flux becomes an easy and reproducible way of quantitatively measuring responses to presence of given doses of agonists. As shown in Fig. 3, the dose response curves of reconstituted
vesicles are identical to those measured with native AcChR-rich
membranes as isolated from electroplax. Furthermore, the desensitization phenomenon is detectable as a flux parameter effector since
preincubation of the reconstituted membrane with carbamylcholine
blocks the response to agonist. Removal of carbamylcholine regenerates a membrane in the resting state as an indicator of the reversibility of the desensitization process (Fig. 3). These results
are a valid testimony to the interpretation that only the 40, 50,
60 and 65,000 subunits of the AcChR molecule are necessary and sufficient for function in ligand recognition, desensitization and
ion channel formation. Reconstitution of this purified receptor with
lipids of known composition shows that asolectin (soybean lipids)
is essentially ineffective when used in the octylglucoside reconstitution procedure. Also, PC alone does not provide a proper environment for function if this function is expressed as the ability to
undergo reversible agonist induced ion flux. On the other hand,
although ion flux responses by agonists can be generated in AcChR
in PE-PC mixtures, we have also detected that the presence of cholesterol is a necessity to yield augmented flux responses (number of
ion/number of moles of carbamylcholine) approaching those values
obtained with complete electroplax lipids. As yet there is no information of the phospholipid fatty acids in AcChR function or structure. The reconstitution results appear to indicate that not only
the type of lipid but the reconstitution procedure and the proportions of the different types of lipids play a role in obtaining a
proper AcChR configuration capable of undergoing optimum agonist
induced influxes. At this time, however, the highest ion flux
responses are obtained when the complete complement found in native
Torpedo lipids is present in reconstitution experiments using octylglucoside purified receptor.

ACKNOWLEDGEMENTS

     Work from our laboratories was supported by the National
Science Foundation (NSF Research Grant BNS 77-24775) and by the
U.S. Public Health Service (NIH Research Grant GM 24885 (MMC) and
NS 17029 (JMGR)).

REFERENCES

1. P. R. Adams, Acetylcholine receptor kinetics, J. Membrane
        Biol. 58: 161 (1981).

2. F. J. Barrantes, Endogenous chemical receptors: some physical aspects, Ann. Rev. Biophys. Bioeng. 8: 287 (1979).

3. J. M. Gonzalez-Ros, A. Paraschos, M. C. Farach and M. Martinez-Carrion, Characterization of acetylcholine receptor isolated from Torpedo californica electroplax through the use of an easily removable detergent, β-D-octylglucoside, Biochim. Biophys. Acta 643: 407 (1981).

4. W. Schiebler and F. Hucho, Reconstitution of active acetylcholine receptor by hybridization of binding site-blocked with ion channel-blocked acetylcholine receptor protein, Biochim. Biophys. Acta 597: 626 (1980).

5. J. M. Wolosin, A. Lyddiatt, J. O. Dolly and E. A. Barnard, Stoichiometry of the ligand-binding sites in the Acetylcholine-receptor oligomers from muscle and from electric organ. Measurement by affinity alkylation with bromoacetylcholine, Eur. J. Biochem. 109: 495 (1980).

6. R. B. Neubig, E. K. Krodel, N. D. Boyd and J. B. Cohen, Acetylcholine and local anesthetic binding to Torpedo nicotinic postsynaptic membranes after removal of nonreceptor peptides, Proc. Nat. Acad. Sci. USA 76: 690 (1979).

7. A. Sobel, T. Heidmann, J. Hofler, J. P. Changeux, Distinct protein components from Torpedo marmorata membranes carry the acetylcholine receptor site and the binding site for local anesthetics and histrionicotoxin, Proc. Nat. Acad. Sci. USA 75: 510 (1978).

8. J. M. Gonzalez-Ros, A. Paraschos and M. Martinez-Carrion, Reconstitution of functional membrane-bound acetylcholine receptor from isolated Torpedo californica receptor protein and electroplax lipids, Proc. Nat. Acad. Sci. USA 77: 1796 (1980).

9. W. C. S. Wu, H-P. Moore and M. A. Raftery, Quantitation of cation transport by reconstituted membrane vesicles containing purified acetylcholine receptor, Proc. Nat. Acad. Sci. USA 78: 775 (1981).

10. J. Lindstrom, R. Anholt, B. Einarson, A. Engel, M. Osame and M. Montal, Purification of acetylcholine receptors, reconstitution into lipid vesicles and study of agonist-induced cation channel regulation, J. Biol. Chem. 255: 8340 (1980).

11. R. L. Huganir, M. A. Schell and E. Racker, Reconstitution of the purified acetylcholine receptor from Torpedo californica, FEBS Lett. 108: 155 (1979).

12. M. J. Ross, M. W. Klymkowski, D. A. Agard and R. M. Stroud, Structural studies on a membrane-bound acetylcholine receptor from Torpedo californica, J. Mol. Biol. 116: 635 (1977).

13. M. W. Klymkowski and R. M. Stroud, Immunospecific identification and three-dimensional structure of a membrane-bound acetylcholine receptor from Torpedo californica, J. Mol. Biol. 128: 319 (1979).

14. D. S. Wise, A. Karlin and B. P. Schoenborn, An analysis by low-angle neutron scattering of the structure of the

acetylcholine receptor from Torpedo californica in detergent solution, Biophys. J. 28: 473 (1979).

15. L. Huang, Transmembrane nature of acetylcholine receptor as evidenced by protease sensitivity, FEBS Lett. 102: 9 (1979).

16. R. Tarrab-Hazdai, B. Geiger, S. Fuchs and A. Amsterdam, Localization of acetylcholine receptor in excitable membrane from the electric organ of Torpedo: evidence for exposure to receptor antigenic sites on both sides of the membrane, Proc. Nat. Acad. Sci. USA 75: 2497 (1978).

17. C. L. Weill, M. G. McNamee and A. Karlin, Affinity-labeling of purified acetylcholine receptor from Torpedo californica, Biochem. Biophys. Res. Commun. 61: 997 (1974).

18. H-P. Moore and M. A. Raftery, Studies of reversible and irreversible interactions of an alkylating agonist with Torpedo californica acetylcholine receptor in membrane-bound and purified states, Biochemistry 18: 1862 (1979).

19. V. Witzeman and M. A. Raftery, Selective photo-affinity labeling of acetylcholine receptor using a cholinergic analogue, Biochemistry 16: 5862 (1977).

20. R. Tarrab-Hazdai, T. Bercovici, V. Goldfarb and C. Gitler, Identification of the acetylcholine receptor subunit in the lipid bilayer of Torpedo electric organ excitable membranes, J. Biol. Chem. 255: 1204 (1980).

21. P. R. Hartig and M. A. Raftery, Lactoperoxidase catalyzed membrane surface labeling of the acetylcholine receptor from Torpedo californica, Biochem. Biophys. Res. Commun. 78: 16 (1977).

22. V. Sator, J. M. Gonzales-Ros, P. Calvo-Fernandez and M. Martinez-Carrion, Pyrene sulfonyl azide: a marker of acetylcholine receptor subunits in contact with membrane hydrophobic environment, Biochemistry 18: 1200 (1979).

23. M. W. Klymkowski, J. E. Heuser and R. M. Stroud, Protease effects on the structure of acetylcholine receptor membranes from Torpedo californica, J. Cell Biol. 85: 823 (1980).

24. M. E. Eldefrawi and A. T. Eldefrawi, Coupling between the nicotinic acetylcholine receptor site and the ionic channel site, Ann. N.Y. Acad. Sci. 358: 239 (1980).

25. M. Raftery, S. Blanchard, J. Elliott, P. Hartig, H-P Moore, U. Quast, M. Schimerlik, V. Witzemann and W. Wu, Properties of Torpedo californica receptor, Adv. in Cytopharmacol. 3: 159 (1979).

26. D. J. Cash, H. Aoshima and G. Hess, Acetylcholine-induced cation translocation across cell membranes and inactivation of the acetylcholine receptor. Chemical kinetic measurements in the millisecond time region, Proc. Nat. Acad. Sci. USA 78: 3318 (1981).

27. S. M. Sine and P. Taylor, The relationship between agonist occupation and the permeability response of the cholinergic receptor revealed by bound Cobra α-toxin, J. Biol. Chem. 255: 10144 (1980).

28. T. J. Andreasen, D. R. Doerge and M. G. McNamee, Effects of phospholipase A_2 on the binding and ion permeability control properties of the acetylcholine receptor, Arch. Biochem. Biophys. 195: 468 (1979).

INTERACTION OF FUSICOCCIN WITH PLANT CELL PLASMA MEMBRANES

Alessandro Ballio

Institute of Biological Chemistry
University of Rome
00185 Rome, Italy

INTRODUCTION

Fusicoccin (FC) is the major phytotoxic metabolite of the fungus _Fusicoccum amygdali_ Del.,[1] the organism responsible of a disease called "canker of almond and peach", causing severe economic losses in the mediterranean area. The role of FC as a chemical determinant of this disease has been well documented[2] and its toxic activity has been thoroughly characterised;[3,4] it is a non-specific but very active phytotoxin.

The structure of FC (Fig. 1) was worked out by Ballio et al.[5] and confirmed by an independent investigation of Barrow et al.[6]; it is the α-glucoside of a highly oxygenated carbotricyclic diterpene.

Studies on the mechanism of action of FC, started in 1964 by Graniti,[7] soon evidenced unexpected effects on several physiological processes in higher plants. Stomatal movement,[8] cell enlargement of different tissues,[9-13] germination of dormant seeds,[14] were stimulated by concentrations as low as $10^{-5} - 10^{-6}$ M in a wide range of plants. All tissues responding to FC showed an almost immediate increase in H^+-extrusion capacity, accompanied by stimulation of K^+ uptake and hyperpolarisation of the transmembrane potential, suggestive of a possible interaction with some membrane component.[15] This suggestion was further strengthened by the finding that the activity of a K^+-dependent ATPase of plasmalemma was stimulated by FC.[16]

Thus, the search for possible FC-specific binding sites on the plasma membrane became of particular interest, since the interaction of FC with them might represent the first step in the biological action of this substance.

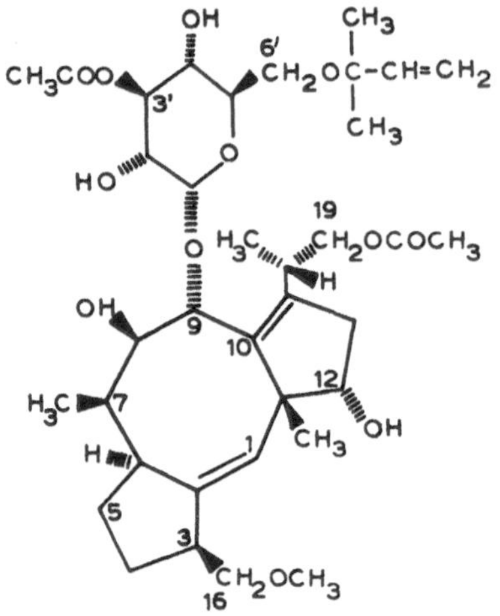

Fig. 1.   Structure of fusicoccin.

FC-BINDING SITES IN MICROSOMAL PREPARATIONS OF PLANT TISSUES

The search for FC-binding sites obviously required the availability of a radioactive ligand. Since [3]H- or ^{14}C-labeled acetate
and mevalonate are poorly incorporated into FC produced by submerged
cultures of F. amygdali, attention was directed to dihydro-FC, a FC
derivative having the double bond of the t-pentenyl unit on glucose
saturated by hydrogen, obtainable by catalytic hydrogenation of the
parent compound.[5] In fact studies on the relationships between
chemical structure and biological activity had shown that dihydro-
FC is as active as FC in tests for phytotoxicity,[17] and in promoting
stomatal opening,[18] plant growth and seed germination.[19,20] Catalytic
addition of [3]H$_2$ to FC afforded dihydro-FC specifically labeled in
the C$_5$ unit on glucose; the yield was quantitative and the specific
activity (62.5 Ci mmole^{-1}) was suitable for the proposed investigation.[21]

Labeled dihydro-FC was found to bind to a microsomal fraction
prepared from homogenates of maize coleoptiles: differential centrifugation and isopycnic sucrose density gradients, with parallel determination of enzymatic markers, indicated that the binding sites
are most probably located at the plasmalemma.[22] The binding is
specific, saturable at about 10^{-6}M, heat-labile and has a pH optimum
at 5.5 and a temperature optimum at 25-26°C.[23,24] Scatchard analysis
of binding data over a large range of FC concentrations gave a
biphasic curve; the higher affinity sites have K$_D$ between 7 and

12×10^{-10} M and a number of sites in the range 0.2-2.4 picomoles g^{-1} fresh tissue.[23,24] The interaction of the ligand with the binding sites is non covalent, since extraction with warm ethanol removes all bound radioactivity as unmodified dihydro-FC.

The recognition in maize coleoptiles of high affinity binding sites for FC allowed to test the competitivity with the radioactive ligand of a large number of compounds structurally related to FC, either fungal metabolites or derivatives obtained from them by chemical modification. The results of this in vitro competition test[25] correlated well with those of in vivo tests,[18,20] thus leaving no doubt about the relevance of the binding to the physiological activity of FC.

The general characteristics of FC-binding have also been investigated in plant tissues different from maize coleoptiles, namely spinach leaves,[21] maize roots[26] and oat roots;[27] except for minor differences an overall similarity with maize coleoptiles was consistently observed.

SOLUBILISATION OF FC-BINDING SITES

FC specifically bound to microsomal preparations from maize coleoptiles can be partially solubilised by Triton X-100, deoxycholate, sodium perchlorate or trypsin in a form still bound to a macromolecular component; on gel filtration this shows a symmetrical peak corresponding to a molecular weight of about 80,000 daltons.[24,28] The solubilised macromolecular complex can be separated by disc electrophoresis from the plasmalemma ATPase activity present in the same preparation:[29] thus the possibility that FC-binding sites are associated with a multimeric transport ATPase remains an open question. Similar results have been obtained by Stout and Cleland with detergent solubilised preparations from oat root plasmalemma-enriched vesicles incubated with radioactive dihydro-FC.[27]

More recently Aducci et al. have reported on the solubilisation of FC-binding sites not complexed with the ligand.[30] The solubilised sites were obtained from acetone-dried microsomal fractions of spinach leaves without use of detergents or chaotropic reagents, and had kinetic properties very close to those of particulate preparations from the same tissue.[21] Notable differences observed with the soluble sites were (i) independence of the binding reaction from pH in the interval 5.5-9.0, as compared to a sharp optimum at 5.5 for microsomal preparations of both maize coleoptiles[24] and spinach leaves;[21] (ii) fast and complete displacement of the radioactive ligand by a chase of cold FC, contrasted by a much slower exchange rate with the particulate fraction from spinach leaves,[21] and by the virtual irreversibility in the case of the particulate fraction from corn coleoptiles.[22,24]

LOCALIZATION OF FC-BINDING SITES

As already mentioned, maximal FC binding activity is associated with plasmalemma markers.[22,24] The subcellular distribution of labeled dihydro-FC fed in vivo to maize coleoptiles is very similar to that of the in vitro binding and, coinciding with that of the K^+-stimulated, diethylstilbestrol-sensitive ATPase, again suggests that plasmalemma is the structure responsible for FC binding.[31]

The availability of a high molecular weight derivative of FC, obtained by conjugation of bovine serum albumin (BSA) with a suitable FC derivative, and of antifusicoccin antibodies, elicited in rabbits in response to the above conjugate,[32] allowed to demonstrate the occurrence of FC binding sites at the outer surface of plasmalemma.[33] In fact when tobacco leaf protoplasts were incubated first with the non permeant BSA-FC conjugate (which retains a very high affinity for FC binding sites) and then with antifusicoccin antibodies, agglutination occurred within 5 minutes. Agglutination was not observed in the absence of the FC-conjugate, or when the conjugate was substituted with BSA, or when rabbit serum was used in place of the antibodies. Up to now the localization of FC-binding sites in structures other than plasmalemma (tonoplast?) cannot be ruled out.

PHYSIOLOGICAL ROLE OF FC-BINDING SITES

The recognition of binding sites specific for FC raises the question of their physiological role. It is difficult to admit that FC represents a new phytohormone. In fact, unlike plant hormones, FC displays effects which are not tissue specific and moreover has a remarkably high metabolic stability in vivo;[34] furthermore, as far as we know, the substance has a very restricted distribution in nature. Therefore it is reasonable to suppose that FC binding sites interact with some widely occurring endogenous compound that under natural conditions elicites in the plant cell a response similar or opposite to that of FC. This hypothesis has led Aducci et al.[26] to detect in maize roots one or more water soluble compounds capable of competing efficiently with FC at the binding sites level (microsomal preparations from roots and coleoptiles of maize and from spinach leaves) and of inhibiting FC-stimulated proton extrusion in vivo. These endogenous ligands, whose structure is under investigation in our laboratory, also inhibit FC binding in soluble preparations from spinach leaves and displace promptly and completely radioactive dihydro-FC bound to the soluble sites,[30] thus ruling out an aspecific effect on the membrane structure.

The inhibition of FC binding with preparations from maize roots and coleoptiles, and from spinach leaves, namely in two tissues of the same plant and in a third tissue of an unrelated plant, suggests for these endogenous ligands an ubiquitous role in higher plants, such as that of modulators of the proton extruding system triggered by FC. This is further supported by the occurrence of equally active compounds in plant tissues other than maize root (unpublished results).

REFERENCES

1. A. Ballio, E. B. Chain, P. De Leo, B. F. Erlanger, M. Mauri and A. Tonolo, Fusicoccin, a new wilting toxin produced by Fusicoccum amygdali Del., Nature, 203: 297 (1964).
2. A. Ballio, V. D'Alessio, G. Randazzo, A. Bottalico, A. Graniti, L. Sparapano, B. Bosnar, C. G. Casinovi and O. Gribanovski-Sassu, Occurrence of fusicoccin in plant tissues infected by Fusicoccum amygdali Del., Physiol. Plant Pathol. 8: 163 (1976).
3. A. Graniti, Qualche dato di fitotossicità della fusicoccina A, una tossina prodotta in vitro da Fusicoccum amygdali Del., Phytopath. mediterr. 3: 125 (1964).
4. E. B. Chain, P. G. Mantle and B. V. Milborrow, Further investigations on the toxicity of fusicoccins. Physiol. Plant Pathol. 1: 495 (1971).
5. A. Ballio, M. Brufani, C. G. Casinovi, S. Cerrini, W. Fedeli, R. Pellicciari, B. Santurbano and A. Vaciago, The structure of fusicoccin A. Experientia 24: 631 (1968).
6. K. D. Barrow, D. H. R. Barton, E. B. Chain, U. F. W. Ohnsorge and R. Thomas, Fusicoccin. Part II. The constitution of fusicoccin. J. Chem. Soc. (C), 1265 (1971).
7. A. Graniti, The role of toxins in the pathogenesis of infections by Fusicoccum amygdali Del. on almond and peach, in "Host-parasite relations in plant pathology", Z. Kiraly and G. Ubriszy, eds. Research Institute for Plant Protection, Budapest (1964).
8. N. C. Turner and A. Graniti, Fusicoccin, a fungal toxin that opens stomata. Nature 223: 1070 (1969).
9. A. Ballio, A. Graniti, F. Pocchiari and V. Silano, Some effects of fusicoccin A on tomato leaf tissues. Life Sci. 7:751 (1968).
10. A. Ballio, F. Pocchiari, S. Russi and V. Silano, Effects of fusicoccin and some related compounds on etiolated pea tissues. Physiol. Plant Pathol. 1: 95 (1971).
11. E. Marrè, P. Lado, F. Rasi-Caldogno and R. Colombo, Fusicoccin as a tool for the analysis of auxin action. Rend. Accad.Naz. Lincei 50: 45 (1971).
12. P. Lado, A. Pennacchioni, F. Rasi-Caldogno, S. Russi and V. Silano, Comparison between some effects of fusicoccin and indole-3-acetic acid on cell enlargement in various plant materials. Physiol. Plant Pathol. 2: 75 (1972).
13. P. Lado, F. Rasi-Caldogno, A. Pennacchioni and E. Marrè, Mechanism of the growth-promoting action of fusicoccin. Interaction with auxin and effects of inhibitors of respiration and

protein synthesis. Planta 110: 311 (1973).

14. P. Lado, F. Rasi-Caldogno and R. Colombo, Promoting effect of
    fusicoccin on seed germination. Physiol. Plant.31: 149
    (1974).

15. E. Marrè, Fusicoccin: a tool in plant physiology. Annu. Rev.
    Plant Physiol. 30: 273 (1979).

16. N. Beffagna, S. Cocucci and E. Marrè, Stimulating effect of
    fusicoccin on K^+-activated ATPase in plasmalemma preparations
    from higher plant tissues. Plant Sci. Lett. 8: 91 (1977).

17. A. Ballio, A. Bottalico, M. Framondino, A. Graniti and G. Ran-
    dazzo, Fusicoccin: structure-phytotoxicity relationships.
    Phytopathol. mediterr.10: 26 (1971).

18. A. Bottalico, A. Graniti and P. Lerario, Further investigation
    on the biological activity of some fusicoccins and cotylenins.
    Phytopathol. mediterr. 17: 127 (1978).

19. A. Ballio, Fusicoccin: structure-activity relationships, in:
    "Regulation of cell membrane activities in plants", E. Marrè
    and O. Ciferri, eds., Elsevier/North-Holland, Amsterdam
    (1977).

20. A. Ballio, M. I. De Michelis, P. Lado and G. Randazzo, Fusicoc-
    cin structure-activity relationships: stimulation of growth
    by cell enlargement and promotion of seed germination.
    Physiol. Plant. 52: 471 (1981).

21. A. Ballio, R. Federico, A. Pessi and D. Scalorbi, Fusicoccin
    binding sites in subcellular preparations of spinach leaves,
    Plant Sci. Lett. 18: 39 (1980).

22. U. Dohrmann, R. Hertel, P. Pesci, S. Cocucci, E. Marrè, G.
    Randazzo and A. Ballio,Localization of "in vitro" binding
    of the fungal toxin fusicoccin to plasma-membrane-rich
    fractions from corn coleoptiles. Plant Sci. Lett. 9: 291
    (1977).

23. A. Ballio, Chemistry and plant growth regulating activity of
    fusicoccin derivatives and analogues, in "Advances in
    pesticide science", H. Geissbühler, ed., Pergamon Press,
    Oxford and New York (1979).

24. P. Pesci, S. Cocucci and G. Randazzo, Characterization of fusi-
    coccin binding to receptor sites on cell membranes of maize
    coleoptile tissues. Plant Cell Environ. 3: 205 (1979).

25. A. Ballio, R. Federico and D. Scalorbi, Fusicoccin structure-
    activity relationships: in vitro binding to microsomal pre-
    parations of maize coleoptiles. Physiol. Plant. 52: 476
    (1981).

26. P. Aducci, R. Federico and A. Ballio, Fusicoccin receptors.
    Evidence for endogenous ligand. Planta 148: 208 (1980).

27. R. G. Stout and R. E. Cleland, Partial characterization of
    fusicoccin binding to receptor sites on oat root membranes.
    Plant Physiol. 66: 353 (1980).

28. P. Pesci, L. Tognoli, N. Beffagna and E. Marrè, Solubilisation
    and partial purification of a fusicoccin-receptor complex
    from maize coleoptiles. Plant Sci. Lett. 15: 313 (1979).

29. L. Tognoli, N. Beffagna, P. Pesci and E. Marrè, On the relation-
     ship between ATPase and fusicoccin binding capacity of crude
     and partially purified microsomal preparations from maize
     coleoptiles. Plant Sci. Lett. 16: 1 (1979).
30. P. Aducci, A. Ballio and R. Federico, Solubilization of fusi-
     coccin binding sites, in "Plasmalemma and tonoplast: their
     functions in the plant cell", D. Marmè, E. Marrè and R.
     Hertel, eds., Elsevier/North-Holland, Amsterdam (1982).
31. N. Beffagna, P. Pesci, L. Tognoli and E. Marrè, Distribution
     of fusicoccin bound in vivo among subcellular fractions
     from maize coleoptiles. Plant Sci. Lett. 15: 323 (1979).
32. C. Pini, G. Vicari, A. Ballio, R. Federico, A. Evidente and
     G. Randazzo, Antibodies specific for fusicoccin. Plant Sci.
     Lett. 16: 343 (1979).
33. P. Aducci, R. Federico and A. Ballio, Interaction of a high
     molecular weight derivative of fusicoccin with plant membra-
     nes. Phytopathol. mediterr. 19: 187 (1980).
34. A. Ballio, R. Federico and D. Scalorbi, Metabolic stability of
     fusicoccin in plant tissues. Rend. Accad. Naz. Lincei 63:
     604 (1977).

STRUCTURE-FUNCTION RELATIONSHIPS IN PYRUVATE

AND α-KETOGLUTARATE DEHYDROGENASE COMPLEXES

Lester J. Reed and Robert M. Oliver

Clayton Foundation Biochemical Institute and
Department of Chemistry
The University of Texas at Austin
Austin, Texas, USA

INTRODUCTION

Enzyme systems that catalyze a lipoic acid-mediated oxidative decarboxylation of pyruvate and α-ketoglutarate have been isolated from microbial and eukaryotic cells as functional units with molecular weights in the millions. Each complex consists of three catalytic components: pyruvate dehydrogenase or α-ketoglutarate dehydrogenase (E_1); dihydrolipoyl transacetylase or dihydrolipoyl transsuccinylase (E_2); and dihydrolipoyl dehydrogenase (E_3), a flavoprotein that is a common component of the two complexes. These three enzymes, acting in sequence, catalyze[1] the reactions shown in Fig. 1. E_1 catalyzes both the decarboxylation of the α-keto acid (reaction 1) and the subsequent reductive acylation of the lipoyl moiety (reaction 2) that is covalently bound[2] to E_2. E_2 catalyzes the transacylation step (reaction 3), and E_3 catalyzes reoxidation of the dihydrolipoyl moiety with NAD^+ as the ultimate electron acceptor (reactions 4 and 5). The pyruvate dehydrogenase complexes from eukaryotic cells also contain small amounts of two regulatory enzymes, a kinase and a phosphatase, that modulate the activity of E_1 by phosphorylation and dephosphorylation, respectively.[3] This report discusses some novel structure-function relationships that underline the active site coupling mechanism in the pyruvate and α-ketoglutarate dehydrogenase complexes from Escherichia coli.

SUBUNIT STRUCTURE OF DIHYDROLIPOYL TRANSACYLASES

Each complex is organized about a core, an oligomer of E_2, to which multiple copies of E_1 and E_3 are joined by noncovalent bonds.

Results of electron microscopic studies,[4,5] x-ray crystallographic studies,[6,7] and sedimentation equilibrium molecular weight determinations[8,9] demonstrated that the E_2 core enzymes of the E. coli pyruvate and α-ketoglutarate dehydrogenase complexes each consists of 24 apparently identical polypeptide chains arranged with 432 molecular symmetry in a cube-like particle. Each transsuccinylase subunit contains one covalently bound lipoyl moiety, whereas each transacetylase subunit contains two lipoyl moieties.[10] Using limited proteolysis with trypsin to probe the subunit structure of the dihydrolipoyl transacetylase, Bleile et al.[11] showed that E. coli dihydrolipoyl transacetylase subunits possess a novel architectural feature (Fig. 2A). Each transacetylase subunit (M_r = 64,500) consists of two principal domains: (1) a compact domain (M_r = 29,600), designated the subunit binding domain, that contains the intersubunit binding sites of E_2, the binding sites for E_1 and E_3 and the catalytic site for transacetylation (reaction 3), and (2) a large flexible extension (M_r = 31,600) that bears the two covalently bound lipoyl moieties (lipoyl domain). The subunit binding domain and the lipoyl domain are apparently connected by a trypsin-sensitive "hinge" region. The E. coli dihydrolipoyl transsuccinilase subunit (M_r = 42,000) (Fig. 2B) also consists of a compact subunit-binding domain (M_r = 28,000) and an extended lipoyl domain (M_r = 11,000) (D. K. McRorie and L. J. Reed, unpublished data). The subunit binding domains form

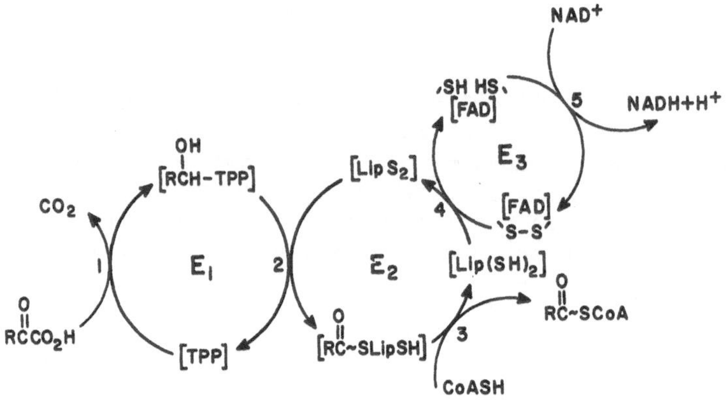

$$RCCO_2H + CoASH + NAD^+ \rightarrow RC{\sim}SCoA + CO_2 + NADH + H^+$$

Fig. 1.   Reaction sequence in α-keto acid oxidation, TPP, thiamin pyrophosphate; $LipS_2$ and $Lip(SH)_2$, lipoyl moiety and its reduced form; CoASH, coenzyme A; FAD, flavin adenine dinucleotide; NAD^+ and NADH, nicotinamide adenine dinucleotide and its reduced form. E_1, pyruvate dehydrogenase or α-ketoglutarate dehydrogenase; E_2, dihydrolipoyl transacetylase or dihydrolipoyl transsuccinylase; E_3, dihydrolipoyl dehydrogenase.

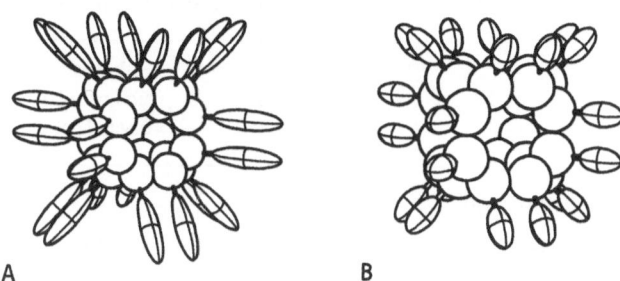

A   B

Fig. 2.   Interpretive models of quaternary structure of dihydrolipoyl
          transacylases. (A) E. coli dihydrolipoyl transacetylase;
          (B) E. coli dihydrolipoyl transsuccinylase. Each of the
          24 subunits is represented by one sphere and its attached
          ellipsoid. The spheres represent the assemblage of subunit
          binding domains and the ellipsoids represent the extended
          lipoyl domains.

a closed structure, the E_2 "inner" core, which determines the stoichiometry and quaternary organization of the complex. This inner core
is responsible for the cube-like appearance of the transacylases on
negative stain electron micrographs. The inner cores of the transacetylase and transsuccinylase are very similar in size and morphology, as demonstrated by electron microscopy and x-ray crystallography.[5,7] It appears that two domains, one compact and the other
extended and bearing the covalently bound lipoyl moiety, is a novel
architectural feature common to the prokaryotic and eukaryotic
dihydrolipoyl transacylases.[2,11,12,13]

SUBUNIT COMPOSITION AND STRUCTURE OF PYRUVATE AND

α –KETOGLUTARATE DEHYDROGENASE COMPLEXES

       Although there is some disagreement regarding the polypeptide
chain stoichiometry of the E. coli pyruvate dehydrogenase complex,[14]
we shall assume that the stoichiometry of the native complex is
that determined by Eley et al.[8] and confirmed by Angelides et al.[15]
i.e., $24E_1:24E_2:12E_3$ (Table 1). The polypeptide chain stoichiometry
of the native α-ketoglutarate dehydrogenase complex[9,16] is $12E_1$:
$24E_2:12E_3$. Space-filling models of the E. coli pyruvate and α-ketoglutarate dehydrogenase complexes are shown in Fig. 3. Fused trimeric
clusters of the subunit binding domain form the morphological units
of the transacetylase and transsuccinylase cores (E_2). Eight of
these units, represented in both models as spheres, are centered at
the vertices of a cube. The lipoyl domains are considered to be
flexible, extended structures that are attached by hinge regions

Table 1.   Subunit Composition of E. coli Pyruvate and
           α-Ketoglutarate Dehydrogenase Complexes

Enzyme	M_r	Subunits		Subunits per molecule of complex
		No.	M_r	
PDC	4,600,000			
E_1	181,000	2	90,500	24
E_2	1,548,000	24	64,500	24
E_3	112,000	2	56,000	12
KGDC	2,500,000			
E_1	190,000	2	95,000	12
E_2	1,000,000	24	42,000	24
E_3	112,000	2	56,000	12

PDC, pyruvate dehydrogenase complex; KGDC, α-ketoglutarate dehydrogenase complex.

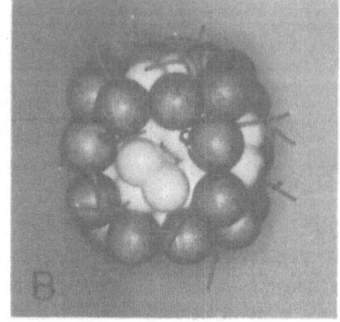

Fig. 3.   Interpretive models of E. coli α-ketoglutarate dehydrogenase
          complex (A) and pyruvate dehydrogenase complex (B).   The
          models are viewed approximately down a fourfold axis of
          the E_2 core.   The eight trimers of E_2 subunit binding do-
          mains (inner core) are represented as spheres and the 24
          lipoyl domains as flexible extensions.   The E_1 subunits are
          bound along the twelve edges and the E_3 dimers are located
          on the six faces of the cube-like E_2 inner core.

to the inner core of subunit binding domains. Pyruvate dehydrogenase
or α-ketoglutarate dehydrogenase subunits (E_1) are uniformly distributed along each of the twelve edges of the inner core, one E_1 subunit on each edge of the transsuccinylase and two E_1 subunits on
each edge of the transacetylase. Crosslinking experiments indicate
that E_3 is present in the complexes as dimers.[17,18] Six E_3 dimers
are thought to be located on the faces of the E_2 inner core of both
complexes, bound to diagonally opposed morphological units of E_2.
The E_3 dimers are assumed to be randomly distributed between the
two equivalent orientations on each of the six faces of the E_2 cores.

These models point up to many basic similarities in the structural features of the two complexes. The major differences appear
to be the presence of 12 additional E_1 subunits and 24 additional
lipoyl moieties in the pyruvate dehydrogenase complex.

ACTIVE SITE COUPLING MECHANISM

The lipoyl moieties in the transacetylase and the transsuccinylase are attached in amide linkage to the ε-amino group of a lysyl
residue.[19] This linkage provides a flexible arm about 14 Å in length,
which we proposed could allow the reactive 1,2-dithiolane ring to
rotate among the catalytic sites of E_1, E_2 and E_3, i.e., a "swinging
arm" active site coupling mechanism. We also pointed out that the
distances between the catalytic sites might be larger than could be
spanned by rotation of a single lipoyllysyl moiety, (~ 28 Å), thus
necessitating interaction between two or more such moieties. These
interactions involve thiol-disulfide interchange and acyl transfer.
The high mobility of the lipoyl moieties revealed by ESR spectroscopy[20,21] supported the swinging arm concept. Fluorescence energy
transfer measurements[16] indicated that the distances between catalytic sites in the E. coli α-ketoglutarate dehydrogenase complex
were consistent with a mechanism involving rotation of a single
lipoyl moiety between catalytic sites of E_1, E_2 and E_3 during a
catalytic cycle. However, the distances between catalytic sites in
the E. coli pyruvate dehydrogenase complex were significantly larger
(at least 40 Å) and apparently could not be spanned by the rotation
of a single lipoyl moiety.[22] Angelides and Hammes[23] proposed that
a series (i.e., tandem) interaction between at least two lipoyl
moieties is required to link the active sites on the pyruvate dehydrogenase complex. This notion seemed compatible with the presence
of two lipoyl moieties on each transacetylase (E_2) subunit and with
the finding that intramolecular transfer of acyl groups and electron pairs between lipoyl moieties can occur under conditions in
which only a few E_1 subunits are functional.[24,25,26] Recent kinetic
data[27,28,29,30] have rendered doubtful the probability that the
normal catalytic mechanism involves transfer of an acetyl group
(and an electron pair) from one lipoyl moiety to a second lipoyl
moiety en route to CoA. However, redox reactions between reduced

and oxidized lipoyl moieties, i.e., thiol-disulfide exchange, are
apparently part of the normal reaction mechanism.

Further insight into the active site coupling mechanism came
from studies on limited proteolysis of the E_2 core of the pyruvate
dehydrogenase complex, demonstrating that the lipoyllysyl moiety is
part of a "super arm", i.e., the lipoyl domain.[11] We proposed that
movement of lipoyl domains and not simply rotation of lipoyl moieties provides the means to span the physical gaps between catalytic
sites on the α-keto acid dehydrogenase complexes.[30] Support for
this proposal has come from our structural and enzymatic studies
and from proton NMR studies of Perham et al.[31] Unexpectedly sharp
lines in the proton NMR spectrum of the E. coli pyruvate dehydrogenase complex are attributed to high internal mobility of the
lipoyl domains.

We have analyzed the active site coupling mechanism in the E.
coli pyruvate and α-ketoglutarate dehydrogenase complexes by means
of reconstitution studies and use of trypsin and lipoamidase to
probe lipoic acid function. Results obtained with the α-ketoglutarate dehydrogenase complex, the simpler of the two complexes, are
summarized below and interpreted in the reaction diagram (Fig. 4).
The diagram depicts four trimeric clusters of subunit binding domains
which form a face of the inner core and the four E_1 subunits bound
to the edges of that face. These components are represented by
circles labeled E_2 and E_1, respectively. The lipoyl domains of an
E_2 trimeric cluster are symmetrically disposed around the threefold
axes of the cubic inner core. Thus, two symmetry related lipoyl
domains may be structurally associated with an edge of the inner
core. In the diagram the lipoyl domains are represented by ellipses.
An E_3 dimer is shown at the center of the face in one of its two
possible orientations. Possible interaction pathways between components are designated by arrows. A transacylation site on an E_2
subunit binding domain is represented by the small, crossed circle.
For clarity, only one of the 24 transacylation sites on E_2 is shown.

Subcomplexes consisting of E_2-E_3 and E_1-E_2 were titrated with
E_1 and E_3, respectively, and overall activity of the reconstituted
complexes was measured. The results show that overall activity is
a linear function of the amount of E_1 added, and that maximum activity is observed at a polypeptide chain ratio of about $12E_1:24E_2$
(J.H. Collins and L. J. Reed, unpublished data). Similar results
were obtained by Angelides and Hammes[16] when the native α-ketoglutarate dehydrogenase complex was inactivated by titration of the E_1
component with the transition state inhibitor, thiamin thiazolone
pyrophosphate. These data indicate that the rate-limiting step in
the overall reaction is catalyzed by E_1 (Fig. 1, reactions 1 and 2).

To determine the relationship between lipoic acid content and
overall activity, α-ketoglutarate dehydrogenase complex containing

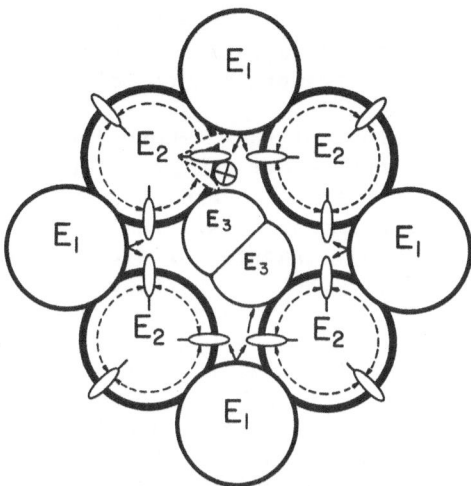

Fig. 4.   Interpretive diagram illustrating active site coupling
          mechanism in E. coli α-ketoglutarate dehydrogenase com-
          plex. The view is along a fourfold axis of the E_2 core.

tritium-labeled lipoyl moieties was subjected to limited digestion
with trypsin and with lipoamidase.[30] Limited tryptic digestion
releases lipoyl domains together with their covalently attached
lipoyl moieties from the E_2 inner core, whereas lipoamidase releases
only the lipoyl moieties, leaving the lipoyl domains otherwise
intact. Plots of loss of overall activity versus loss of lipoyl
moieties or lipoyl domains were very similar and curvilinear. After
half of the lipoyl moieties were removed, the α-ketoglutarate dehydrogenase complex still exhibited about 75% of its maximum activity. These results indicate that each E_1 subunit may be serviced
by more than one lipoyl moiety. Assuming random removal of lipoyl
domains or lipoyl moieties, calculations show that the best fit to
the experimental data is obtained based on the assumption that each
E_1 is serviced by two lipoyl moieties. Because two lipoyl domains
are identically positioned with respect to each edge of the E_2 inner
core, we propose that service to a particular E_1 is provided only
by this pair of lipoyl domains, as depicted in Fig. 4. Although
the α-ketoglutarate dehydrogenase complex possesses the functional
capacity to transfer succinyl groups (and electron pairs) between
the 24 lipoyl moieties of E_2 in the absence of CoA,[25] it has not
been demonstrated that such intramolecular transfers are on the
main catalytic pathway.[16] It should be noted that the geometry for
the lipoyl moiety and transsuccinylation site interactions is fixed
by the oligomer structure of E_2 and is identical for all lipoyl
moieties. We propose that after any lipoyl moiety is charged with
a succinyl group by interaction at E_1 (Fig. 1, reactions 1 and 2),

rotation of the lipoyl domain brings the S-succinyldihydrolipoyl
moiety to an adjacent transsuccinylation site on the E_2 inner core
where the succinyl group is transferred to CoA (Fig. 1, reaction 3).

Because of the structure of the α-ketoglutarate dehydrogenase
complex (Fig. 2), all of the lipoyl domains of E_2 cannot be identically positioned with respect to E_3, and it is unlikely that all of
the dihydrolipoyl moieties can be reoxidized by direct interaction
with E_3 (Fig. 1, reaction 4). Transfer of electron pairs between
reduced and oxidized lipoyl moieties, particularly between those
common to a morphological unit, appears to be an essential property
of the complex. The data obtained by titrating E_1-E_2 subcomplex with
E_3 yields curvilinear plots,[9] indicating the presence of multiple,
interconnected pathways from E_1 to E_3. Binding studies indicate that
when 50% of maximum activity has been reconstituted, an average of
only one of six E_3 dimers has been bound in the complex (P. F. Davis,
R. M. Oliver, and L. J. Reed, unpublished data). This level of activity requires the participation of six E_1 subunits (because the steps
catalyzed by E_1 are rate limiting) and, presumably, six different
lipoyl domains. These two sets of six components are conceived to
be those associated with the two morphological units to which the
E_3 dimer is bound. The geometries with respect to an E_3 dimer are
different for each of the three lipoyl moieties common to a morphological unit, and we consider it unlikely that all six lipoyl moieties have access to the same E_3 dimer. Therefore, we propose that
only two of the set of six lipoyl moieties (one from each morphological unit) have direct access to E_3 when only one E_3 dimer is bound
to the E_2 inner core. The other four dihydrolipoyl moieties are
reoxidized by thiol-disulfide exchange with lipoyl moieties that
have direct access to E_3. These possible pathways are designated on
the reaction diagram (Fig. 4) by dashed arrows. Due to the symmetrical geometry of the system, the possible orientations assumed by an
E_3 dimer on binding to the E_2 inner core will have no effect on the
overall activity of the system. In the native complex twelve of the
lipoyl moieties have a common geometry with and direct access to the
E_3 dimers. Consideration of the possible arrangements of E_3 dimers
in the complex shows that only one thiol-disulfide exchange is ever
needed to transfer an electron pair to E_3.

Preliminary analysis of the data on the E. coli pyruvate dehydrogenase complex indicates that the active site coupling mechanism
in that complex is similar to the model proposed for the α-ketoglu-
tarate dehydrogenase complex, with some modification to accommodate
12 additional E_1 subunits and 24 additional lipoyl moieties in the
former complex.

This work was supported in part by Grant GM06590 from the U.S.
Public Health Service.

REFERENCES

1. L. J. Reed, Multienzyme complexes, Acc. Chem. Res. 7: 40 (1974).
2. D. M. Bleile, M. L. Hacker, F. H. Pettit, and L. J. Reed, Subunit structure of dihydrolipoyl transacetylase component of pyruvate dehydrogenase complex from bovine heart, J. Biol. Chem. 256: 514 (1981).
3. T. C. Linn, F. H. Pettit, and L. J. Reed, Regulation of the activity of the pyruvate dehydrogenase complex from beef kidney mitochondria by phosphorylation and dephosphorylation, Proc. Natl. Acad. Sci. USA 62: 234 (1969).
4. L. J. Reed, and R. M. Oliver, The multienzyme α-keto acid dehydrogenase complexes, Brookhaven Symp. Biol. 21: 397 (1968).
5. R. M. Oliver, and L. J. Reed, Multienzyme complexes, in: "Electron microscopy of proteins," Vol. 2, J. R. Harris, ed., Academic Press, London, in press.
6. D. J. DeRosier, R. M. Oliver, and L. J. Reed, Crystallization and preliminary structural analysis of dihydrolipoyl trans-succinylase, the core of the 2-oxoglutarate dehydrogenase complex, Proc. Natl. Acad. Sci. USA 68: 1135 (1971).
7. C. C. Fuller, L. J. Reed, R. M. Oliver, and M. L. Hacker, Crystallization of a dihydrolipoyl transacetylase–dihydrolipoyl dehydrogenase subcomplex and its implications regarding the subunit structure of the pyruvate dehydrogenase complex from Escherichia coli, Biochem. Biophys. Res. Commun. 90: 431 (1979).
8. M. H. Eley, G. Namihira, L. Hamilton, P. Munk, and L. J. Reed, α-Keto acid dehydrogenase complexes. XVIII. Subunit composition of the Escherichia coli pyruvate dehydrogenase complex, Arch. Biochem. Biophys. 152: 655 (1972).
9. F. H. Pettit, L. Hamilton, P. Munk, G. Namihira, M. H. Eley, C. R. Willms, and L. J. Reed, α-Keto acid dehydrogenase complexes. XIX. Subunit structure of the Escherichia coli α-ketoglutarate dehydrogenase complex, J. Biol. Chem. 248: 5282 (1973).
10. R. H. White, D. M. Bleile, and L. J. Reed, Lipoic acid content of dihydrolipoyl transacylases determined by isotope dilution analysis, Biochem. Biophys. Res. Commun. 94: 78 (1980).
11. D. M. Bleile, P. Munk, R. M. Oliver, and L. J. Reed, Subunit structure of dihydrolipoyl transacetylase component of pyruvate dehydrogenase complex from Escherichia coli, Proc. Natl. Acad. Sci. USA 76: 4385 (1979).
12. G.-B. Kresze, and H. Ronft, Bovine kidney pyruvate dehydrogenase complex. Limited proteolysis and molecular structure of the lipoate acetyltransferase component, Eur. J. Biochem. 112: 589 (1980).
13. R. N. Perham, and A. O. M. Wilkie, Inner core and domain structure of the pyruvate dehydrogenase multienzyme complex of Bacillus stearothermophilus, Biochem. Int. 1: 470 (1980).

14. M. J. Danson, G. Hale, P. Johnson, R. N. Perham, J. Smith and
     P. Spragg, Molecular weight and symmetry of the pyruvate
     dehydrogenase multienzyme complex of Escherichia coli, J.
     Mol. Biol. 129: 603 (1979).
15. K. J. Angelides, S. K. Akiyama, and G. G. Hammes, Subunit
     stoichiometry and molecular weight of the pyruvate dehydro-
     genase multienzyme complex from Escherichia coli, Proc. Natl.
     Acad. Sci. USA 76: 3279 (1979).
16. K. J. Angelides, and G. G. Hammes, Structural and mechanistic
     studies of the α-ketoglutarate dehydrogenase multienzyme
     complex from Escherichia coli, Biochemistry 18: 5531 (1979).
17. L. J. Reed, F. H. Pettit, M. H. Eley, L. Hamilton, J. H.
     Collins, and R. M. Oliver, Reconstitution of the Escherichia
     coli pyruvate dehydrogenase complex, Proc. Natl. Acad. Sci.
     USA 72: 3068 (1975).
18. J. R. Coggins, E. A. Hooper, and R. N. Perham, Use of dimethyl
     suberimidate and novel periodate-cleavable bis(imido esters)
     to study the quaternary structure of the pyruvate dehydro-
     genase multienzyme complex of Escherichia coli, Biochemistry
     15: 2527 (1976).
19. H. Nawa, W. R. Brady, M. Koike, and L. J. Reed, Studies on the
     nature of protein-bound lipoic acid, J. Am. Chem. Soc., 82:
     896 (1960).
20. M. C. Ambrose, and R. N. Perham, Spin-label study of the mobi-
     lity of enzyme-bound lipoic acid in the pyruvate dehydro-
     genase multienzyme complex of Escherichia coli, Biochem. J.
     155: 429 (1976).
21. H. J. Grande, H. J. Van Telgen, and C. Veeger, Symmetry and
     asymmetry of the pyruvate dehydrogenase complexes from
     Azotobacter vinelandii and Escherichia coli as reflected
     by fluorescence and spin-label studies, Eur. J. Biochem.
     71: 509 (1976).
22. G. B. Shepherd, and G. G. Hammes, Fluorescence energy transfer
     measurements in the pyruvate dehydrogenase multienzyme
     complex from Escherichia coli with chemically modified lipoic
     acid, Biochemistry 16: 5234 (1977).
23. K. J. Angelides, and G. G. Hammes, Mechanism of action of the
     pyruvate dehydrogenase multienzyme complex from Escherichia
     coli, Proc. Natl. Acad. Sci. USA 75: 4877 (1978).
24. D. L. Bates, M. J. Danson, G. Hale, E. A. Hooper, and R. N.
     Perham, Self-assembly and catalytic activity of the pyru-
     vate dehydrogenase multienzyme complex of Escherichia coli,
     Nature 268: 313 (1977).
25. J. H. Collins, and L. J. Reed, Acyl group and electron pair
     relay system: A network of interacting lipoyl moieties in
     the pyruvate and α-ketoglutarate dehydrogenase complexes
     from Escherichia coli, Proc. Natl. Acad. Sci. USA 74: 4223
     (1977).
26. R. L. Cate, T. E. Roche, and L. C. Davis, Rapid intersite
     transfer of acetyl groups and movement of pyruvate dehydro-

genase complex, J. Biol. Chem. 255: 7556 (1980).

27. P. A. Frey, B. H. Ikeda, G. R. Gavino, D. C. Speckhard, and S. S. Wong, Escherichia coli pyruvate dehydrogenase complex. Site coupling in electron and acetyl group transfer pathways, J. Biol. Chem. 253: 7234 (1978).

28. M. C. Ambrose-Griffin, M. J. Danson, W. G. Griffin, G. Hale, and R. N. Perham, Kinetic analysis of the role of lipoic acid residues in the pyruvate dehydrogenase multienzyme complex of Escherichia coli, Biochem. J. 187: 393 (1980).

29. S. K. Akiyama, and G. G. Hammes, Elementary steps in the reaction mechanism of the pyruvate dehydrogenase multienzyme complex from Escherichia coli: kinetics of acetylation and deacetylation, Biochemistry 19: 4208 (1980).

30. L. R. Stepp, D. M. Bleile, D. K. McRorie, F. H. Pettit, and L. J. Reed, Use of trypsin and lipoamidase to study the role of lipoic acid moieties in the pyruvate and α-ketoglutarate dehydrogenase complexes of Escherichia coli, Biochemistry 20: 4555 (1981).

31. R. N. Perham, and A. O. M. Wilkie, Inner core and domain structure of the pyruvate dehydrogenase multienzyme complex of Bacillus stearothermophilus, Biochem. Int. 1: 470 (1980).

FACILITATION OF ENZYME CATALYSIS

BY FORMATION OF MULTIENZYME COMPLEXES

Costantino Salerno and Paolo Fasella

Institute of Biological Chemistry and C.N.R. Center of
Molecular Biology, University of Rome, 00185 Rome, Italy

INTRODUCTION

There has been considerable discussion whether enzymes, generally identified as constituent of 'soluble cellular fraction', may
be organized into multienzyme complexes.[1,2] Much attention has been
paid to the possibility of complex formation between enzymes which
catalyze successive reactions in the same metabolic pathway[3-6].

Enzyme clusters, in contrast to the soluble system of independent enzymes, have the potential of compartmentalizing the intermediates of a pathway. It might be imagined that intermediates could
be transferred from one active site to the next by conformational
changes in the proteins or, alternatively, could be confined to an
unmixed layer in the enzyme surface or be contained in a microcom-
partment within the protein structure of the enzyme cluster. Such
multienzyme complex could provide a suitable device in the control
of metabolic cross-roads, preventing the conversion of intermediate
substrates by other enzymes. Studies with labelled substrates led to
the identification of distinct segregated pools of intermediate
metabolites within the cell.[7]

Another potential advantage of aggregation between soluble
enzymes catalyzing subsequent steps in a metabolic pathway is an
increased efficiency in the catalysis of the reaction sequence
through the decrease of transit time for the passage of metabolites
between the enzymes. The relevance of this potential advantage may

sometimes be difficult to estimate, essentially because free diffusion in water may not be a realistic model of substrate transfer
between active sites of enzymes localized in hydrophobic regions of
the cell or associated with membranous structures in a highly hetero-
geneous environment. The reduction of transit time could be advantageous only if the enzyme-bound steps are all so fast that the rate of
transit is limiting. Hess[3] has argued that, at least in glycolysis,
transit times are so short, because of high cellular enzyme concentration, that their reduction is unimportant.

The effect of substrate diffusion on the efficiency of enzymic
reactions has been examined comparing the kinetic properties of free
and matrix-bound enzymes. For instance, aspartate aminotransferase
exhibits a lower affinity for oxaloacetate but a higher affinity for
glutamate as a result of its attachment on the surface of collagen
membrane.[8] Even though diffusional limitations between bulk solution
and film surface are negligible for glutamate, the apparent affinity
of the enzyme for this substrate is increased as a result of diffusional limitations for the other substrate. A 40-fold decrease in
the apparent Km for pyruvate and a 6-fold increase in the apparent
Km for NADH after immobilization of lactate dehydrogenase on glass
beads have been reported.[9] Under the experimental conditions employed,
diffusional limitations are expected to be more important for NADH
than for pyruvate and thus may indeed be responsible[8] for the apparent decrease and increase in NADH and pyruvate affinities, respectively.

In the living cell, interactions between heterologous proteins
could be quite weak. In fact, considering that the intracellular
concentration of many enzymes is of the order of 10-100 μM,[3] interaction between enzymes catalyzing subsequent steps in a metabolic
pathway could play a role even if the equilibrium dissociation
constant of the heterologous enzyme complexes was of the order of
1 mM, i.e. large enough to make aggregation quite difficult to detect
in vitro.

INTERACTION BETWEEN TRANSAMINASE AND DEHYDROGENASE

In order to discuss the experimental difficulties encountered
and the results obtained in the study of complexes between heterologous weakly-interacting enzymes, in the following pages we will
summarize the physico-chemical evidence for the interaction between

aspartate aminotransferase and glutamate dehydrogenase. Mitochondrial aspartate aminotransferase and glutamate dehydrogenase are mainly present in the soluble sub-mitochondrial fraction of liver cells.[10] They are involved in important metabolic processes such as amino acid dehydrogenation, ammonium production, and urea synthesis. This enzyme system is endowed with several features which recommend it for accurate study. Both enzymes are available in high yield and purity and a great amount is known concerning the relationship between structure and function for both enzymes as they act separately.[11,12] The high degree of homology among tissues and species, observed for both enzymes, substantiates the physiological relevance of the study of their interaction.

Gel Filtration Chromatography

The interaction between pig liver mitochondrial aspartate aminotransferase and pig liver glutamate dehydrogenase has been studied by gel filtration chromatography[13] using the procedure described by Hummel and Dreyer.[14] A 2 ml-volume solution containing 1.4 µM dehydrogenase and 2.8 µM transaminase was placed on a Sephadex G-100 column (100 x 0.5 cm) pre-equilibrated with the same transaminase solution as was used to dissolve dehydrogenase. The column was thereafter eluted with the transaminase solution and the concentration of transaminase in the eluate was measured. As the dehydrogenase peak emerged at the excluded volume of the column, the total amount of transaminase rose above the equilibrium level; correspondly, at some point after the dehydrogenase peak, the concentration of transaminase eluate was decreased below the base-line to form a trough which extended to the elution volume predicted for transaminase.

Several unsuccessful attempts were made to determine the molecular weight of the heterologous enzyme complex by filtration through columns of gels which include both proteins. It was reported that, when the two enzymes were chromatographed on Sephadex G-200, a small amount of transaminase was eluted with the beginning of the dehydrogenase fraction as a peak separate from the major peak of transaminase activity.[15] This effect was not confirmed by the technique of Hummel and Dreyer: equilibration of a Sephadex G-200 column with dehydrogenase had no appreciable effect on the elution profile of transaminase.[16] Likewise, the preequilibration of a Sepharose 6B column with transaminase did not alter the elution profile of dehydrogenase.[16] The last negative result is not surprising[16] because

of the expected small change in the apparent molecular weight of
dehydrogenase and the poor discrimination between small molecular
weight changes by Sepharose 6B chromatography.

In order to determine the affinity constant for the heterologous
enzymes, gel filtration equilibrium experiments were performed using
Sephadex G-200 beads. This insoluble matrix was chosen because dehydrogenase is excluded from the gel[17] whereas transaminase can pene-
trate the insoluble phase. A problem with this method was that at
least 2 hours were required for equilibration and mitochondrial
aspartate aminotransferase was not stable under the experimental
conditions employed.[18] The interaction between bovine liver glutamate dehydrogenase and the more stable pig heart cytoplasmic aspartate aminotransferase has been studied by Fahien et al. using this
technique.[17] Glutamate dehydrogenase, especially in the presence of
NADH plus ammonium ions or NAD plus aspartate decreased the distribution coefficient of transaminase in Sephadex G-200. The apparent
equilibrium dissociation constant (16-67 µM) for this unphysiologi-
cal heterologous enzyme complex was calculated[17] from the dehydrogenase concentration in the aqueous phase times the ratio of the
'free' to the 'transaminase-bound' fraction.

Divalent Cross-linkers

The divalent cross-linker, dimethyl 3,3'-dithiobis-propionimi-
date, was used to stabilize the heterologous enzyme complex.[19] When
bovine liver mitochondrial aspartate aminotransferase and glutamate
dehydrogenase were incubated with the bifunctional reagent, an
anionic, high molecular weight (greater than 200,000) form of transaminase was produced which could be eluted with the dehydrogenase
from DEAE and Sephadex G-200 columns. If dehydrogenase was omitted,
then transaminase was eluted from the columns as the native enzyme
which is cationic and has a molecular weight of about 100,000. These
results suggest[19] that (i) glutamate dehydrogenase and mitochondrial
aspartate aminotransferase form an enzyme-enzyme complex, (ii) the
cross-linker prevents dissociation of this complex by forming a
covalent bond between the two proteins, (iii) the heterologous macromolecular aggregate has a high molecular weight and is anionic
because these are the properties of glutamate dehydrogenase.

In another series of experiments, transaminase was treated with
sodium borohydride in the presence of excess of 14-C glutamate or
14-C aspartate and then dialyzed versus buffer.[19] Since this proce-

dure had little effect on the amount of transaminase cross-linked to
dehydrogenase when compared to transaminase reduced with sodium borohydride alone, it was concluded[19] that a bond is apparently formed
between the two enzymes which is distal from their active sites. Mattingly convincingly questionated this interpretation pointing out[16]
that the 14-C pyridoxyl derivatives were not covalently bound to transaminase[20] and, therefore, could be removed in a large extent during
the dialysis of the enzyme.

Coprecipitation in Polyethylene Glycol

When 1.1 nmol bovine liver mitochondrial aspartate aminotransferase was incubated with 0.15 nmol bovine liver glutamate dehydrogenase
in 20 mM phosphate buffer, pH 7.0, containing 14% polyethylene glycol,
a significant amount of both enzymes precipitated.[21] The maximum
ratio of the two enzymes in the precipitate approached three transaminase dimer per one dehydrogenase hexamer. Under the experimental
conditions employed, the enzymes did not precipitate to a significant extent when incubated alone. Coprecipitation appears to be
quite specific since other proteins (bovine serum albumin, citrate
synthase, cytoplasmic malate dehydrogenase) did not coprecipitate
with either glutamate dehydrogenase or aspartate aminotransferase.[21]
Coprecipitation occurred when polyethylene glycol was added to solutions containing glutamate dehydrogenase and mitochondrial malate
dehydrogenase.[21]

These results were interpreted in terms of the formation of
multienzyme complexes insoluble in polyethylene glycol. However alternative explanations cannot be excluded. It is known that addition
of proteins to polyethylene glycol causes a destabilization of the
solvent system because of an unfavourable interaction between polyethylene glycol and the proteins.[22] The addition of dehydrogenase
in the experiments reported above could serve to alter the state
of the solvent so as to reduce the enzyme solubility[16]; thus some
of the proteins could precipitate, the solvent being the only in-
termediary of the interaction between transaminase and dehydrogenase.
Moreover, assuming that the transaminase-dehydrogenase complex is
present and precipitates in polyethylene glycol, it is not yet known
how to estimate the extent of enzyme association. If the heterologous enzyme complex is completely insoluble in polyethylene glycol,
then the less concentrated protein must completely precipitate as
constituent of the heterologous enzyme complex. This was not the

case since, in the experiments reported above, only 28% of transaminase and 66% of dehydrogenase precipitated. Therefore, we must assume that the heterologous enzyme complex is partially soluble in polyethylene glycol or that its precipitation is a slow process.

The coprecipitation of transaminase and dehydrogenase was markedly decreased by increasing either the pH and/or ionic strength[23] or by addition of ligands such as malate plus NADPH, NADH or GTP.[21] On the contrary, ornithine and, to major extent, aspartate plus NADP+ enhanced coprecipitation.[21,23] These results could be attributed to a variation of concentration of the enzyme-enzyme complex as well as to variation of solubility or velocity of precipitation of the protein aggregates in polyethylene glycol.

Fluorescence Polarization

Experiments were performed by covalently labeling transaminase with fluorescent dye and then monitoring the change in nanosecond and/or steady-state fluorescence polarization of the labeled enzyme upon addition of dehydrogenase.

Pig liver mitochondrial aspartate aminotransferase was labeled with fluorescein isothiocyanate.[13] The free fluorescent dye was removed from the protein-conjugated dye by exhaustive dialysis against buffer and by gel filtration chromatography. The latter procedure removed all traces of the free fluorescent compound as demonstrated by polyacrylamide gel electrophoresis. The degree of labeling was 2-3 mol of dye per mol of dimeric enzyme. The labeled transaminase retained about all the original specific activity.

After addition of pig liver glutamate dehydrogenase (ranging from 0 to 1 μM) to a solution containing 20 mM phosphate buffer, pH 7.8, and 81 nM labeled transaminase, the fluorescence polarization increased slowly with time approaching a limiting value within 2 hours at 20°C. The excitation and emission spectrum and the intensity of fluorescence were not affected by the presence of dehydrogenase as high as 1 μM. No inactivation of either enzyme occurred during the incubation. After 2-hour incubation of the two enzymes, the fluorescence polarization of labeled transaminase at equilibrium with increasing concentrations of dehydrogenase was a direct function of dehydrogenase concentration and approached a limiting value in the presence of excess of dehydrogenase. In order to test the reversibility of the phenomenon, a mixture of dehydrogenase

and labeled transaminase in a molar ratio of 1:3 with an initial
transaminase concentration of 0.13 µM was diluted with buffer and
subjected to fluorescence polarization measurements. The results obtained by measuring the fluorescence polarization of the mixture
after each dilution were identical (within the experimental error)
to those observed when separate solutions of dehydrogenase and labeled transaminase were first diluted and then mixed prior to measurements of polarization. The fluorescence polarization of mixtures
of labeled transaminase and dehydrogenase approached, upon dilution,
that of the transaminase-conjugated dye.

These observations seem to exclude the possibility of transfer
of any non-covalently bound fluorescein from transaminase to dehydrogenase and/or a binding of transaminase to dehydrogenase through
the fluorescent moiety. The above considerations suggest that the
observed increase in fluorescence polarization of labeled transaminase upon dehydrogenase addition is related to binding of transaminase to dehydrogenase. The apparent stoichiometry of the interaction
was measured by a continuous variation study by mixing stock solutions of dehydrogenase and labeled transaminase in various proportions but keeping the total volume of each mixture constant.[13] The
apparent stoichiometry of the complex is two hexameric dehydrogenase
molecules per dimeric labeled transaminase molecule. The same value
was obtained by an appropriate analysis of the polarization titration data.[13]

Caution must be used when interpreting these data since they
could be influenced by various factors, such as the aggregation of
dehydrogenase at the higher concentrations[12] and/or the dissociation
of labeled transaminase into monomeric chains upon binding to dehydrogenase. In this connection it must be pointed out that the apparent affinity between the two proteins decrease at the higher concentrations.[13] A plot of the dilution titration data according to
Rawitch and Weber[24] suggests that negative cooperativity occurs
(n = 0.5) in the binding process. An average dissociation constant
of 40 nM was determined for protein concentrations ranging from 10
to 100 nM.

Similar results were obtained by Churchich[25] using beef liver
mitochondrial aspartate aminotransferase and glutamate dehydrogenase.
Transaminase was labeled with l-dimethylaminonaptalene-5-sulphonyl
chloride (1.1 - 1.4 mol of dye per mol of enzyme). The fluorescence
polarization of 3 µM labeled transaminase increased from 0.22 to

to 0.25 after addition of 20 µM dehydrogenase, This change in the
degree of polarization was not accompanied by any change in the
fluorescence lifetime of the dansyl chromophore, Emission anisotropy
decay measurements indicated the presence of macromolecular species
characterized by a rotational correlation time of 175 nsec which is
longer than that of free transaminase (43 nsec). Although this finding suggests that the two enzymes form a macromolecular aggregate,
it must be pointed out that transaminase could be not rigidly
trapped by dehydrogenase since the rotational correlation time determined from mixtures of the two enzymes was shorter than the value
expected from macromolecular species of molecular weight larger than
300,000 (about 300 nsec). An alternative explanation is that dehydrogenase dissociates into monomeric peptide chains upon binding to
transaminase.

Similar results have been reported for glutamate dehydrogenase
and aspartate aminotransferase from pig brain.[26] In these experiments,
the transaminase labeled with the sulfhydryl reagent (N-iodoacetyla-
minoethyl)-5-naphtylamine-1-sulfonic acid was sensible to substrate
binding in its fluorescence emission spectrum. Dehydrogenase elicited
a similar response and determined an increase of the rotational correlation time of the solution containing the two proteins. Mattingly
[16] pointed out that the mitochondrial transaminase would be expected
to react sluggishly with the sulfhydryl reagent and that the fluorophore could be associated with the transaminase by a non-covalent
force and displaced under some circumstances. Beef liver glutamate
dehydrogenase did not change the fluorescence polarization of beef
liver mitochondrial aspartate aminotransferase covalently labeled
with the sulfhydryl reagent N-(1-pyrene)-maleimide.[16]

The binding of glutamate dehydrogenase appears non-specific for
the mitochondrial isoenzyme of aspartate aminotransferase. An unphy-
siological complex between bovine liver glutamate dehydrogenase and
pig heart cytoplasmic aspartate aminotransferase has been demonstrated by fluorescence polarization measurements.[27] The cytoplasmic
isoenzyme of aspartate aminotransferase seems to interact also with
cystathionase[28] and with glyceraldehyde-3-phosphate dehydrogenase.[4]
On the other hand, it has been demonstrated that bovine serum
albumin, γ-globulin, ovoalbumin, and lactate dehydrogenase did not
alter the nanosecond and/or steady state fluorescence polarization
of labeled mitochondrial aspartate aminotransferase.[25]

CONCLUSION

Despite the large body of information and the early appearance of the proposed multienzyme complex between aspartate aminotransferase and glutamate dehydrogenase, little quantitative information concerning its structure has appeared. Moreover, no indication of interactions between the two enzymes has been obtained by sedimentation equilibrium experiments[16,18] or using electron spin resonance probes.[16]

Tempting though they be, extrapolations about the possible physiological significance of the results reported above should be made with great caution. Even if some indication of the existence of a complex between the two weakly-interacting proteins has been obtained in vitro, there is no assurance that undiscovered cellular components would not invalidate our extrapolation from the laboratory to the cell.

REFERENCES

1. G.R. Welch, On the role of organized multienzyme systems in cellular metabolism: a general synthesis, Prog. Biophys. Mol. Biol. 72:4218 (1977).
2. F.H. Gaertner, Unique catalytic properties of enzyme clusters, TIBS 3:63 (1978).
3. B. Hess and A. Boiteux, Heterologous enzyme-enzyme interactions, in: "Protein-Protein Interactions", R. Jaenicke and E. Helmreich, eds., Springer Verlag, Berlin (1972).
4. J. Ovadi, C. Salerno, T. Keleti, and P. Fasella, Physicochemical evidence for the interaction between aldolase and glyceraldehyde-3-phosphate dehydrogenase, Eur. J. Biochem. 90:499 (1978).
5. F. Gavilanes, C. Salerno, and P. Fasella, Heterologous enzyme-enzyme complex between D-fructose-1,6-bisphosphate aldolase and triosephosphate isomerase from Ceratitis capitata, Biochim. Biophys. Acta 660:154 (1981).
6. C.F.A. Bryce, D.C. Williams, R.A. John, and P. Fasella, The anomalous kinetics of coupled aspartate aminotransferase and malate dehydrogenase, Biochem. J. 153:571 (1976).
7. R. MacNab, V. Moses, and J. Mowbray, Evidence for metabolic compartmentation in Escherichia coli, Eur. J. Biochem. 34:15 (1973).
8. J.M. Engasser, P.R. Coulet, and D.C. Gautheron, Kinetics of soluble and collagen-bound aspartate aminotransferase: diffusional effects with a two-substrate enzymatic reaction, J. Biol. Chem. 252:7919 (1977).

9. I.C. Cho and H. Swaisgood, Surface-bound lactate dehydrogenase: preparation and study of the effect of matrix microenvironment on kinetic and structural properties, Biochim. Biophys. Acta 334:243 (1974).

10. R. Marco, A. Pastana, J. Sebastian, and A. Sols, Oxaloacetate metabolic crossroads in liver: enzyme compartmentation and regulation of gluconeogenesis, Mol. Cell. Biochem. 3:53 (1974).

11. P. Fasella and C. Turano, Structure and catalytic role of the functional groups of aspartate aminotransferase, Vitam. Horm. 28:157 (1970).

12. H.F. Fisher, Glutamate dehydrogenase-ligand complexes and their relationship to the mechanism of the reaction, Adv. Enzymol. 39:369 (1973).

13. C. Salerno, J. Ovadi, J. Churchich, and P. Fasella, Interaction between transaminases and dehydrogenases, Proc. FEBS Meet. 32: 147 (1975).

14. J.P. Hummel and W.J. Dreyer, Measurement of protein binding phenomena by gel filtration, Biochim. Biophys. Acta 63:530 (1962).

15. L.A. Fahien and S.E. Smith, The effect of transaminases on reactions catalyzed by glutamate dehydrogenase, Arch. Biochem. Biophys. 135:136 (1969).

16. J.R. Mattingly, An investigation of the plausibility of a metabolically significant interaction between glutamate dehydrogenase and transaminases, Ph. D. Thesis, Department of Chemistry, Notre Dame, Indiana (1981).

17. L.A. Fahien and S.E. Smith, The enzyme-enzyme complex of transaminase and glutamate dehydrogenase, J. Biol. Chem. 249:2696 (1974).

18. C. Salerno, unpublished work.

19. L.A. Fahien, A.E. Ruoho, and E. Kmiotek, A study of glutamate dehydrogenase-aminotransferase complexes with a bifunctional imidate, J. Biol. Chem. 253:5745 (1978).

20. F. Riva, P. Vecchini, C. Turano, and P. Fasella, Chemical study of enzyme substrate intermediates in aspartic aminotransferase, Int. Cong. Biochem. Abstr. 140:329 (1964).

21. L.A. Fahien and E. Kmiotek, Precipitation of complexes between glutamate dehydrogenase and mitochondrial enzymes, J. Biol. Chem. 254:5983 (1979).

22. J.C. Lee and L.L.Y. Lee, Preferential solvent interactions between proteins and polyethylene glycols, J. Biol. Chem. 256: 625 (1981).

23. C. Salerno, P. Fasella, and L.A. Fahien, Interaction of transa-

minase with other metabolically linked enzymes, in: "Transaminases", P. Christen and D. Metzler eds'., John Wiley & Sons, New
York (1982).

24. A.B. Rawitch and G. Weber, The reversible association of lysozyme
and thyroglobulin: cooperative binding by near-neighbor interactions, J. Biol. Chem. 247:680 (1972).

25. J.E. Churchich and Y-H. Lee, Nanosecond emission anisotropy of
interacting enzymes aspartate aminotransferase glutamate dehydrogenase, Biochem. Biophys. Res. Commun. 68:409 (1976).

26. J.E. Churchich, Interaction between brain enzymes glutamate
dehydrogenase and aspartate aminotransferase, Biochem. Biophys.
Res. Commun. 83:1105 (1978).

27. C. Salerno, J.E. Churchich, and P. Fasella, Fluorescence polarization studies on the binding between glutamate dehydrogenase
and cytoplasmic aspartate aminotransferase, It. J. Biochem.
24:351 (1975).

28. J.E. Churchich and K-J. Oh, Fluorescence studies on the interaction between pyridoxal phosphate enzymes, J. Biol. Chem.
249:5623 (1974).

COFACTOR-DEPENDENT ENZYMES

SOME COMPARATIVE ASPECTS OF PYRIDOXAL PHOSPHATE AND PYRUVOYL-DEPENDENT AMINO ACID DECARBOXYLASES

Esmond E. Snell

Departments of Microbiology and
Chemistry, The University of
Texas at Austin, Austin, Texas
U.S.A.

INTRODUCTION

The conversion of amino acids to the corresponding amines by loss of the carboxyl group was among the first degradative reactions of amino acids to be discovered. Our understanding of the catalytic mechanism of this reaction, however, is still incomplete. Among the early milestones leading to our present understanding of these reactions was the discovery of a series of inducible bacterial amino acid decarboxylases by Gale and his colleagues (Fig. 1) and their discovery that an unidentified, heat stable coenzyme also was required for the reaction.[1] A similar unidentified coenzyme, required for the bacterial degradation of tryptophan to indole, was also reported by Usawa.[2] Identification of this coenzyme followed two developments: (a) discovery in our laboratory of pyridoxal as one of two previously unrecognized forms of vitamin B_6 that, unlike pyridoxine, were highly active in promoting growth of lactic acid bacteria,[3] and (b), the discovery by Gunsalus et al.[4] (Fig. 2) that Streptococcus faecalis failed to produce tyrosine decarboxy-

$$RCHNH_2COOH \longrightarrow RCH_2NH_2 + CO_2$$
(Gale 1940 - 1945; R = tyr, lys, arg, glu, his, orn)

$$Tryptophan \longrightarrow Indole + CH_3COCOOH + NH_3$$
(Uzawa, 1942)

Fig. 1. An unidentified coenzyme for reaction of amino acids.

257

lase when grown in a medium lacking vitamin B_6. Addition of pyridoxal to such cells resulted in formation of an active enzyme; addition of pyridoxine did not. In dried cells, pyridoxal was effective only when added together with ATP, a finding which clearly pointed to a phosphorylated pyridoxal as the coenzyme and led to its eventual identification as pyridoxal 5'-phosphate (PLP).

PYRIDOXAL-P DEPENDENT DECARBOXYLASES

How does PLP facilitate the decarboxylation reaction? Our early investigations showed that pyridoxal promoted a number of reactions of amino acids in non-enzymatic systems that were closely analogous to known enzymatic reactions of amino acids, e.g. transamination, serine dehydration, etc.[5] We concluded that all of these reactions proceeded via Schiff's bases formed between amino acid and pyridoxal, as shown in Scheme 1. In such complexes, labilization of bonds a, b, or c of the amino acids as a result of the strongly electrophilic character of the heterocyclic nitrogen of pyridoxal, leads to the enhanced reactivity of each of the three groups surrounding the α carbon atom. Many of these non-enzymatic reactions were stimulated by appropriate di- or trivalent metal ions and appeared to proceed through a metal ion-stabilized Schiff's base such as I (Scheme 1). Model decarboxylation reactions, however, were slightly inhibited by the presence of such metal ions, indicating that loss of the carboxyl group from the chelate complex as CO_2 proceeded less readily than from a Schiff's base such as II (Scheme 1) that contains the free carboxyl group.[6] We postulated that the coenzyme in pyridoxal phosphate-dependent enzymes functioned through formation of similar enzyme-bound complexes. Braunstein and Shemyakin independently came to similar conclusions.[7]

PLP·E
(1) Tyrosine ——————→ Tyramine + CO_2

    (Gunsalus, Bellamy & Umbreit, 1944;
    Baddiley & Gale, 1945).

(2) Tryptophan + H_2O $\xrightarrow{\text{PLP·E}}$ Indole + Pyruvate NH_3

    (Wood, Gunsalus & Umbreit, 1947).

(3) Glu + oxaloacetate $\rightleftharpoons$ α-Ketoglutarate + Asp

    (Schlenk & Snell, 1945; Lichstein, Gunsalus &
    Umbreit, 1945).

Fig. 2. Identification of PLP as a coenzyme.

Scheme   1

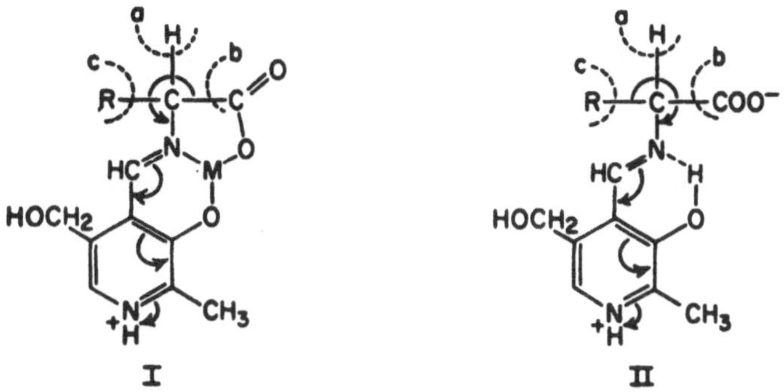

I   II

Does decarboxylation in PLP enzymes result from direct labilization of bond b, or is it a secondary consequence of labilization
of the α hydrogen atom (bond a)? For the simple decarboxylases
this question was settled when it was shown by Mandeles and others[8]
that only one deuterium atom from D_2O was incorporated into tyramine
during the enzymatic decarboxylation of tyrosine, and that tyrosine
decarboxylase also catalyzed the slow decarboxylation of α-methyl-
tyrosine (Fig. 3). Direct labilization of the carboxyl group also
occurs in pyridoxal-catalyzed, non-enzymatic decarboxylation of
amino acids. Thus when α-aminoisobutyrate is heated with pyridoxal
we observed[6] two reactions: (a) formation of CO_2 and isopropylamine
(pyridoxal remains unchanged), and (b) the production of CO_2, acetone and pyridoxamine (Scheme 2). Under the same conditions transamination between pyridoxal and isopropylamine does not occur, i.e.,
acetone and pyridoxamine could not have formed by that route. Isopropylamine formation from α-aminoisobutyrate must occur, therefore, by direct labilization of the α-carboxyl group (Scheme 2).
The postulated mechanism of this reaction is strengthened by·the
appearance of acetone and pyridoxamine, which are formed when the
extra electron pair, instead of localizing on carbon 1 of the incipient amine, localize on the 4' carbon atom of pyridoxal; hydrolysis
of the resultant Schiff's base gives a decarboxylation-dependent
transamination reaction.[6] At the time we first observed this reaction, no corresponding enzymatic reactions were known; shortly thereafter, however, studies of the bacterial degradation of α-aminoisobutyrate showed that the initial attack on this amino acid (Fig. 4)
proceeds via (a) the decarboxylation-dependent transamination of
α-aminoisobutyrate with a PLP-enzyme to yield acetone, carbon
dioxide, and the PMP-enzyme, followed by (b) transamination between
the PMP-enzyme and pyruvate to regenerate the PLP-enzyme with for-

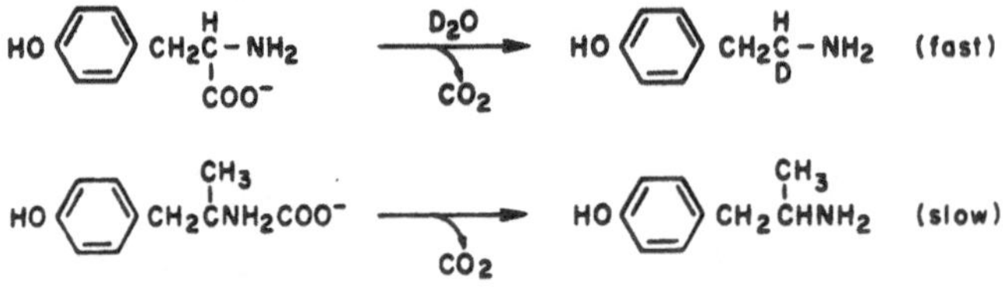

Fig. 3.   Tyrosine decarboxylase: direct labilization of carboxyl group.

Scheme 2

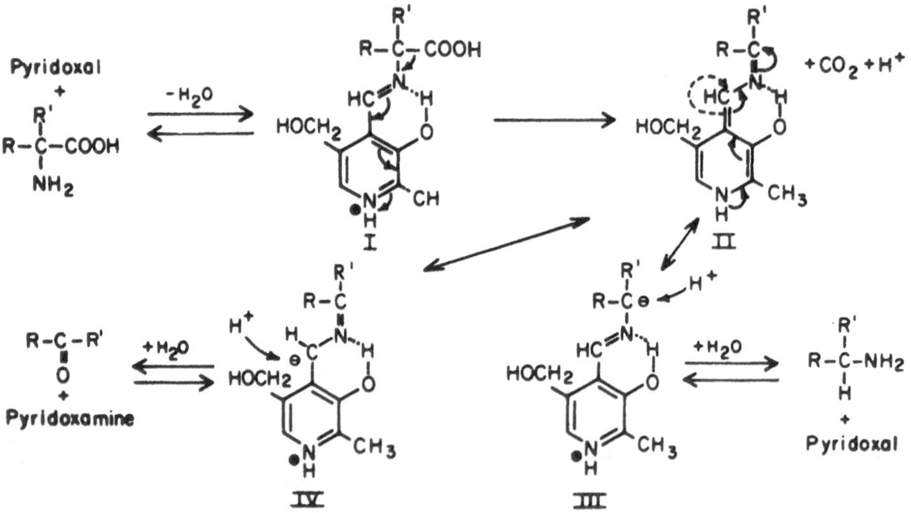

Pyridoxal
+
$R-\underset{\underset{NH_2}{|}}{\overset{\overset{R'}{|}}{C}}-COOH$

$R-\underset{\overset{||}{O}}{\overset{|}{C}}-R'$ $+H_2O$

Pyridoxamine

mation of alanine. Summation of the two subreactions gives the overall reaction catalyzed by this interesting enzyme.[9,10]

In the non-enzymatic system, isopropylamine and acetone are formed in similar amounts. It is therefore of interest to enquire how, in enzymatic systems, the position at which the electron pair localizes is controlled. Formation of the two products requires that a proton be placed on either carbon atom 1 of the incipient amine, for simple decarboxylation, or upon the 4'-carbon of the pyridoxal residue for decarboxylation-dependent transamination (Scheme 2). It appears likely that the position this proton assumes is determined with remarkable accuracy by the spatial positioning of a proton-donating group on the protein. However, control is not absolute.

Almost every PLP-dependent decarboxylase so far examined (Fig. 5) undergoes a slow substrate-dependent inactivation, and this inactivation has been traced to the occasional occurrence of a decarboxylation-dependent transamination reaction that leads to an inactive PMP enzyme.[11,12] Occasionally, it has been postulated that such inactivation may provide a physiological control mechanism for decarboxylase activity. Such a role appears to me less likely than that inactivation represents slightly less than 100% control of spe-

(1) $(CH_3)_2CNH_2COO^- + \text{Pyruvate} \longrightarrow CH_3\overset{O}{\overset{\|}{C}}-CH_3 + CO_2 + \text{Alanine}$

(a) $(CH_3)_2CNH_2COO^- + PLP \cdot E \longrightarrow CH_3COCH_3 + CO_2 + PMP \cdot E$

(b) $PMP \cdot E + \text{Pyruvate} \rightleftharpoons PLP \cdot E + \text{Alanine}$

(2) Similar reactions with isovaline have also been observed.

Fig. 4.   Enzymatic decarboxylation-dependent transamination reactions (Aaslestad, Bouis, Phillips & Larson; Bailey & Dempsey).

1. Glutamate decarboxylase (Huntley & Metzler, 1968)

   a. $\alpha\text{-MeGlu} \xrightarrow{\text{PLP} \cdot \text{E}} CO_2 + \gamma\text{-NH}_2 \cdot \text{valerate}$

   b. $\alpha\text{-MeGlu} + PLP \cdot E \longrightarrow CO_2 + PMP \cdot E + \gamma\text{-oxovalerate}$

   c. Also occurs with Glu as substrate (?)

2. Dopa decarboxylase (O'Leary & Bahn, 1975)

   a. $\alpha\text{-Me-Dopa} \xrightarrow{\text{PLP} \cdot \text{E}} CO_2 + 3,4\text{-diOH}\phi CH_2CH_2NH_2$

   b. $\alpha\text{-Me-Dopa} + PLP \cdot E \longrightarrow CO_2 + PMP \cdot E + 3,4\text{-diOH}\phi CH_2CHO$

   c. Also occurs with L-Dopa as substrate

3. Ornithine decarboxylase (O'Leary & Herreid, 1978)

4. Arginine decarboxylase (O'Leary, Herreid, & Wittenberg, 1978)

Fig. 5.   Inactivation of α -decarboxylases by decarboxylation-dependent transamination.

cificity by the enzyme. <u>In vivo</u>, such inactivation should have little physiological significance, since dissociation of the inactive PMP-enzyme can occur readily to PMP and apoenzyme, and the active holo-enzyme can be regenerated in the presence of PLP.

A final type of PLP-dependent, α-decarboxylation reaction is shown in Scheme 3. The reaction is exemplified by the synthesis of the sphingosine precursor, 3-ketodihydrosphingosine, from serine and palmityl CoA,[13] the synthesis of δ-aminolevulinic acid from glycine and succinyl CoA,[14] and similar reactions involved in the biosynthesis of biotin and perhaps other products. The reaction may occur by direct labilization and loss of the carboxyl group of the amino acid; the carbanion thus formed attacks the electron-deficient carbonyl group of the acyl CoA with displacement of CoA. In the case of δ-aminolevulinic acid synthetase, however, studies of deuterium labilization indicate that an α-hydrogen atom of glycine is first labilized, and that the resulting carbanion adds to succinyl CoA to displace CoA.[14] The enzyme-bound β-ketocarboxylic acid thus formed is doubly activated for loss of the carboxyl group by the β-keto group and by the azomethine group.

Thus, three types of PLP-dependent α-decarboxylation reactions of amino acids are known, as illustrated in Fig. 6. Although the role of PLP in these reactions is rather well understood, we still have no definite information concerning the identity of functional groups in the specific apoenzymes that contribute to the catalytic action.

PYRUVOYL-DEPENDENT DECARBOXYLASES

The first representative of a new type of amino acid decarboxy-lase that does not require PLP was isolated in my laboratory in 1965.[15] We were attracted to this enzyme, an inducible histidine decarboxylase from <u>Lactobacillus</u> 30A, by a report from Rodwell[16] that crude preparations showed no evidence of a PLP requirement. Homogeneous preparations of the enzyme proved in fact to have the spectrum of a simple protein but were none the less inhibited by carbonyl reagents such as phenylhydrazine and cyanide[15,17] and were inactivated by reduction with sodium borohydride.[17] To determine the nature of the carbonyl group, we first reduced with tritiated NaBH$_4$, then after thorough dialysis hydrolyzed the protein and fractionated over a Dowex 50 column. The radioactive derivative was not absorbed to the column and showed the chromatographic characteristics of lactic acid. It was identified as lactate by preparing its p-bromo-phenacyl ester in the presence of excess unlabelled carrier lactate and recrystallizing to constant specific radioactivity. In a second approach, we used ^{14}C labelled phenylhydrazine to form the enzyme phenylhydrazone, digested the product with chymotrypsin, and isolated the phenylhydrazone of pyruvoylphenylalaline.[17] The active center of

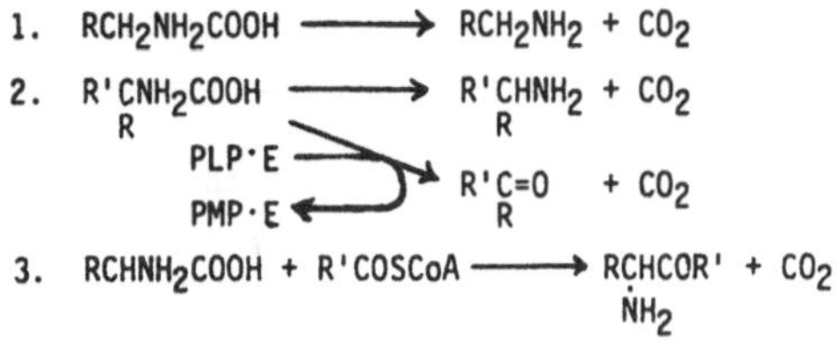

1. $RCH_2NH_2COOH \longrightarrow RCH_2NH_2 + CO_2$

2. $R'CNH_2COOH \longrightarrow R'CHNH_2 + CO_2$
    with R subscripts

$$PLP \cdot E \searrow$$
$$PMP \cdot E \longleftarrow \quad R'C{=}O \; + CO_2$$

3. $RCHNH_2COOH + R'COSCoA \longrightarrow RCHCOR' + CO_2$ (with NH_2)

Fig. 6.   PLP-dependent α-decarboxylations.

Scheme 3

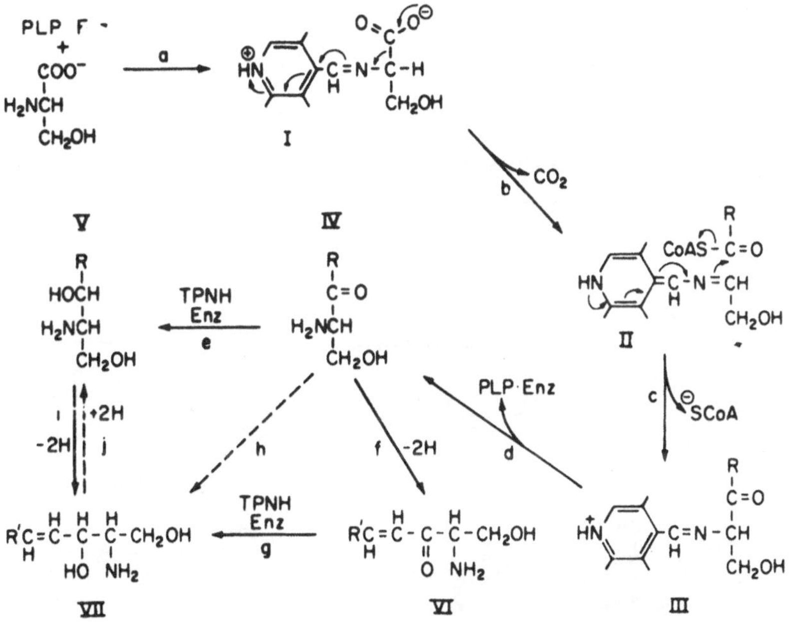

the enzyme thus contains a pyruvoyl group linked as an amide to a phenylalanine residue of the protein. During this same time period, Hodgins and Abeles[18] showed that crude preparations of proline reductase from Clostridium sticklandii also contained a catalytically essential, covalently bound pyruvate residue. Since these early studies several additional pyruvoyl-dependent decarboxylases, listed in Table 1, have been discovered.

These findings raise a number of questions: (a) How does the

Table 1. List of piruvoyl-dependent enzymes

Enzymes		Mol.wt.x10^{-3}		Pyruvate per Enzyme
	Subunits	Subunits	Native	
I. Decarboxylases				
A. Histidine (Lactobacillus 30a)		$28(\alpha)$, $9(\beta)$	$208(\alpha_6\ \beta_6)$	6
" (Micrococcus sp.n.)			$102(\alpha_3\ \beta_3)$	3
" (C. perfringens)		–	–	–
B. S-Adenosylmethionine (E. coli)		$17(\alpha)$	$108(\alpha_6)$	6
" (S. cerevisiae)		$41(\alpha)$	$88(\alpha_2)$	2
" (rat liver)		$42(\alpha)$	$155(\alpha_4)$	4
C. Phosphatidylserine (E. coli)		$36(\alpha)$	$?(\alpha_x)$	x
D. α-Aspartate (E. coli)		$(11.8, 9.8, 6.4)$	58	1
II. D-Proline reductase (C. sticklandii)		$31(\alpha)$	$300(\alpha_{10})$	10

Scheme 4

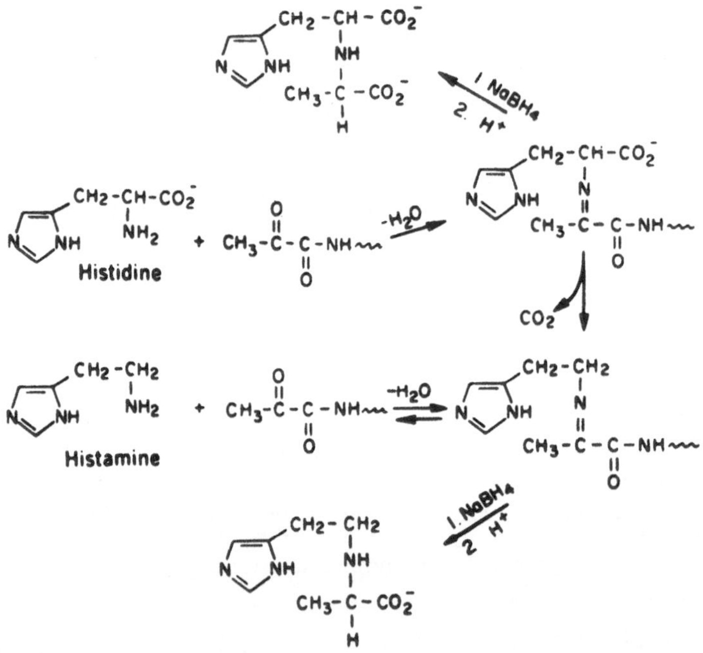

pyruvoyl group function in the decarboxylation reaction? (b) What is the origin of the pyruvate in these proteins? (c) How does the pyruvoyl group become covalently attached to protein? By analogy to the role of the coenzyme in PLP-dependent decarboxylases we suspected that histidine decarboxylase, too, would act by a Schiff's base mechanism as shown in Scheme 4. We tested this postulate by adding ^{14}C-histidine and trapping the predicted intermediates of the reaction by reduction with sodium borohydride. From acid hydrolysates of the reduced reaction mixture we isolated both small amounts of the predicted histidine adduct (Top, Scheme 4) and much larger quantities of the histamine adduct (Bottom, Scheme 4). The latter adduct was also formed when large amounts of histamine were added to the free enzyme, reduced with borohydride, and acid hydrolyzed; the amount formed was stoichiometric with the amount of pyruvate obtained from acid hydrolysates of the free enzyme.[19] It is clear, therefore, that decarboxylases of this class like the PLP-dependent decarboxylases, utilize a Schiff's base mechanism. Unfortunately, this is as far as our mechanistic information currently takes us. In the case of the PLP-dependent decarboxylases there are two strongly electrophilic groups in the intermediate Schiff's base, i.e., a protonated azomethine nitrogen and the protonated pyridinium nitrogen, both placed appropriately for their function as electron withdrawing groups. It is unclear how the pyruvoyl-dependent decarboxylases achieve a similar degree of electrophilicity. The pH profile of histidine decarboxylase shows an essentially constant V_{max} between pH 4 and pH 7.8, demonstrating that functional groups with pK values within this pH range cannot be concerned in the catalytic events of decarboxylation, although such groups are involved in binding the substrate.[19]

We turn now to the origin of pyruvate within the protein. Since Lactobacillus 30A produces pyruvate as an intermediate in glucose fermentation, we thought pyruvate would be derived from glucose. However, neither the amino acid residues of the enzyme nor the pyruvate residue were labelled when the organism was grown with ^{14}C-glucose. When grown with ^{14}C-serine, however, both the serine residues of the enzyme protein and the pyruvoyl residue become labelled without dilution (Table 2); none of the other amino acids in the protein were labelled.[20] Thus the pyruvoyl group arises from serine. The precursor serine could, in principle, be part of the primary sequence of a precursor protein; alternatively, it could be attached to a precursor protein either before or after its conversion to an activated pyruvoyl derivative. Each of these mechanisms for inserting the pyruvoyl residue requires the existence of an inactive precursor enzyme. We therefore raised rabbit antibodies to homogeneous histidine decarboxylase and used these antibodies in an attempt to demonstrate the presence of an enzymatically inactive precursor protein in wild-type Lactobacillus 30A under various conditions of growth. Initial attempts in this direction were unsuccessful. However, by use of nitrosoguanidine as a mutagen we were able to isolate several

Table 2. Comparison of specific radioactivities of
          serine and pyruvate

Compound analysed	Specific activity (dpm/nmol)
L-Serine added to medium	1320
Serine isolated from enzyme	1300
Pyruvate in $CH_3\overset{\displaystyle NNC_6H_5}{\underset{\displaystyle}{C}}COPhe$ fragment from enzyme	1320
Lactate extracted from medium	19

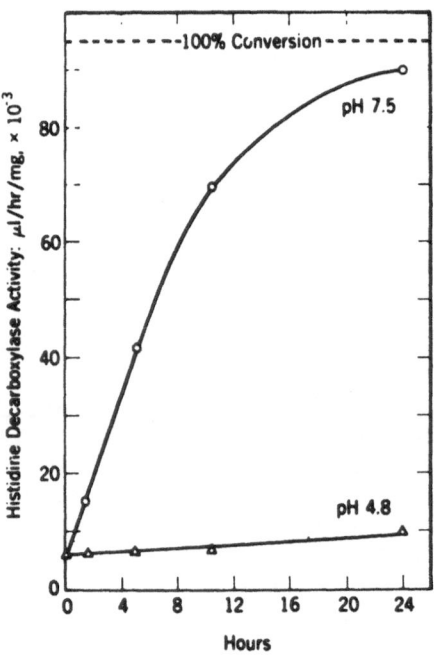

Fig. 7. Comparative rates of activation of prohistidine decarboxy-
         lase from mutant 3 of <u>Lactobacillus</u> 30 A. For experimental
         details see[21].

histidine decarboxylaseless mutants of <u>Lactobacillus</u> 30A which contained proteins that were precipitated by the antiserum. From mutant
3 we isolated a homogeneous protein which contained only 3 to 5
percent of the activity of equal amounts of the wild-type enzyme.

Most interestingly, although stable at pH 4.8, upon storage at pH 7.6 it gradually gained activity, eventually reaching almost the same level shown by the wild-type enzyme (Fig. 7). This cross-reactive protein, therefore, acts as a prohistidine decarboxylase[21], and it became essential to compare its molecular properties with those of the wild type enzyme. Results of gel electrophoresis of wild type enzyme, mutant proenzyme, and activated mutant proenzyme are shown in Fig. 8. No differences in migration rates were detected. When the proteins were denatured with sodium dodecyl sulfate (SDS), and subjected to SDS gel electrophoresis, a quite different picture emerged: both wild type enzyme (Fig. 8, Track 1.B) and activated proenzyme (Track 3.B) showed the presence of two subunits, whereas the proenzyme before activation contains only a single major peptide larger than either of the subunit peptides of the active enzyme (Track 2.B).

Activation is thus an unusual and interesting reaction: cleavage of a peptide chain occurs under extremely mild conditions of temperature and pH coincident with conversion of a serine residue of the precursor chain to a pyruvoyl group in one daughter chain.[21] We still know very little about the mechanism of this activation reaction. Each proenzyme molecule appears to activate itself in an intramolecular reaction. Monovalent cations are required for the conversion (Fig. 9), potassium being most effective; the conversion is first order with respect to monovalent cations.[22] This half order dependency on monovalent cations, which is not due to ionic strength, is likewise not understood.

The structural relationship of proenzyme subunits to the two daughter subunits is shown in Scheme 5. Cleavage occurs between two serine residues to generate the pyruvate-free β-chain and the larger pyruvoyl-containing α-chain. Nothing is split out except ammonia. The two daughter chains can be separated by chromatography in guanidine solutions; reconstitution of enzyme activity requires both chains.[22] Both wild-type and mutant enzymes have an $\alpha_6 \beta_6$ structure that reversibly dissociates at neutral pH (but not at the optimal pH of 4.8) and low ionic strength to an $\alpha_3 \beta_3$ structure.[23]

Recently, we have been able to demonstrate the presence of a proenzyme in wild-type cells as well as in mutant cells. The activation reaction is similar in both proteins, although faster in the wild type proenzyme.[24] Although covering one of two -SH groups of the enzyme by reaction with DTNB inactivates the enzyme, inactivation appears to result from spacial distortion about the active site since replacement of the large TNB group by CN regenerates an active enzyme.[24] We are currently conducting a careful structural comparison of this enzyme and its mutational variants in the hope of throwing additional light on the nature of both the activation reaction and the mechanism of the decarboxylation reaction.

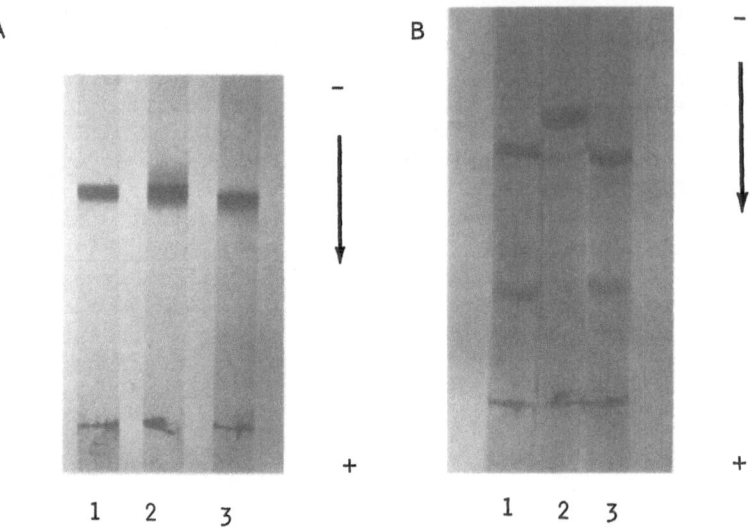

Fig. 8.   Polyacrylamide disc gel electrophoresis of wild-type histi-
          dine decarboxylase, prohistidine decarboxylase and activated
          prohistidine decarboxylase in the absence (Tracks 1A, 2A and
          3A, respectively) and in the presence (Tracks 1B,2B and 3B,
          respectively) of SDS.

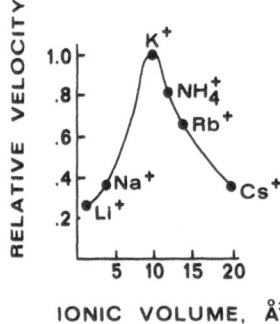

Fig. 9.   Comparative rates of activation of prohistidine decarboxy-
          lase in the presence of various cations. For experimental
          details see[22].

Scheme 5

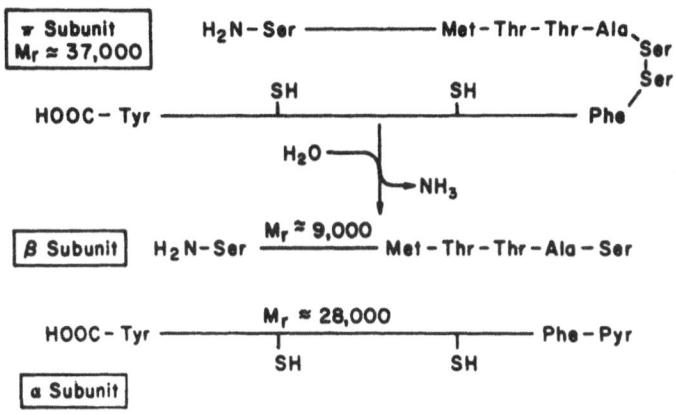

In summary, all of the known decarboxylases that act on amino acids contain carbonyl or azomethine groups, and all of the decarboxylation reactions proceed by a Schiff's base mechanism. Two major groups of such decarboxylases exist, those dependent upon the dissociable coenzyme, pyridoxal phosphate, and those dependent upon a covalently bound, non-dissociable pyruvoyl group; several representatives of both types are now known. The reason for this duality in structure is not yet known, and aside from the catalytic advantages attributable to Schiff's base formation in the ES complex, little is known of the catalytic group provided by the enzyme protein in either class of decarboxylase. The stereochemistry of decarboxylation has been examined with a number of different enzymes of both types (Table 3); it proceeds with retention of configuration in all cases.[25] Finally, it should be observed that while many of the decarboxylases appear to have only degradative functions, all are polysubunit enzymes for reasons that are not understood.

I hope it is clear from this brief review that a number of intriguing questions concerning the mechanism of action of these enzymes and their biogenesis remain to be answered. Conversion of the proenzyme of histidine decarboxylase to the active enzyme, in particular, is most unusual: it represents an apparently non-proteolytic cleavage of a peptide chain coincident with formation of the active site pyruvoyl residue of the daughter enzyme. It seems unlikely that such a mechanism is restricted to this small group of enzymes, and it is not known whether the pyruvoyl group can also serve as a blocking group in proteins that lack an identified enzymatic activity. If this mode of processing is common to the pyruvoyl enzymes listed in Table 1, it is obvious from their subunit composition that the pyruvate-free fragment derived during

Table 3.   Stereochemistry of decarboxylation catalyzed by amino
acid decarboxylases[a]

Decarboxylase	Source	Stereochemistry	α-Proton exchange
A. Pyruvoyl			
Histidine	Lactobacillus	Retention	No
Histidine	C.perfringens	Retention	
S-Adenosyl methionine	E. coli	Retention	No
B. PLP			
Tyrosine	S. faecalis	Retention	Yes
Glutamate	E. coli	Retention	No
Lysine	B. cadaveris	Retention	Yes

[a]Modified from[25].

the activation process must frequently be discarded, rather than being retained, as it is in the histidine decarboxylases.

Work from this laboratory cited herein was supported in part by Grants AM 19898 and AI 13940 from the U.S. Public Health Service.

REFERENCES

1. E. F. Gale, The bacterial amino acid decarboxylases, Advances in Enzymology, 6: 1 (1946).
2. S. Usawa, Formation of indole from L-tryptophan (VII). Separation of apo- and co-tryptophanase, J. Osaka Med. Assoc., 42: 1637 (1943).
3. E. E. Snell, The vitamin activities of pyridoxal and pyridoxamine, J. Biol. Chem., 154: 313 (1944).
4. I. C. Gunsalus, W. D. Bellamy, and W. W. Umbreit, A phosphorylated derivative of pyridoxal as coenzyme of tyrosine decarboxylase, J. Biol. Chem., 155: 685 (1944).
5. E. E. Metzler, M. Ikawa, and E. E. Snell, A general mechanism for vitamin B_6-catalyzed reactions, J. Amer. Chem. Soc., 76: 648 (1954).
6. G. D. Kalyankar and E. E. Snell, Pyridoxal-catalyzed decarboxylation of amino acids, Biochemistry, 1: 594 (1962).
7. A. E. Braunstein, and M. M. Shemyakin, A theory of the process of the amino acid exchange catalyzed by pyridoxal enzymes, Biokhimiya, 18: 393 (1953).

8. S. Mandeles, R. Koppelman, and M. E. Hanke, Deuterium studies
    on the mechanism of enzymatic amino acid decarboxylation,
    J. Biol. Chem., 209: 327 (1954).
9. G. B. Bailey, and W. B. Dempsey, Purification and properties of
    an α -dialkylamino acid transaminase, Biochemistry, 6: 1526
    (1967).
10. H. G. Aaslestad and A. D. Larson, Bacterial metabolism of 2-
    methylamine, J. Bacteriol., 88: 1296 (1964).
11. M. H. O'Leary, and R. M. Herreid, Mechanism of inactivation of
    ornithine decarboxylase by α-methylornithine, Biochemistry,
    17: 1010 (1978).
12. M. H. O'Leary and R. L. Baughn, Decarboxylation-dependent tran-
    samination catalyzed by 3,4-dihydroxyphenylalanine decarbo-
    xylase, J. Biol. Chem., 252: 7168 (1977).
13. E. E. Snell, S. J. DiMari, and R. N. Brady, Biosynthesis of
    sphingosine and dihydrosphingosine by cell-free systems from
    Hansenula ciferri, Chem. Phys. Lipids, 5:116 (1970).
14. P. M. Jordan, and D. Shemin, δ-Aminolevulinic acid synthetase,
    in The Enzymes, 3rd Ed., Vol. VII: 339, P.D. Boyer, ed.,
    Academic Press, N.Y. (1972).
15. J. Rosenthaler, B. M. Guirard, G. W. Chang, and E. E. Snell,
    Purification and properties of histidine decarboxylase from
    Lactobacillus 30a, Proc. Nat. Acad. Sci. U.S.A., 54: 152
    (1965).
16. A. W. Rodwell, The histidine decarboxylase of a species of
    Lactobacillus: Apparent dispensability of pyridoxal phospha-
    te as coenzyme, J. Gen. Microbiol., 8: 233 (1953).
17. W. D. Riley and E. E. Snell, Histidine decarboxylase of
    Lactobacillus 30a. IV. The presence of covalently bound
    pyruvate as the prosthetic group, Biochemistry, 7: 3520
    (1968).
18. D. S. Hodgins and R. H. Abeles, Studies on the mechanism of
    action of D-proline reductase: The presence of covalently
    bound pyruvate and its role in the catalytic process. Arch.
    Biochem. Biophys., 130: 274 (1969).
19. P. A. Recsei and E. E. Snell, Histidine decarboxylase of
    Lactobacillus 30a. VI. Mechanism of action and kinetic pro-
    perties, Biochemistry, 9: 1492 (1970).
20. W. D. Riley and E. E. Snell, Histidine decarboxylase of
    Lactobacillus 30a. V. Origin of enzyme-bound pyruvate and
    separation of non-identical subunits, Biochemistry, 9: 1485
    (1970).
21. P. A. Recsei and E. E. Snell, Prohistidine decarboxylase from
    Lactobacillus 30a. A new type of zymogen, Biochemistry,
    12: 365 (1973).
22. P. A. Recsei and E. E. Snell, Bacterial prohistidine decarbo-
    xylase: kinetics of conversion to the active enzyme, in
    Metabolic Interconversion of Enzymes 1980, H. Holzer ed.,
    Springer-Verlag, Heidelberg (1981).
23. M. L. Hackert, W. E. Meader, R. M. Oliver, J. B. Salmon, P. A.

Recsei, and E. E. Snell, Crystallization and subunit structure of histidine decarboxylase from <u>Lactobacillus</u> 30a, <u>J.
Biol. Chem.</u>, 256: 687 (1981).

24. P. A. Recsei and E. E. Snell, Histidine decarboxylase of
<u>Lactobacillus</u> 30a. Comparative properties of wild-type and
mutant proenzymes and their derived enzymes, <u>J. Biol. Chem.</u>,
in press.

25. R. A. Allen and J. P. Klinman, Stereochemistry and kinetic
isotope effects in the decarboxylation of S-adenosylmethionine catalyzed by the pyruvyl enzyme, S-adenosylmethionine
decarboxylase, <u>J. Biol. Chem.</u>, 256: 3233 (1981).

STRUCTURAL STUDIES OF ASPARTATE AMINOTRANSFERASE ISOZYMES

Donatella Barra, Francesco Bossa and Shawn Doonan°

Istituto di Chimica Biologica e Centro di Biologia Molecolare del CNR, Università di Roma, Città Universitaria
00185 Roma, Italy and
° Department of Biochemistry, University College, Cork,
Ireland

INTRODUCTION

Enzymes which depend for their activity on the cofactor pyridoxal
5'-phosphate have been the object of extensive study over the last
thirty or more years. One of the main reasons for this has been that
the chemistry of the cofactor is well understood, and knowledge of
the chemistry of the cofactor gives direct insights into the mechanisms of action of the enzymes in this group. Probably the most
studied and still the best understood of these enzymes is aspartate
aminotransferase which catalyses the reaction:

$$
\begin{array}{ccccccc}
\text{COOH} & & \text{COOH} & & \text{COOH} & & \text{COOH} \\
| & & | & & | & & | \\
\text{CO} & + & \text{CHNH}_2 & \rightleftharpoons & \text{CHNH}_2 & + & \text{CO} \\
| & & | & & | & & | \\
\text{CH}_2 & & \text{CH}_2 & & \text{CH}_2 & & \text{CH}_2 \\
| & & | & & | & & | \\
\text{CH}_2 & & \text{COOH} & & \text{CH}_2 & & \text{COOH} \\
| & & & & | & & \\
\text{COOH} & & & & \text{COOH} & &
\end{array}
$$

Braunstein and Shemyakin[1] and Metzler, Ikawa and Snell[2] independently
suggested that the overall reaction proceeds by two half-reactions
as shown in Fig. 1. In the first half reaction, an amino acid substrate reacts with the pyridoxal 5'-phosphate form of the enzyme
(represented as E-CHO) to give a Schiff base intermediate which then
undergoes prototropic rearrangement followed by hydrolysis to yield
the oxoacid product and the pyridoxamine 5'-phosphate form of the

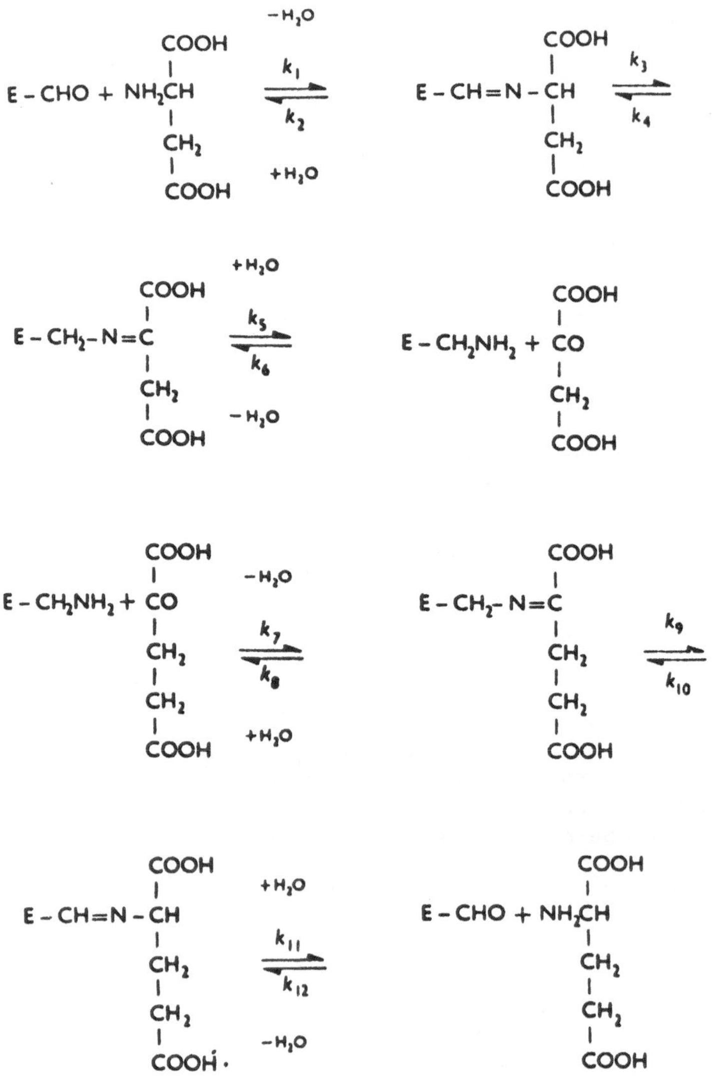

Fig. 1. A minimal mechanism for the action of aspartate aminotransferase isozymes. E-CHO and E-CH$_2$NH$_2$ represent the forms
of the holoenzyme containing pyridoxal 5'-phosphate and
pyridoxamine 5'-phosphate respectively. The aldehydic form
of the cofactor is linked via a Schiff base to a lysine
residue in the protein; hence the first and last reactions
in the sequence should properly be shown as transaldiminat-
ions. Similarly the prototropic rearrangements are more
complex than shown here.

enzyme ($E-CH_2NH_2$). A symmetric process with the oxoacid substrate completes the reaction. The sequence in Fig. 1 represents a minimal reaction mechanism; it is now known that the prototropic rearrangement proceeds in more than one step, but the details need not concern us here.

Of particular importance in the study of aspartate aminotransferases was the realisation that, since pyridoxal alone can catalyse the formation of oxoacids from amino acids in a process entirely analogous to the enzyme-catalysed reaction[2], then a direct comparison could, at least in principle, be made of the enzymic and of the model systems and insights thereby gained into the role of the protein part of the enzyme. In a detailed study Banks et al.[3] showed that the natural cofactors and substrates of aspartate aminotransferase underwent reactions identical to those shown in Fig.1 but with co-factors in place of the enzyme-cofactor complex. Moreover, they were able to obtain values for nine of the twelve rate constants specified in the Figure and show that, as would be expected, the prototropic rearrangements (described by k_3, k_4, k_9 and k_{10}) are the slow steps in the model system for transamination. Two different approaches have been taken to obtain values for the corresponding rate constants in the enzyme-catalysed reaction in both cases using the cytosolic enzyme from pig heart. One such approach made use of steady state kinetics with amino acid substrates deuterated on the α-carbon atoms; on the assumption that this deuteration would affect the rates only of those reactions in which the C_α-D bond is broken (k_3 and k_{10} in Fig. 1) then it was possible to obtain values for all four of the rate constants describing the prototropic rearrangements. In a second approach, Fasella and Hammes[4] used fast-reaction kinetic methods to obtain values for the same rate-constants. The values obtained by the two groups were in excellent agreement and showed that the protein part of the holoenzyme catalyses the slow steps of transamination by factors of $10^8 - 10^9$ fold; this was one of the earliest measurements of the catalytic power of the protein part of an enzyme in terms of the rate constants for unit steps of the reaction in the presence and absence of the catalytic protein.

The results obtained in the studies outlined above gave a quantitative measure of the degree of catalysis exerted by the apoprotein of aspartate aminotransferases but left the origin of the catalysis unexplained. It seemed clear that the slow step in transamination involves a process of proton abstraction from the Schiff base intermediate, bond rearrangement and reprotonation, the steps being either concerted or sequential. In either case there must be amino acid side chains in the active site of the enzyme one of which acts as a strong base and the other as a strong acid. A minimum account of the factors responsible for catalysis by the enzyme would require not only identification of these amino acids but also description of the chemical and physical environments of their side chains which result in their enhanced acidity or basicity with respect to

the enzyme - Schiff base intermediates.[5]   Information of this type
is available only from X-ray diffraction analyses of the protein.
With this ultimate objective in mind we embarked on determination
of the primary structure of the cytosolic aspartate aminotransferase from pig heart. Subsequently the work was extended to the
mitochondrial isozyme and to the isozymes from other organisms. The
results obtained are described below.

PRIMARY STRUCTURES OF AMINOTRANSFERASES FROM PIG HEART

     The cytosolic aspartate aminotransferase from pig heart is a
dimer of two identical subunits each with 412 amino acid residues;
the primary structure of the subunit is shown in Fig. 2. This structure was determined essentially by classical methods of fragmentation, peptide isolation, and sequence analysis using the dansyl-Edman
procedure. Because of the large size of the polypeptide a multiplicity of digestion methods had to be used to obtain the complete
sequence. These included digestion with trypsin and pepsin[6] and with
thermolysin and elastase.[7] The thermolytic digest was particularly
useful since it allowed 375 of the 412 amino acids to be identified
in peptides. The results obtained were also of value in defining
more clearly the specificity of this enzyme which had not previously
been much used in sequence studies; cleavage mainly occurred, in
order of decreasing frequency, at isoleucine, phenylalanine, leucine
and valine.[7] The combined information from these four digests, plus
a small amount of extra data from cleavage with cyanogen bromide
and with chymotrypsin, allowed us to assemble 10 composite fragments
containing the bulk of the residues in the protein.[7] Of central importance for the completion of the sequence of the cytosolic isozyme
was the use of a newly-isolated protease from the basidiomycete
Armillaria mellea.[8] Preliminary studies by the discoverers of the
enzyme had suggested that it was specific for lysine residues and
moreover that it cleaved on the N-terminal side of lysine. Application of the protease to digestion of cytosolic aspartate aminotransferase confirmed and extended knowledge of the specificity of the
enzyme[9] as well as allowing completion of the sequence. Of the 19
lysine residues in cytosolic aspartate aminotransferase, complete
cleavage occurred at 12 but cleavage was either restricted (4) or
absent (3) at the remaining residues; the reasons for restricted
cleavage at these bonds is still unclear. Interestingly, cleavage
was also observed on the C-terminal side of three arginine residues;
in each case the arginine residue was followed in the sequence by
leucine or isoleucine. Unfortunately the protease from A. mellea has
not become commercially available and hence it has not been widely
used as a sequencing tool. Consequently there are still several unan-
swered questions on the specificity of this enzyme.

     Knowledge of the primary structure of cytosolic aspartate ami-
notransferase did not, of course, contribute greatly to understanding
of the mechanism of action of the enzyme. Advances on this front will

```
                                                                    35
A P P S V F A E V P Q A Q P V L V F K L I A D F R E D P D P R K V N L
  S S W W A H V E M G P P D P I L G V T E A F K R D T N S K K M N L

                                                                    70
G V G A Y R T D D C Q P W V L P V V R K V E Q R I A N N S S L N H E Y
G V G A Y R D D N G K P Y V L P S V R K A E A Q I A A K   N L D K E Y

                                                                   105
L P I L G L A E F R T C A S R L A L G D D S P A L Q E K R V G G V Q S
L P I G G L A E F C K A S A E L A L G E N N E V L K S G R Y V T V Q T

                                                                   140
L G G T G A L R I G A E F L A R W Y N G T N N K D T P V Y V S S P T W
I S G T G A L R I G A N F L Q R F F K     F S R D     V F L P K P S W

                                                                   175
E N H N G V F T T A G F K D I R S Y R Y W D T E K R G L D L Q G F L S
G N H T P I F R D A G M Q   L H S Y R Y Y D P K T C G F D F T G A L E

                                                                   210
D L E N A P E F S I F V L H A C A H N P T G T D P T P E Q W K Q I A S
D I S K I P A Q S V I L L H A C A H N P T G V D P R P E Q W K E M A T

                                                                   245
V M K R R F L F P F F D S A Y Q G F A S G N L E K D A W A I R Y F V S
L V K K N N L F A F F D M A Y Q G F A S G D G N K D A W A V R H F I E

                                                                   280
E G F E L F C A Q S F S K N F G L Y N E R V G N L T V V A K E P D S I
Q G I N V C L C Q S Y A K N M G L Y G E R V G A F T V V C K D A E E A

                                                                   315
L R V L S Q M Q K I V R V T W S N P P A Q G A R I V A R T L S D P E L
K R V Q S Q L K I L I R P M Y S N P P V N G A R I A S T I L T S P D L

                                                                   350
F H E W T G N V K T M A D R I L S M R S E L R A R L E A L K T P G T W
R Q Q W L Q E V K G M A D R I I S M R T Q L V S N L K K E G S S H N W

                                                                   385
N H I T D Q I G M F S F T G L N P K Q V E Y L I N Q K H I Y L L P S G
Q H I V D Q I G M F C F T G I K P E Q V E R L T K E F S I Y M T K D G

R I N M C G L T T K N L D Y V A T S I H E A V T K I Q
R I S V A G V T S G N V G Y L A H A I H Q   V T K
```

Fig. 2.   Comparison of the amino acid sequences of cytosolic (upper
           line) and mitochondrial (lower line) aspartate aminotran-
           sferases from pig heart. Identical residues in the two se-
           quences are underlined. The mitochondrial isozyme is eleven
           residues shorter than the cytosolic form and hence gaps
           have been left to maximise the homology.

depend on determination of the complete three-dimensional structure
of the enzyme; this topic will be returned to in the next section.
It had been known for some time, however, that aspartate aminotransferase, in common with a small number of other enzymes, exists as
distinct molecular forms in the cytosol and the mitochondria of
eukaryotic cells.[10] In no case were the structural relationships
between such a pair of isozymes known and hence we decided to extend
our studies to determination of the primary structure of the mito-

chondrial aspartate aminotransferase from pig heart. The approach was basically similar to that taken with the cytosolic isozyme in that digestions with trypsin, thermolysin, chymotrypsin and pepsin were used. Digestions with thermolysin and chymotrypsin[11] were highly successful in that the peptides isolated contained a total of 98.5% of the amino acid residues in the protein, but many of these peptides were small and could not be positioned unambiguously in the sequence. In the light of this, and based on our experience with the cytosolic isozyme, digestion with pepsin was carried out under mild conditions[12] in an attempt to obtain larger peptides but still retain a soluble digest. This approach was successful in that peptides containing all but two of the amino acids in the protein were isolated. Moreover many of the peptides arose from partial cleavage of susceptible bonds and hence the digest contained families of overlapping peptides such that parts of the sequence of the protein could be reconstructed from the peptic digest alone. This advantage had to be paid for in the sense that the digest was extremely complex, containing more than 120 peptides, and hence isolation was correspondingly difficult.

Use was again made of the A. mellea protease for studies of the mitochondrial isozyme.[13] In one digest, the protein substrate was trifluoroacetylated and S-aminoethylated to block cleavage at lysine and produce cleavage at modified cysteine; this method of producing specific enzymic hydrolysis at cysteine residues had previously been reported.[14] This digest gave some useful information but, because of the small number (7) of cysteine residues in the protein, several of the polypeptide fragments were large and could not be isolated; this limited the information obtained from the digest. Finally, some use was also made of the glutamic-acid-specific protease from S. aureus. We did not attempt to isolate all the peptides produced, but some valuable information was obtained particularly for distinguishing between glutamic acid and glutamine.[12]

The complete amino acid sequence of mitochondrial aspartate aminotransferase from pig heart is shown in Fig. 2 aligned with that of the cytosolic form.[15,16] It is immediately apparent that, if a small number of gaps are left in the sequence of the mitochondrial isozyme to accomodate its smaller size (401 residues) compared with that of the cytosolic form (412 residues), then there is a substantial degree of sequence identity between the two isozymes. In fact the sequence identity is 47% with the structures aligned as in Fig. 2; the identical residues (194) are underlined. Of the remainder of the residues, 128 substitutions could have arisen by a single base mutation. The sequence identities are obviously distributed unequally along the proteins. For example, it is very low up to residue 33 but residues 34 to 41 are identical in both proteins; this region includes tyrosine-40 which is at or near the active site of the cytosolic isozyme.[17] Similarly, the lysine residues to which the cofactor pyridoxal 5'-phosphate is attached (258) are in a

region of highly conserved structure. Other highly conserved residues
are presumably important for enzymic function but again three-dimensional structures will be required for a complete interpretation of
the distribution of conserved and variable regions.

The results given in Fig. 2 represent the most complete structural comparison available for any pair of cytosolic and mitochondrial isozymes and show clearly their genetic origins as distinct,
but evolutionarily related, gene products. Many questions about the
evolution of the isozymes remain outstanding, however. For example,
have they evolved at equal rates or has the necessity for uptake of
the mitochondrial isozyme from its site of synthesis in the cytosol
into the mitochondrion imposed extra constraints on its evolution?
Indirect approaches to this problem have given conflicting results.
Immunochemical comparisons[18,19] suggest closer relationships between
mitochondrial than between cytosolic isozymes from different organisms whereas statistical comparisons of amino acid compositions do
not support this conclusion.[20] Our view is that the isozymes have
evolved at roughly equal rates at the level of overall amino acid
sequence but that certain restricted regions of the mitochondrial
isozymes have been particularly highly conserved compared with the
same regions in the cytosolic forms; these highly conserved regions
may be involved in the process of uptake of the mitochondrial isozyme
into the organelles. The N-terminus appears to be one such region
and is more highly conserved in mitochondrial than in cytosolic
isozymes.[21] Complete sequences of aspartate aminotransferases from
a variety of sources will be required to elucidate further the evolutionary history of these isozymes; such studies are in progress
in our and other laboratories but lack of space precludes further
discussion here.

THREE DIMENSIONAL STRUCTURES OF THE ISOZYMES

As pointed out above further understanding of the mechanism of
catalysis by aspartate aminotransferases and of other aspects of their
function requires knowledge of the three dimensional structures. This
final section gives a brief review of progress to date in this area.

Work is actively underway on the crystal structures of the cytosolic isozyme from pig heart,[22] the cytosolic[23] and mitochondrial[24]
isozymes from chicken heart and the mitochondrial isozyme from beef
heart.[25] Results available to date suggest that the three dimensional
structures of all these proteins are very similar.[24] This is consistent with our own results on prediction of the secondary structures
of the pig heart isozymes where considerable similarities were observed.[16] Hence it seems that the relationships observed between
isozymes of aspartate aminotransferase at the level of primary structure are reflected in similarities in their higher order structures.

Of the three forms of the enzyme mentioned above, studies of

that from chicken heart mitochondria have so far been published at
highest resolution (0.28 nm).[24] A complete description of the results
obtained is beyond the scope of this article but some factors relevant to catalysis by the enzyme may be briefly reviewed. Amino acid
side chains involved in positioning of the cofactor pyridoxal 5'-
phosphate, in addition to lysine-258, probably include tryptophan-
140, histidines -143, -189 and -193, aspartic acid -222, tyrosine
-225, alanine -224, arginine -266 and serine -255; all of these
residues are conserved in the structures shown in Fig. 2. Of considerable interest is the observation that each of the active sites
of the enzyme is formed by amino acid residues from both subunits.
Candidates for substrate binding are arginine -386 and asparagine
-194 for the α-carboxyl group of the substrate but arginine -292[‡]
(where ‡ represents an amino acid from the other subunit) for the
ω-carboxyl group. As was pointed out in the Introduction, catalysis by aspartate aminotransferase requires an amino acid side chain
to act as a strong acid to promote the prototropic rearrangement
which constitutes the slow step of transamination. The most probable
groups involved here are tyrosine -70[‡] and lysine -258. It is interesting that no histidine side chains are observed in the appropriate positions to act as proton donors or acceptors; involvement of
histidine had been predicted on other grounds.[26] Clearly it is not
yet possible to give a complete account of catalysis by aspartate
aminotransferase; this must await more detailed results from X-ray
diffraction studies. Nevertheless, our original aim in embarking
on primary structure analysis of the enzyme, that is to provide the
necessary preliminary information for the crystallographic work,
is obviously now paying dividends and a rich harvest of mechanistic
information can be expected within the next few years. In addition,
the scope of the collaboration has become widened to include a study
of the genetic origins and evolutionary history of the aspartate
aminotransferases which we hope will throw light on such fascinating
problems as the mechanisms by which the isozymes are selectively
compartmentalised in eukaryotic cells.

REFERENCES

1. A. E. Braunstein and M. M. Shemyakin, A theory of amino acid
   metabolic processes catalysed by pyridoxal-dependent enzymes, Biokhimja 18: 393 (1953).
2. D. E. Metzler, M. Ikawa and E. E. Snell, A general mechanism
   for vitamin B_6-catalyzed reactions. J. Amer. Chem. Soc. 76:
   648 (1954).
3. B. E. C. Banks, M. P. Bell, A. J. Lawrence and C. A. Vernon,
   A model system for aspartate aminotransferase in "Pyridoxal
   Catalysis: Enzymes and Model Systems", E.E. Snell, A. E.
   Braunstein, E. S. Severin and Yu. M. Torchinsky, eds., Interscience, New York (1968).
4. P. Fasella and G. G. Hammes, A temperature jump study of aspartate aminotransferase. A reinvestigation. Biochemistry, 6:
   1798 (1967).

5. S. Doonan, C. A. Vernon and B. E. C. Banks, Mechanisms of enzyme action. Prog. Biophys. Molec. Biol. 20: 247 (1970).

6. S. Doonan, H. J. Doonan, F. Riva, C. A. Vernon, J. M. Walker, F. Bossa, D. Barra, M. Carloni and P. Fasella, The primary structure of aspartate aminotransferase from pig heart muscle. Partial sequences determined by digestion with pepsin and trypsin. Biochem. J.130: 443 (1972).

7. F. Bossa, D. Barra, M. Carloni, P. Fasella, F. Riva, S. Doonan, H. J. Doonan, R. Hanford, C. A. Vernon and J. M. Walker, The primary structure of aspartate aminotransferase from pig heart muscle. Partial sequences determined by digestion with thermolysin and elastase. Biochem. J. 133: 805 (1973).

8. P. L. Walton, R. W. Turner and D. Broadbent, Brit. Patent, 1263956 (1972).

9. S. Doonan, H. J. Doonan, R. Hanford, C. A. Vernon, J. M. Walker, L. P. da S. Airoldi, F. Bossa, D. Barra, M. Carloni, P. Fasella and F. Riva, The primary structure of aspartate aminotransferase from pig heart muscle. Digestion with a proteinase having specificity for lysine residues. Biochem. J. 149: 497 (1975).

10. H. Wada and Y. Morino, Comparative studies on glutamic oxalacetic transaminases from the mitochondrial and soluble fractions of mammalian tissues. Vit. Horm. 22: 411 (1964).

11. D. Barra, R. Petruzzelli, F. Martini, F. Bossa and S. Doonan, The primary structure of mitochondrial aspartate aminotransferase from pig heart: peptides obtained by cleavage with thermolysin and chymotrypsin. Ital. J. Biochem. 28: 456 (1979).

12. D. Barra, M. R. Savi, R. Petruzzelli, F. Bossa and S. Doonan, The primary structure of mitochondrial aspartate aminotransferase from pig heart. Peptides obtained by cleavage with pepsin and with Staphylococcus aureus protease. Ital. J. Biochem. 28: 478 (1979).

13. S. Doonan, H. M. A. Fahmy, G. J. Hughes, D. Barra and F. Bossa, The primary structure of mitochondrial aspartate aminotransferase from pig heart: peptides obtained by cleavage at basic residues. Ital. J. Biochem. 28: 441 (1979).

14. S. Doonan and H.M. A. Fahmy, Specific enzymic cleavage of polypeptides at cysteine residues. Eur. J. Biochem. 56: 421 (1975).

15. D. Barra, F. Bossa, S. Doonan, H. M. A. Fahmy, G.J. Hughes, K. Y. Kakoz, F. Martini and R. Petruzzelli, The structure of mitochondrial aspartate aminotransferase from pig heart and comparison with that of the cytoplasmic isozyme. FEBS Lett. 83: 241 (1977).

16. D. Barra, F. Bossa, S. Doonan, H. M. A. Fahmy, G. J. Hughes, F. Martini, R. Petruzzelli and B. Wittmann-Liebold, The cytosolic and mitochondrial aspartate aminotransferases from pig heart. A comparison of their primary structures, predicted secondary structures and some physical properties. Eur. J. Biochem. 108: 405 (1980).

17. O. L. Polyanovsky, T. V. Demidkina and C. A. Egorov, The position of an essential tyrosine residue in the polypeptide chain of aspartate transaminase. FEBS Lett. 23: 262 (1972).
18. P. Sonderegger and P. Christen, Comparison of the evolution rates of cytosolic and mitochondrial aspartate aminotransferases. Nature 275: 157 (1978).
19. P. B. Porter, S. Doonan and F. L. Pearce, Interspecies comparisons of aspartate aminotransferases based on immunochemical methods. Comp. Biochem. Physiol. 69B: 761 (1981).
20. S. Doonan, D. Barra, F. Bossa, P. B. Porter and S. M. Wilkinson, Interspecies comparisons of aspartate aminotransferases based on amino acid compositions. Comp. Biochem. Physiol. 69B: 747 (1981).
21. F. Bossa, D. Barra, F. Martini, E. Schininà, S. Doonan and K. M. C. O'Donovan, Interspecies comparisons of aspartate aminotransferases based on terminal and active site sequences. Comp. Biochem. Physiol. 69B: 753 (1981).
22. A. Arnone, P. H. Rogers, J. Schmidt, C-N.Han, C. M. Harris and D. E. Metzler, Preliminary crystallographic study of aspartate: 2-oxoglutarate aminotransferase from pig heart. J. Mol. Biol. 112: 509 (1977).
23. V. V. Borisov, S. N. Borisova, N. I. Sosfenov and B. K. Vainshtein, Electron density map of chicken heart cytosol aspartate transaminase at 3.5 Å resolution. Nature 284: 189 (1980).
24. G. C. Ford, G. Eichele and J. N. Jansonius, Three-dimensional structure of a pyridoxal-phosphate-dependent enzyme, mitochondrial aspartate aminotransferase. Proc. Natl. Acad. Sci. USA 77: 2559 (1980).
25. S. Capasso, A. M. Garzillo, G. Marino, L. Mazzarella, P. Pucci and G. Sannia, Mitochondrial bovine aspartate aminotransferase. Preliminary sequence and crystallographic data, FEBS Lett. 101: 351 (1979).
26. D. L. Peterson and M. Martinez-Carrion, The mechanism of transamination. Function of the histidyl residue at the active site of supernatant aspartate transaminase. J. Biol. Chem. 245: 806 (1970).

RECENT ADVANCES IN THE STUDY OF COENZYME BINDING TO ASPARTATE

APOAMINOTRANSFERASES

Carlo Turano, Francesca Riva and Anna Giartosio

Institutes of Biological Chemistry of the Universities
of Rome, Camerino and Cagliari and CNR Center of Molecular
Biology, University of Rome

Aminotransferases perform their catalytic function directing
and increasing the ability of their coenzyme to form, tautomerize
and process Schiff bases with aminoacids[1,2].

The coenzyme, pyridoxal-5'-phosphate, is tightly bound to the
holoenzyme, but in most cases it can be easily removed in conditions
mild enough to obtain a catalytically inactive apoenzyme, which in
turn can be completely reactivated by addition of new coenzyme.

Aminotransferases offer therefore the possibility of testing
which features of the protein molecule are responsible for the catalytic power of an enzyme. It is no wonder that the binding of
coenzyme to apoenzyme in aspartate aminotransferase has been the
subject of many studies with a number of different techniques.[3-5]

The role of the different groups of pyridoxal phosphate in binding has been investigated mainly by interaction of coenzyme analogs
with the apoprotein derived from the two isozymes from pig heart.

Apart from the obvious importance of the imine forming aldehydic group, it has been shown that i) the 2-methyl group, which is
dispensable in non enzymic reactions[1], is also dispensable in the
holoenzyme, and different degrees of alkyl substitution, if not too
bulky, are well tolerated;[3,6] ii) methyl substitution at the 6 posi-

tion has little effect on binding or on enzymic activity;[6,7] iii)
the 3-hydroxyl is required for catalytic activity,[1] and appears to
be important in inducing a slow conformational change associated
with the binding;[7] iv) the role of the phosphate group is very important and still poorly understood. Phosphate binding has been
suggested to be the first step in the formation of apoenzyme-
coenzyme complex.[7]

Protein groups of the cytosolic isozyme at or near the active
site that have been recognized by chemical modification or by spectroscopic methods are Lys 258 which binds pyridoxal phosphate in
iminic linkage,[8-9] Tyr,[10] His,[11] Trp[12] and Cys[13] residues.

X-ray crystallography data[14] on chicken heart mitochondrial
aspartate aminotransferase (active site structure, as will be discussed later, appears to be very similar in crystals of enzymes
differing in source or localization) agree with these data and
suggest that the coenzyme at the active site lies with the A face
flat against strands of a β sheet, while the B face is partially
exposed to the solvent, partially covered by a Trp residue. The
coenzyme phosphate group is anchored in a positive pocket and
interacts with Arg 266. The ring nitrogen might be hydrogen bonded
to Asp 222 and the phenolic hydroxyl in position 3 of the coenzyme
is within hydrogen bonding distance with Tyr 225. A very interesting
feature is that the active site is composed by elements of both
subunits.

Pyridoxal phosphate is therefore in contact with both monomers
of the dimeric protein and this may well account for the greater
stability of the holoenzyme in respect to the apoenzyme. This well
known empirical observation has recently been confirmed by calorimetric data on the cytosolic isozyme:[15] the denaturation temperature is noticeably higher for the holoenzyme than for the apoenzyme,
the increased stability is ascribed mainly to the aldiminic bond
and to the interaction between protein and phosphate group, while
the pyridine ring does not seem to play a major role in this respect.

The binding of cytosolic apoaminotransferase to pyridoxal phosphate has been analyzed also by Vergé and Arrio-Dupont in a recent
stopped flow study.[16] These authors describe the interaction as an
initial reversible binding step, followed by one or more isomeriza-
tion steps, during which conformational changes in the protein are
likely to occur.

We have tackled the problem by a different approach, i.e. the calorimetric determination of the heat of binding at different temperatures and in a wide range of pH values.[17]

Fig. 1 shows ΔH for the binding of pyridoxal-5'-phosphate to cytosolic aspartate apoaminotransferase as a function of pH at 19° and 25°C.

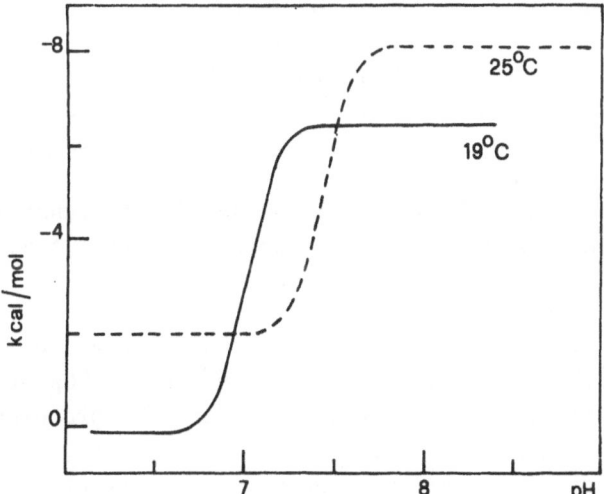

Fig. 1. Enthalpy change in apoenzyme-coenzyme interaction. 3×10^{-7} mol pyridoxal-5'-phosphate and 8×10^{-8} mol aspartate apoaminotransferase, both in 0.1 M buffer (cacodylate, PIPES or triethanolamine), were mixed at definite pH in the LKB batch calorimeter at the indicated temperature. Complete reactivation of the apoenzyme was checked at the end of each test. ΔH is calculated in kcal/mol of apoenzyme after correction for dilution heats.

The same isotherms were determined using buffers with different ionization heat, so that it must be concluded that no buffer protonation accompanies the reaction. This is an unexpected result, in view of the known pK values of the ionizing groups of coenzyme, apoenzyme and holoenzyme, and up to now unexplained.

The heat of binding at each temperature is pH dependent: Δ (Δ H) from pH 7 to pH 8 at 25°C, and from pH 6.5 to pH 7.5 at 19°C is equal to about 6 kcal/mol. At pH 6.5 or 8.0 ΔH values are more negative at higher temperature, showing that the heat capacity change in the pH independent tracts of the isotherms is negative and of the order of -350 cal/mol °K.

While the curves at 19°C and at 25°C have similar features, they are not only temperature dependent in respect to ΔH values, but also in respect to the midpoint (pH 7 at 19°C, pH 7.5 at 25°C) of the pH dependent transition. Van't Hoff equation may be applied to the data, yielding ΔH = +30 kcal/mol for the pH dependent transition. Such a value is too positive to be attributed to the protonation of one or few aminoacidic residues. To interpret these data we propose the existence of two states of apoenzyme, both capable of binding pyridoxal phosphate, but differing in ionization state. A similar hypothesis was put forward by Fonda and Auerbach[18] to explain the higher rate of recombination at lower pH.

Nothing can be said about the pH dependent transition at the moment: future X-ray diffraction studies on crystals grown in different conditions will perhaps solve the problem.

The negative heat capacity change can instead be more easily interpreted. It is currently assumed[19,20] that ΔCp of this sign and order of magnitude is attributable for many proteins to an increase in hydrophobic contacts and to a loss of degrees of freedom by apoenzyme and coenzyme, summing up to a contraction of the protein structure. Such a "tightening" of the apoenzyme upon binding pyridoxal-5'-phosphate is in accord with the enhanced stability of the holoenzyme in respect to the apoenzyme noticed by Relimpio et al.[15] and with the reaction dependent conformational change suggested by a number of authors.[16,21,22]

Coenzyme binding studies on the mitochondrial isozyme of pig heart aspartate aminotransferase are less abundant than those on

the cytosolic form, but the general picture is similar. In fact,
the similarities between the two binding sites are such that re-
searchers are often tempted to forget the dissimilarities. This
has been particularly true since the first crystallographic data
have been published: the active site features, as already mentioned,
are supposed to be conserved in all the aspartate aminotransferases
so far examined,[23,24] and no relevant difference on this point
has been noticed in the X-ray diffraction data of the two isozymes
from chicken heart.[14]

Subtle structural differences between isozymes have however been
found in solution by several approaches and may account for functional differences connected with the location and metabolic requirements within the cell.

The pyridoxal phosphate chromophore is a built-in probe for
spectroscopic studies and becomes optically active upon its binding
to the apoenzyme. Both holo and apoenzyme exhibit dichroic effects
in correspondence of their absorption spectra.[6,21] It is noticeable that substrate binding induces different dissymmetry factors
for the two isozymes in the 330 nm band related to the ketimine
formation.[25]

Differences are also reported in the phosphate binding site.
^{31}P NMR of the coenzyme phosphate provided information about this
group of critical importance in the apoenzyme-coenzyme interaction.
Experimental evidence suggests that the coenzyme phosphate group
is bound as dianion in the cytosolic enzyme and as monoanion in
the mitochondrial enzyme.[18,26,27]

Chemical approaches have been widely used to compare the active site structure. By reaction with iodoacetate, the aminic group
of enzyme bound pyridoxamine-5'-phosphate has been shown to have
pK = 8.3 in the cytosolic and pK = 9.1 in the mitochondrial
enzyme.[28]

Recently we have thoroughly investigated the reaction of aspartate transaminases with a series of coenzyme analogs provided
with a reactive function in a modified 4'-substituent on the
pyridine ring. In particular, one of these analogs, 4'-fluorodi-
nitrophenyl-pyridoxamine phosphate binds with great affinity to
the active site of both apoenzymes, but inhibits irreversibly only

the cytosolic isozyme.[29] It has been demonstrated that this irreversible inhibition is due to a covalent binding of the dinitrophenyl moiety of the compound to the ε-amino group of lysine 258,
which in the holoenzyme forms an aldiminic bond with pyridoxal
phosphate. The reaction can be conveniently followed by CD measurements, which show similar spectra for the initial inhibitor-
apoenzyme complexes for both isozymes.[30,31] The CD spectra of the
initial complex with the cytosolic enzyme is slowly changed to one
characteristic of the complex in which the covalent bond has been
formed, and which has a different asymmetry.

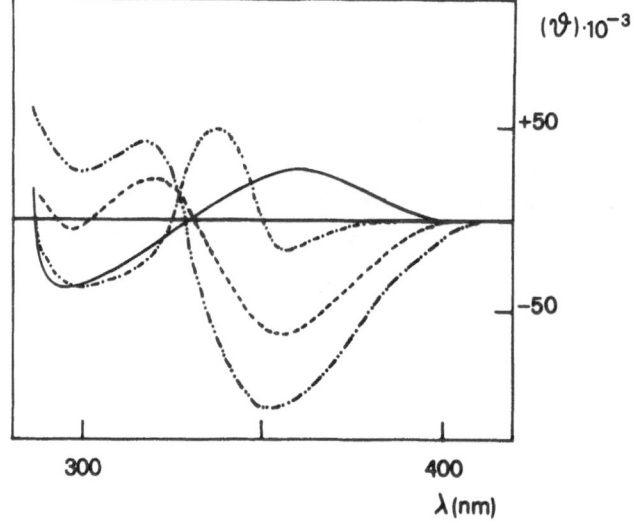

Fig. 2. CD Spectra of the cytosolic and mitochondrial apoenzymes
(3.5x10^{-5}M) reacted with 4'-fluorodinitrophenyl-pyridoxa-
mine-5'-phosphate (3.5x10^{-5}M) in 0.05 M Tris buffer, pH 8,
20°C. (---) cytosolic after 3' of reaction; (-·--·-) cytosolic after 24 h of reaction; (-··--··) mitochondrial after 3' of
reaction; (———) mitochondrial after 24 h of reaction.

The final spectrum of the mitochondrial enzyme after 24 h of reaction, as will be explained later, corresponds instead to the spectrum
of the holoenzyme in the aldehydic form.

These differences of behaviour of the two isozymes imply a
small structural difference, possibly consisting in a different mobility of the lysine side chain. It can be noted that this extreme
sensitivity of chemical methods in detecting subtle structural features in proteins makes them still useful for integrating crystallographic data.

One point of interest is the reproducibility of these results
in the crystal.[32,33]

Previous results have shown that the events occurring in the
crystal essentially parallel those occurring in solution.[34]

The inhibitor is an unnatural and bulky molecule: its binding
in the crystal is not predictable on the basis of previous knowledge of the interactions taking place in solution. The fact that in
the crystal it discriminates between the two active sites as it
does in solution indicates a definite structural difference in the
two isozymes, not abolished by packing forces present in the crystal
lattice. It is to be noticed that the same forces are held responsible for changing the functional state of one of the two active
sites in the dimeric molecule.[35,36]

In the course of the experiments carried out in solution a
second reaction was found to occur, identical for both isozymes,
leading to a slow regeneration of pyridoxal-5'-phosphate from 4'-
fluorodinitrophenyl-pyridoxamine phosphate bound to the apoenzyme
(see fig. 2). This reaction was better investigated with a compound
lacking the fluorine on the dinitrophenyl ring and thus incapable
of forming covalent bonds. (Table 1). Lys 258 appears to be involved in the process since upon carbamylation of its ε-amino group
the same initial complex of the inhibitor with apoenzyme appears
to be formed, as monitored by CD measurements, but no pyridoxal-
5'-phosphate is produced.[30,31] The mechanism of this reaction and
its products, except pyridoxal-5'-phosphate, are still unknown.

This reaction however deserves to be better investigated, since it represents a good example of labilization of a stable cova-

Table 1. % reactivation of the apoisozymes after 24 h reaction with
4'-fluorodinitrophenyl-pyridoxamine-5'-phosphate (A) or its
analog lacking the fluorine atom (B).

Molar ratio Apoenzyme: inhibitor	Addition of coenzyme (°)	% Activity (°°)	
		Cytosolic enzyme	Mitochondrial enzyme
A			
1:1	−	15	30
1:1	+	15	95
1:5	−		71
1:5	+		95
B			
1:1	−	26	30
1:5	−	115	97

(°)   The coenzyme was added after 24 h of reaction.

(°°) Activity is expressed as percent of the activity of the untrea-
ted apoenzyme after addition of the coenzyme.

lent bond of a compound quite different from the substrate, occurring upon binding to an enzymic active site. It is highly probable
that this labilization reflects either a strain imposed on the
groups substituted in position 4', or a particular microenvironment at the active site, which could be of some importance in the
normal catalytic process in the presence of the real substrates.

REFERENCES

1. E. E. Snell, Chemical structure in relation to biological acti-
    vities of Vitamin B$_6$, Vitam. Horm. 16: 77 (1958).
2. A. Meister,"Biochemistry of the Amino Acids" II ed., Academic
    Press, New York (1965).
3. P. Fasella and C. Turano, Structure and catalytic role of the
    functional groups of aspartate aminotransferase, Vitam.
    Horm. 28: 157 (1970).

4. M. Martinez-Carrion, S. Cheng, M. J. Stankewicz and A. Relimpio,
    Spectroscopic probes of enzyme-coenzyme-substrate complexes
    of aspartate transaminases, in: "Isozymes", C. L. Markers,
    ed., Academic Press, New York (1975).

5. A. E. Braunstein, Amino group transfer, in: "The Enzymes" IX B,
    P. D. Boyer, ed., Academic Press, New York (1973).

6. V. L. Florentiev, V. I. Ivanov and M. Ya. Karpeisky, Synthesis
    and physicochemical and coenzyme properties of alkyl-substi-
    tuted analogs of the B_6 vitamins and pyridoxal phosphate,
    Methods Enzymol. 18 A: 567 (1970).

7. M. H. O'Leary and S. W. Koontz, Coenzyme binding site of gluta-
    mate decarboxylase, Biochemistry 19: 3400 (1980).

8. S. Doonan, H. J. Doonan, H. Hanford, C. A. Vernon, J. M. Walker,
    F. Bossa, D. Barra, M. Carloni, P. Fasella, F. Riva and P.
    L. Walton, The primary structure of aspartate aminotransfe-
    rase from pig heart muscle determined in part using a pro-
    tease with specificity for lysine, FEBS Lett. 29: 31 (1974).

9. R. C. Hughes, W. T. Jenkins and E. H. Fischer, The site of
    binding of pyridoxal-5'-phosphate to heart glutamic-aspartic
    transaminase, Proc. Natl. Acad. Sci. U.S.A. 48: 1615 (1962).

10. C. Turano, A. Giartosio, F. Riva, D. Barra and F. Bossa, Studies
    on the active site of aspartate aminotransferase, in: "Sympo-
    sium on pyridoxal enzymes", Maruzen Co., Tokio (1968).

11. M. Martinez-Carrion, C. Turano, F. Riva and P. Fasella, Eviden-
    ce of a critical histidine residue in soluble aspartic ami-
    notransferase, J. Biol. Chem. 242: 1426 (1967).

12. M. Arrio-Dupont, Fluorescence of aromatic amino acids in a py-
    ridoxal phosphate enzyme: aspartate aminotransferase, Eur.
    J. Biochem. 91: 369 (1978).

13. W. Birchmeier, K. J. Wilson and P. Christen, Syncatalytic modi-
    fication of cytoplasmic aspartate aminotransferase: identi-
    fication of a peptide containing the modified cysteinyl re-
    sidue, FEBS Lett.26: 113 (1972).

14. G. C. Ford, G. Eichele and J. N. Jansonius, Three-dimensional
    structure of a pyridoxal-phosphate-dependent enzyme, mito-
    chondrial aspartate aminotransferase, Proc. Natl. Acad. Sci.
    U.S.A. 77: 2559 (1980).

15. A. Relimpio, A. Iriarte, J. F. Chebowski and M. Martinez-Carrion,
    Differential scanning calorimetry of cytoplasmic aspartate
    transaminase, J. Biol. Chem. 256: 4478 (1981).

16. D. Vergé and M. Arrio-Dupont, Interactions between apoaspartate
    aminotransferase and pyridoxal-5'-phosphate. A stopped-flow
    study, Biochemistry 20: 1210 (1981).

17. A. Giartosio, C. Salerno, F. Franchetta and C. Turano, A calorimetric study of the interaction of pyridoxal-5'-phosphate with aspartate apoaminotransferase and model compounds. Manuscript in preparation.

18. M. L. Fonda and S. B. Auerbach, The interaction of pyridoxal phosphate with aspartate apoaminotransferase, Biochim. Biophys. Acta 422: 38 (1976).

19. C. Tanford, Protein denaturation, Adv. Prot. Chem. 24: 1 (1970).

20. H. J. Hinz, D. D. F. Shiao and J. M. Sturtevant, Calorimetric investigation of inhibitor binding to rabbit muscle aldolase, Biochemistry 10: 1347 (1971).

21. P. Fasella and G. G. Hammes, An optical rotatory dispersion study of aspartic aminotransferase, Biochemistry 3: 530 (1964).

22. L. V. Abaturov, O. L. Polyanovsky, Yu. M. Torchinsky and Ya. M. Varshavsky, Studies on hydrogen deuterium exchange in aspartate transaminase, in: "Pyridoxal catalysis: Enzymes and model systems," E. E. Snell, A. E. Braunstein, E. S. Severin and Yu. M. Torchinsky, eds., Wiley, New York (1968).

23. A. Arnone, P. H. Rogers, J. Schmidt, C. N. Han, C. M. Harris and D. E. Metzler, Preliminary crystallographic study of aspartate: 2-oxoglutarate aminotransferase from pig heart, J. Mol. Biol. 112: 509 (1977).

24. V. V. Borisov, S. N. Borisova, G. S. Kachalova, N. I. Sasfenov, B. K. Vainshtein, Yu. M. Torchinsky, and A. E. Braunstein, Threedimensional structure at 5 Å resolution of cytosolic aspartate transaminase from chicken heart, J. Mol. Biol. 125: 275 (1978).

25. M. Martinez-Carrion, D. C. Tiemeier and D. L. Peterson, Conformational properties of the isoenzymes of aspartate transaminase and the enzyme-substrate complexes, Biochemistry 9: 2574 (1970).

26. M. Martinez-Carrion, ^{31}P nuclear-magnetic-resonance studies of pyridoxal and pyridoxamine phosphates. Interaction with cytoplasmic transaminase, Eur. J. Biochem. 54: 39 (1975).

27. E. M. Mattingly, J. R. Mattingly and M. Martinez-Carrion, Does the phosphate of PLP function differently in the isozymes of porcine aspartate aminotransferase? National Symposium of Pyridoxal, Knoxville (1981).

28. Y. Morino, M. Okamoto, and S. Tanase, Selective inactivation of pyridoxamine form of aspartate aminotransferase by iodoacetate, J. Biol. Chem. 253: 6026 (1978).

29. F. Riva, D. Carotti, D. Barra, A. Giartosio and C. Turano, Different reactivity of mitochondrial and cytoplasmic aspartate aminotransferases toward an affinity labeling reagent analog of the coenzyme, J. Biol. Chem. 255: 9230 (1980).

30. F. Riva, F. Ascoli, D. Carotti, R. Santucci, Different reactivity of aspartate aminotransferase isozymes towards an analog of the coenzyme, followed by CD spectra, 14th FEBS Meeting, Edinburgh, Abst. 330P (1981).

31. D. Carotti, F. Riva, F. Ascoli, R. Santucci, P. Fasella, Circular dichroism study of the interaction of aspartate amino transferase isozymes with a coenzyme analog, manuscript in preparation.

32. A. Mozzarelli, S. Ottonello, D. Carotti e F. Riva, Reattività dell'aspartato aminotransferasi allo stato cristallino con un analogo di coenzima, 27° Congresso Nazionale della Società Italiana di Biochimica, Parma (1981).

33. G. L. Rossi, A. Mozzarelli, S. Ottonello, D. Carotti, F. Riva, Reactivity of aspartate aminotransferase in crystal toward a coenzyme analog, manuscript in preparation.

34. A. Mozzarelli, S. Ottonello, G. L. Rossi and P. Fasella, Catalytic activity of aspartate aminotransferase in the crystal. Equilibrium and kinetic analysis, Eur. J. Biochem. 98: 173 (1979).

35. A. Arnone, P. O. Briley, P. H. Rogers, G. G. Hyde, C. M. Metzler and D. E. Metzler, Crystallographic studies showing domain movement in cytoplasmic aspartate aminotransferase, National Symposium of Pyridoxal, Knoxville (1981).

36. H. Kirsten, H. Gehring and P. Christen, Catalytic properties of crystalline mitochondrial aspartate aminotransferase, lattice-induced functional asymmetry of the two subunits, 14th FEBS Meeting, Edinburgh, Abst. 323P (1981).

FLAVOPROTEINS: CORRELATION OF STRUCTURE AND FUNCTION

Vincent Massey

Department of Biological Chemistry
University of Michigan
Ann Arbor, Michigan 48109, U.S.A.

The differences in properties among different flavoproteins and their differences from the free coenzymes have long intrigued workers in this field. For example, in the oxidized state, free flavins have an intense greenish yellow fluorescence, which is mostly quenched, sometimes partially, and often completely, on incorporation into a particular flavoprotein. In the reduced state, free flavins react rapidly with oxygen, through a complex series of reactions involving flavin hydroperoxide, flavin radical and superoxide radical, yielding finally oxidized flavin and H_2O_2. Some flavoproteins retain this ability to react rapidly with O_2, others lose it almost completely. Free flavins also show a thermodynamic stabilization of the radical state, characterized by a dismutation-comproportionation equilibrium, and a pK between neutral and anionic semiquinone of ~ 8.5:

$$Fl_{ox} + Fl_{red}H_2 \rightleftharpoons 2FlH° \tag{1}$$

$$FlH° \xrightleftharpoons{\text{pK 8.5}} Fl°^- + H^+ \tag{2}$$

In the free system, the equilibrium of equation 1) lies predominantly to the left. Among flavoproteins, a great variety of responses is found. Many flavoproteins show a considerable increase in the thermodynamic stabilization of the radical state, so that the equilibrium of equation 1) now lies heavily to the right. In yet a number of other cases, considerable amounts of radical may be found on one-electron reduction of the flavoprotein, but now the stabilization is kinetic in nature, not thermodynamic. This is not difficult to envisage, since the only way a flavoprotein can discharge the radical state is to make contact with an electron acceptor; if

the only acceptor available is another flavoprotein radical there
may be quite severe kinetic barriers to dismutation, especially with
the anion radicals. In many cases of flavoprotein radicals cited
in the literature, it is not clear whether radical stabilization is
thermodynamic or kinetic in nature. As we shall see later, this is
an important differentiation in terms of correlation of protein
structure and biological function of flavoproteins. What is clear
is that in most cases the protein environment causes a shift in pK
of the ionization of equation 2), so that over the entire range of
stability of the protein, either the blue neutral semiquinone is
found (pK raised) or the red anionic semiquinone is found (pK lowered).

It is obvious that the properties described above, and others
which we will consider later, must be modulated by the environment
of the protein enclosing the flavin prosthetic group in each particular flavoprotein. The concept that the particular environment
might be similar within a particular class of flavoproteins began
to be formulated about 10 years ago, when it was found in a survey
of different flavoproteins, that those which react rapidly with O_2,
the conventional "oxidases", have a number of properties in common,
which are not shared by the conventional "dehydrogenases", those
enzymes which fail to react rapidly with O_2[1,2]. Thus, the oxidases
were found to stabilize the red anionic flavin radical, to stabilize
the adduct of sulfite at the flavin N(5)-position, and on reaction
of the reduced form with O_2 to produce H_2O_2 and oxidized flavoprotein without detectable intermediates. On the other hand, most
"dehydrogenases" stabilized the blue neutral semiquinone, failed to
form sulfite adducts, and on reaction of the reduced form with O_2
gave the blue flavin semiquinone and the 1-electron reduction product of oxygen, O_2^-. It was also found that in general the oxidases
failed to react rapidly with 1-electron acceptors such as ferricyanide or cytochrome c, whereas the "dehydrogenases" reacted rapidly
with most 1-electron acceptors. Thus those flavoenzymes whose
natural function was to react with obligatory 1-electron acceptors
such as iron-sulfur proteins or cytochromes, also treated O_2 as a
1-electron acceptor, whereas the oxidases treated it apparently
strictly as a 2-electron acceptor.

Although the conventional classification of flavoenzymes as
"oxidases" or "dehydrogenases" had to be expanded to include the
group of "monooxygenases" which catalyse the incorporation of one
atom of the oxygen molecule into a second substrate and convert the
other into H_2O, it was not until recently that Hemmerich began to
emphasize the inadequacy and inconsistency of the old classification[3]. This arises from the fact that the vast majority of flavoproteins function catalytically by alternate reduction and reoxidation, and that in most cases the reduction of the flavin is a
2-electron reduction accomplished by dehydrogenation of its reducing
substrate. In some cases the reoxidation of the flavin is

accomplished by a 2-electron reaction also involving hydrogen transfer, in other cases by a 2-electron reduction of O_2, and in yet other
cases by two 1-electron transfers to the oxidant. Thus a more logical and accurate classification has been proposed, with five main
classes, as follows[4,5]:

Class 1   Transhydrogenases, which carry out dehydrogenation ($2e^-$-
oxidation) of one substrate and "rehydrogenation" ($2e^-$-reduction) of
another. This group can be further subdivided according to whether
the centers involved in the hydrogen transfer are carbon, nitrogen
or sulfur atoms.

Class 2   Dehydrogenase-oxidase, carrying out the reoxidation of the
$2e^-$-reduced flavin by molecular oxygen with the production of H_2O_2.
This is the classical oxidase, and it is convenient to retain the
simple term, "oxidase".

Class 3   Dehydrogenase-oxygenase. These enzymes carry out the
splitting of the oxygen molecule, with one atom being reduced to
H_2O and the other inserted into a second substrate. This group of
enzymes can be further subdivided depending on whether the substrate
into which oxygen is inserted also provides the reducing equivalents
to produce the reduced flavin required for oxygen activation (the
"internal" monooxygenases) or whether a third substrate (a reduced
pyridine nucleotide) is required for that function (the "external"
monooxygenases). The internal monooxygenases appear in fact to
belong better to Class 2, the oxygenase function being an adventi-
tious one, due to the H_2O_2 product being slow to leave the reoxidized
enzyme, and reacting with the dehydrogenated product while both are
still enzyme-bound[6]. The true monooxygenases then all use NADH or
NADPH as reducing agent, and incorporate an atom of oxygen (the
second substrate) into a third molecule. The nature of this third
substrate provides a further differentiation within this class.
Thus we have the aromatic or phenolic hydroxylases, where a phenolic
compound is further hydroxylated in a reaction involving the flavin
C(4a)-hydroperoxide as a nucleophile, and a second group, so far
without a suitable name, where the flavin hydroperoxide appears to
behave as an electrophile. Examples of this latter group include
cyclohexanone monooxygenase, where the ring is expanded by oxygen
insertion[7] and the microsomal N(S)-monooxygenase which forms N- and
S-oxides with a large number of amines and sulfur-containing compounds[8,9].

Class 4   Dehydrogenase-electron transferase enzymes, involving a
two-electron equivalent dehydrogenation, and reoxidation in $1e^-$
steps with obligatory 1-electron acceptors, such as iron-sulfur
proteins and cytochromes. This class also includes enzymes which
work in the opposite direction, such as ferredoxin-NADP reductase,
where the flavin is reduced by single electron steps to the fully
reduced form, which then is reoxidized in a 2-electron equivalent
step at the expense of reduction of $NADP^+$.

Class 5   Pure electron transferases, where both reduction and oxidation of the flavin appear to be accomplished in single electron
transfer steps. The flavodoxins are examples of this class.

Our initial correlations of properties and catalytic function
may now be refined in light of this more logical classification.
Despite a few exceptions, which may have very interesting explanations, the initial correlations seem to have stood the test of time
very well, with the caveat that for the original "dehydrogenase"
we now read the electron transferases of Classes 4 and 5, and for
oxidases those enzymes of Class 2 (and not the oxygenases of Class
3, which although reacting rapidly with O_2, have their own very
distinctive properties, quite different from those of the oxidases).
Unfortunately the least well characterized of the flavoenzymes are
the transhydrogenases of Class 1, which so far appear to have rather
variable properties, producing either blue or red semiquinone (most
of which, however, appear only to be stabilized kinetically, not
thermodynamically) and which have rather variable oxygen reactivities[5]. It should be noted also that some of the more complex flavoproteins are rather difficult to fit neatly into a single category.
An example is milk xanthine oxidase, which although it is a true
oxidase in the sense that its fully reduced flavin reacts rapidly
with O_2, is also an electron transferase, where all three redox constituents, flavin, molybdenum and iron-sulfur centers, are in rapid
equilibrium[10]. In accordance with this classification is its stabilization of the blue neutral flavin semiquinone and its lack of
formation of a flavin N(5)-sulfite adduct. A widely quoted example
of an oxidase which gives a stable red semiquinone but no sulfite
adduct, is putrescine oxidase[11]. This lack of correlation has been
taken as evidence undermining the oxidase classification[12], but the
explanation is probably rather trivial. Putrescine is a dication
and there is evidence for a protein anionic residue being involved
in substrate binding[13]; this negatively-charged residue is probably
located in the region of the flavin N(5)-position and simply repels
the negatively-charged sulfite[13]. A similar explanation may hold
for the Old Yellow Enzyme, which forms a red semiquinone (which
however is thermodynamically unstable) but no sulfite adduct[1]. In
view of previous misconceptions, it seems important to state at this
point that in general, oxidases stabilize thermodynamically the red
flavin anion radical and the N(5)-sulfite adduct; however not every
flavoprotein which yields a red radical is an oxidase, and not
every red radical flavoprotein will form a sulfite-adduct.

Another much-emphasized exception to the above classification-
correlation is that of yeast lactate dehydrogenase or flavocytochrome b_2. This enzyme in the Hemmerich-Massey classification clearly
belongs in Class 4, yet it forms both a red semiquinone and a sulfite
adduct[12]. Rather than destroy the basis of the classification,

however, these properties probably are a consequence of the carbanion mechanism of the dehydrogenation reaction, which appears to
correlate well with formation of flavin anion radical and N(5)-
sulfite adduct[5,14].

Another interesting correlation which has recently come to
light is that all flavoproteins so far examined which contain flavin
covalently linked to the protein, give on partial reduction the red
semiquinone anion, irrespective of whether the enzyme reacts rapidly
with O_2, or not[15]. This may be a consequence of the biosynthesis
of the covalent linkage, which is always at the flavin 8-α-position
or the 6-position. These positions should be activated for attack
by a protein nucleophile (cf ref 16) when a positively-charged group
of the protein is located at the N(1)-C(2α) locus of the flavin.

This type of charge/hydrogen bonding effect is believed to be
responsible for stabilization of the red anionic semiquinone and the
flavin N(5)-sulfite adduct[5,14]. Such an interaction may therefore
be very important in the biosynthesis of covalently-bound flavins.

Active Site Probes of Flavoproteins

The possible explanation for the correlations discussed in the
previous section received experimental support recently from studies
with artificial flavins which act as "indicator" groups for the protein environment surrounding the flavin. It was first suggested by
Ghisla and colleagues[17-19] that because of their existence in spectrally distinct tautomeric and mesomeric forms, 6-hydroxy and
8-hydroxy flavins might serve as probes of the active sites of
flavoproteins. In their anionic state these flavins exist

predominantly in the benzoquinoid mesomeric form, with the negative
charge localized in the N(1)-C(2α) region of the flavin. The pK's
for ionization are 7·1 for 6-hydroxyflavin[18] and 4·8 for 8-hydroxy-
flavin[19]. In both cases, a positively-charged protein residue in
the neighborhood of the flavin N(1) position would stabilize the
benzoquinoid form, and would result in a lowering of the pK of the
protein-bound flavin. The benzoquinoid form could also be favored
by a negatively-charged residue in the vicinity of the flavin
6-position or of the 8-position, but in these cases the pK of the
protein-bound form would actually be raised.

To these can now be added an even more versatile probe,
8-mercaptoflavin[14].

(A) $\lambda_{max} \sim 470nm$ pK 3.8 | +H⁺   (B) $\lambda_{max} \approx 560nm$

(C)   (D)

In this case the pK is very low, pH 3.8, and in free solution
the preferred form of the anion is the 8-thiolate (form C), rather
than the N(1)-ionized benzoquinoid form[14]. From the spectral
changes which occur on binding to proteins it can be seen that two
forms with dramatically different absorption spectra are stabilized.
One, attributable to the predominant thiolate, is deep red in color,
with λ max in the 520-550 nm region, and only a single well-marked
absorption band in the visible region. The other, attributable to
the benzoquinoid form, is bright blue in color, with λ max in the
region 570-610 nm, and with three distinctive absorption bands in
the visible-near UV (Fig. 1). In the case of the 8-mercaptoflavin
anion it is only when there is a positive charge in the protein in
the vicinity of the flavin N(1)-position, or a negative charge in
the neighborhood of position-8, that the blue benzoquinoid form
would be stabilized. Hence a study of the spectral properties of a
flavoprotein in which the native flavin has been removed and replaced

by 8-mercaptoflavin, coupled with the pK-behavior of the 6-hydroxy-
and 8-hydroxy flavin enzymes, should allow definite conclusions to
be made.   Such studies have shown that with the enzymes studied of
the dehydrogenase-oxidase class, the benzoquinoid forms are stabilized in all cases, indicating the existence of a protein positive
charge interacting with the N(1)-C(2α) locus of the flavin.   As discussed in more detail elsewhere[5,14], we take these results as providing strong evidence for a common type of flavin-protein interaction among all the enzymes of this class, and that this interaction
with the reduced flavin in some way directs its high reactivity with
O_2.

Significantly, all of the enzymes of this class also exhibit in
the reduced form the spectrum typical of that of the anion, FlH^{-}[5,14].
This structural feature of a protein positive charge in the vicinity
of the flavin N(1)-position thus serves in the oxidized state to
stabilize the benzoquinoid form of 6-hydroxy-, 8-hydroxy- and 8-mer-
captoflavin, and to promote the attack of sulfite at the N(5)-
position and stabilize the resulting adduct.   In the semiquinoid
form, the same protein positive charge stabilizes the anion form of
the native flavin, with its negative charge at the N(1)-position.
Similarly, the anion form of fully reduced flavin, again with the

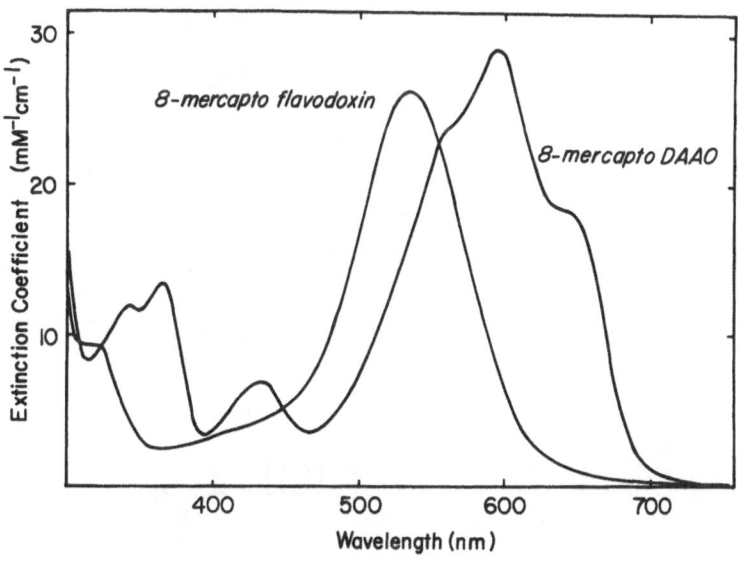

Fig. 1.   Spectra of 8-mercaptoflavoproteins in the 8-thiolate form
          (flavodoxin) and in the benzoquinoid form (D-amino acid oxi-
          dase)

negative charge at the N(1)-position, is stabilized. This leads to
yet another prediction, which seems to be borne out by the limited
results so far available, that enzymes of this class should also
have redox potentials more positive than that of the free coenzyme.
This follows from the fact that a positive charge in the region of
the N(1)-position of the flavin would lead to a higher binding
affinity for the reduced flavin, compared to that for the oxidized
form. Thus, from the Nernst equation, the redox potential in such
cases should be more positive than that of free flavin.

The existence of a protein interaction with the flavin
N(1)-C(2α) locus would favor formation of a C(10a) hydroperoxide
when the reduced flavin reacts with O_2, whereas hydrogen bonding
interaction with the flavin N(5) position would favor formation and
stabilization of a C(4a)-hydroperoxide. It is thus tempting to
speculate that enzymes of classes 4 and 5, the electron transferases,
which mostly stabilize the blue neutral radical, have protein interactions with the flavin N(5)H, and react with O_2 to form the C(4a)-
hydroperoxide[4,5] as shown in the accompanying Scheme.

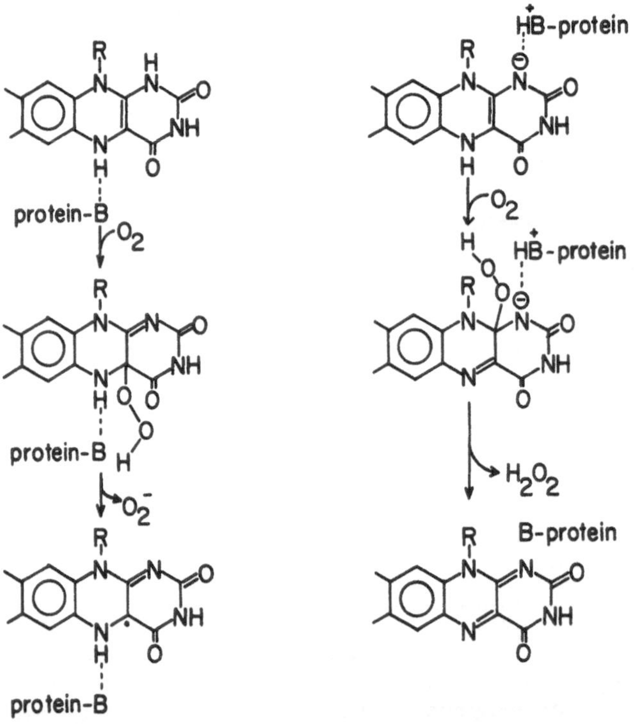

Flavin N(5)-Interactions   Flavin N(I)-Interactions

By homolytic cleavage this species would yield O_2^- and the stabilized blue neutral radical, in keeping with the behavior observed with
enzymes of these two classes[1,2]. Enzymes of Class 2, the oxidases,
presumably react with O_2 to form the C(10a)-hydroperoxide. In this
case the preferred route of breakdown would appear to be by hetero-
lytic cleavage, promoted by the negative charge of the flavin, to
yield H_2O_2 and oxidized flavin without intermediate formation of
O_2^- and flavin radical, again in keeping with observations with
this group of enzymes[1,2].

Chemically Reactive Flavins as Probes of Flavoprotein Topography

The facile nucleophilic displacement of the halo-substituent
from 8-halogeno-flavins by thiolates and sulfide, and the resulting
spectral changes[14] provide very sensitive methods for testing whether
the benzene subnucleus of enzyme-bound flavin is exposed to solvent.

Many 8-chloroflavin-substituted flavoproteins have been tested
in this way[5,14,20]. While reaction rates of sulfide or thiophenol
with 8-chloroflavoproteins as fast or faster than those with the
free 8-chloroflavins clearly indicate exposure of the flavin
8-position to the solvent, low rates of reaction are more difficult
to interpret, since they may be due to geometric constraints imposed
by the protein on formation of the tetrahedral intermediate which
is presumably formed in these reactions. This difficulty in interpretation may be overcome by making use of the chemical reactivity
of 8-mercaptoflavins, which react readily with alkylating agents
such as iodoacetic acid and iodoacetamide, and with methylmethane-
thiolsulfonate[20].

The disulfide formed in the latter reaction is readily reduced
by reagents such as dithiothreitol, giving back the starting
8-mercaptoflavin. In this way it is possible to tell whether the
8-mercaptoflavin is still bound to the enzyme, or whether it has
been dissociated as a result of modification of protein thiol residues.

A survey of flavoproteins along the above lines has revealed
the interesting fact that while in many the benzene subnucleus is
buried in the protein (eg lactate oxidase, glucose oxidase, egg
white-riboflavin-binding protein) in all cases so far examined where
pyridine nucleotide is a cosubstrate, the flavin 8-position is quite
exposed (melilotate hydroxylase[20], p-hydroxybenzoate hydroxylase[20],
phenol hydroxylase[21], putidaredoxin reductase[22], NADP-ferredoxin
reductase[23], Old Yellow Enzyme[20], glutathione reductase[24]). While
more examples need to be obtained, it is tempting to speculate that
all pyridine nucleotide-linked flavoproteins may have common structural features, and that the pyridine nucleotide gains access to
the active site by a channel across the benzenoid ring of the protein-bound flavin. This hypothesis is certainly consistent with
the information on the three-dimensional structure of glutathione
reductase[25] and p-hydroxybenzoate hydroxylase[26] available from X-ray
diffraction analysis. The results demonstrate clearly the value of
such artificial flavins as probes of the protein environment surrounding the bound flavin.

Another useful probe which we have investigated recently is
one where the oxygen at the 2-position is replaced by sulfur,
2-thioflavin[27]. This flavin has been found to have a pK of 9.8,

and in its ionized form reacts rapidly with methylmethanethiolsul-
fonate, with pronounced changes in absorption spectrum[28]:

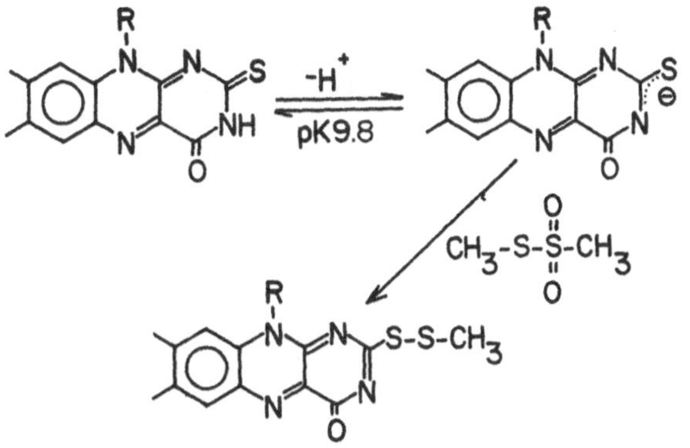

Hence, a study of the ionization properties of the appropriate
2-thioFMN or 2-thioFAD bound to a particular protein, and its reactivity with methylmethane thiolsulfonate, yields information
about the protein topography in this region of the flavin. No class
correlations have yet emerged from such studies, although some very
interesting information about individual proteins has emerged. For
example, it has been found that the flavin 2-position in p-hydroxybenzoate hydroxylase is quite exposed to solvent, but becomes buried
on complex formation with the substrate p-hydroxybenzoate[28]. This
finding is in accord with the X-ray crystallographic information,
which is only available for the enzyme-substrate complex[26]. However,
the latter shows that the substrate is located in such a way that a
conformational change burying the substrate must have occurred as a
result of its binding to the protein. The results with 2-thioFAD-
enzyme provide a striking independent demonstration of this phenomenon.

These examples illustrate the type of information available
from such probes. There are several more potential probes already
available, but awaiting testing for their possible utility. These
include 4-thioflavins, synthesized many years ago by Hemmerich and
coworkers[29], which would be expected to behave in a similar fashion
to 2-thioflavins. Walsh and colleagues[30] have described the potential utility of 3-deaza flavins as active site probes, due to the
favorable pK of 5.8 and the large differences in spectra of the
neutral and anion forms.

Thus, in analogy with the results described for 8-mercapto and 6-hydroxyflavins, where a protein positive charge in the vicinity of the flavin N(1) position appears to stabilize the benzoquinoid anion form, a protein positively-charged group in the vicinity of the flavin position 3 would be expected to stabilize the anion form of 3-deazaflavin, and lower its pK in the protein-bound form. Conversely, a negatively-charged residue in the protein around this position would be expected to stabilize the neutral form, and so raise the pK. Such a stabilization of the neutral form of 3-deazaflavin by egg white riboflavin-binding protein has been observed[30] and is consistent with the preference of binding neutral flavins generally observed for this protein[30-33,14]. Walsh et al. ascribe this effect to the flavin binding in a hydrophobic pocket[30]. However, the suggestion of a protein negative charge in this region of the flavin could explain the results equally well, and be consistent with the conclusions of McCormick et al.[33] and those from our laboratory with 2-thioriboflavin[28] that the pyrimidine subnucleus is exposed to solvent.

REFERENCES

1. V. Massey, F. Müller, R. Feldberg, M. Schuman, P. A. Sullivan, L. G. Howell, S. G. Mayhew, R. G. Matthews and G. P. Foust, The reactivity of flavoproteins with sulfite. Possible relevance to the problem of oxygen reactivity, J. Biol. Chem. 244: 3999 (1969).

2. V. Massey, S. Strickland, S. G. Mayhew, L. G. Howell, P. C. Engel, R. G. Matthews, M. Schuman, and P. A. Sullivan, The production of superoxide anion radicals in the reaction of reduced flavins and flavoproteins with molecular oxygen, Biochem. Biophys. Res. Commun.36: 891 (1969).

3. P. Hemmerich, V. Massey and H. Fenner, Flavin and 5-deazaflavin: a chemical evaluation of 'modified' flavoproteins with respect to the mechanisms of redox biocatalysis, FEBS Letters 84: 5 (1977).

4. P. Hemmerich and V. Massey, The role of apoprotein in directing pathways of flavin catalysis, in: 3rd International Symposium on Oxidases, T. E. King, H. S. Mason and M. Morrison, eds., University Park Press, Baltimore, still in press.

5. V. Massey and P. Hemmerich, Active site probes of flavoproteins, Biochem. Soc. Transactions 8: 246 (1980).

6. M. S. Flashner and V. Massey, Flavoprotein oxygenases in: "Molecular mechanisms of oxygen activation", O. Hayaishi, ed., Academic Press, New York (1974).

7. C. Walsh, F. Jacobson and C. Ryerson, in: "Biomimetic Chemistry" D. Dolphin, ed., American Chemical Society, Washington, D. C. (1979).

8. L. L. Poulsen and D. M. Ziegler, The liver microsomal FAD-containing monooxygenase, spectral characterization and kinetic studies, J. Biol. Chem. 254: 6449 (1979).

9. N. B. Beaty and D. P. Ballou, The oxidative half-reaction of liver microsomal FAD-containing monooxygenase, J. Biol. Chem. 256: 4619 (1981).

10. J. S. Olson, D. P. Ballou, G. Palmer and V. Massey, The mechanism of action of xanthine oxidase, J. Biol. Chem. 249: 4363 (1974).

11. R. DeSa, Putrescine oxidase from Micrococcus rubens. Purification and properties of the enzyme, J. Biol. Chem. 247: 5527 (1972).

12. F. Lederer, Sulfite binding to a flavodehydrogenase, cytochrome b_2 from baker's yeast, Eur. J. Biochem. 88: 425 (1978).

13. W. F. Swain and R. DeSa, Mechanism of action of putrescine oxidase. Binding characteristics of the active site of putrescine oxidase from Micrococcus rubens, Biochim. Biophys. Acta 429: 331 (1976).

14. V. Massey, S. Ghisla and E. G. Moore, 8-mercaptoflavins as active site probes of flavoenzymes, J. Biol. Chem. 254: 9640 (1979).

15. D. E. Edmondson, B. A. C. Ackrell and E. B. Kearney, Identification of neutral and anionic 8α substituted flavin semiquinones in flavoproteins by electron spin resonance spectroscopy, Archiv. Biochem. Biophys. 208: 69 (1981).

16. C. Walsh, Flavin coenzymes: at the crossroads of biological redox chemistry, Acc. Chem. Res. 13: 148 (1980).

17. S. Ghisla, V. Massey and S. G. Mayhew, Studies on the active centers of flavoproteins: binding of 8-hydroxy-FAD and 8-hydroxy-FMN to apoproteins, in: "Flavins and Flavoproteins" T. P. Singer, ed., Elsevier, Amsterdam (1976).

18. S. G. Mayhew, C. D. Whitfield, S. Ghisla and M. Schuman-Jorns, Identification and properties of new flavins in electron-transferring flavoproteins from Peptostreptococcus elsdenii and pig-liver glycolate oxidase, Eur. J. Biochem. 44: 579 (1974).

19. S. Ghisla and S. G. Mayhew, Identification and properties of 8-hydroxyflavin-adenine dinucleotide in electron-transferring flavoprotein from Peptostreptococcus elsdenii, Eur. J. Biochem. 63: 373 (1976).

20. L. M. Schopfer, V. Massey and A. Claiborne, Active site probes of flavoproteins. Determination of the solvent accessibility

of the flavin position 8 for a series of flavoproteins,
J. Biol. Chem. 256: 7329 (1981).

21. K. Detmer and V. Massey, unpublished results.

22. G. Wagner, I. C. Gunsalus and V. Massey, unpublished results.

23. G. Zanetti and V. Massey, unpublished results.

24. H. Schirmer and S. Ghisla, personal communication.

25. G. Schulz and E. Pai in: "Flavins and Flavoproteins", V. Massey
     and C. H. Williams, eds., Elsevier-North Holland, in press.

26. R. K. Wierenga, R. J. deJong, K. H. Kalk, W. G. J. Hol and J.
     Drenth, Crystal structure of p-hydroxybenzoate hydroxylase,
     J. Mol. Biol. 131: 55 (1979).

27. P. Hemmerich, S. Fallab and H. Erlenmeyer, Synthesen in der
     Lumiflavin Reihe, Helv. Chim. Acta 39: 1242 (1956).

28. A. Claiborne, L. M. Schopfer, P. Fitzpatrick, and V. Massey,
     J. Biol. Chem. , in press (1981).

29. F. Müller and P. Hemmerich, Thione, Imine, Oxime und Azine des
     Riboflavins. Nucleophile Substitutionsreaktionen am Flavin-
     kern, Helv. Chim. Acta 49: 2352 (1966).

30. C. Walsh, J. Fisher, R. Spencer, D. W. Graham, W. T. Ashton,
     J. E. Brown, R. D. Brown and E. F. Rogers, Chemical and
     enzymatic properties of riboflavin analogs, Biochemistry
     17: 1942 (1978).

31. J. E. Becvar, Ph. D. Dissertation, University of Michigan (1973).

32. G. Blankenhorn, Riboflavin binding in egg-white flavoprotein:
     the role of tryptophan and tyrosine, Eur. J. Biochem. 82:
     155 (1978).

33. J. D. Choi and D. B. McCormick, The interaction of flavins
     with egg white riboflavin-binding protein, Archiv. Biochem.
     Biophys. 204: 41 (1980).

ROLE OF FLAVIN AND IRON SULFUR CENTERS IN THE TRANSITION OF SUCCINATE DEHYDROGENASE FROM THE ACTIVATED TO THE NON-ACTIVATED FORM

Franco Bonomi, Silvia Pagani, and Paolo Cerletti

Department of General Biochemistry
University of Milan
I-20133 Milano, Italy

The present report deals mainly with the aspects of succinate dehydrogenase which were developed by recent studies in our laboratory, i.e. the properties of the redox active centers of the enzyme in relation to the molecular mechanism of activation and the catalytic cycle.

Mammalian succinate dehydrogenase (succinate: (acceptor) oxido-reductase, E.C. 1.3.3.99) is a flavoprotein of the inner mitochondrial membrane. Its molecular weight is 97,000 daltons (1, 2) and it contains two different subunits, one having a molecular weight of 70,000 daltons the other one 27,000. The larger subunit contains the flavin, FAD, which is covalently bound through the methyl group in position 8 to a histidine residue of the peptide (3) and two tetrahedral 2Fe2S clusters (4), called centers S_1 and S_2, whereas the cubic 4Fe4S cluster, center S_3, is situated in the small subunit (4).

The enzyme is subjected to rigorous regulation, adjusting its activity with respect to the metabolic state of the mitochondria (5, 6). This regulation, the so called activation process, involves the reversible shift of the enzyme between a fully active form, the activated dehydrogenase, and one devoid of catalytic activity, the deactivated enzyme. It is achieved by a single negative modulator oxaloacetate (7-9) which was recognized first by Wojtczack et al. in 1969 (10), and by a number of positive effectors, substrate and substrate analogs (9, 11), anions (7, 9) reduced ubiquinone (12), ATP, ITP and IDP (5, 13, 14), and reduction (15, 16). In all cases, the non-active enzyme is identified with the enzyme-oxalo-acetate complex. The role of the activators is to form stable complexes with the active form of the enzyme, which prevent the enzyme from reacting with oxaloacetate.

The first activating agents and their alledged mechanism were described by Kearney as early as 1957 (11). In two decades much information has been gathered on the process (for review see ref. 18) and a number of reaction schemes were proposed to detail the action of different types of positive effectors namely substrate and substrate analogs, monovalent anions, reduction (18). However no direct evidence was available of the molecular mechanism underlying activation. Gutman proposed a model whereby in succinate dehydrogenase the flavin lies in a crevice of the protein which distorts its planar form (fig. 1) (18). The free oxidized flavin is planar (19, 20) and the reduced coenzyme is bent along the N_{10}-N_5 axis (19). Redox potentials of free FAD and 8α histidyl flavin are respectively -209 mV (21) and -167 mV (16). This potential is too negative to allow reduction by succinate (E'_O=+30 mV (21)) though the flavin of the activé enzyme is reduced by this substrate (1, 17). Bending will destabilize the oxidized form and favour the reduced form, raising the potential to a value allowing interaction with succinate. The deactivating agent, oxaloacetate, widens the jaws around the flavin and allows it to assume the stable planar form of the oxidized state: thereby the flavin recovers a potential close to that of free flavins and cannot be anymore reduced by succinate. In the reduced enzyme the reduced flavin is spontaneously

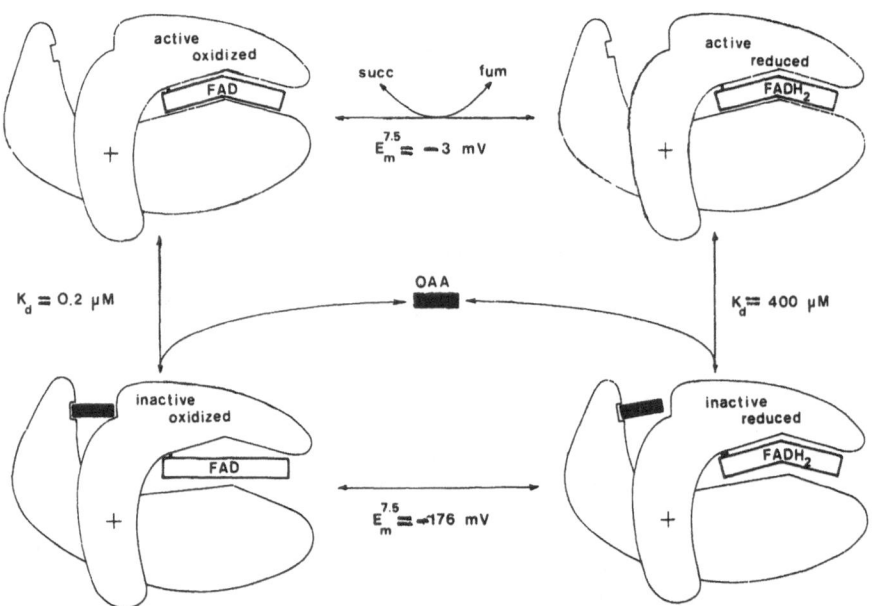

Fig. 1. Modulation of flavin redox potential driven by changes in protein conformation in the transition of succinate dehydrogenase from the activated to the non-activated form. Redrawn after Gutman (18).

bent and does not favour the open jaws conformation of the protein
as the planar oxidized flavin does: this tends to expel oxaloacetate from the regulatory site and explains the decreased affinity
for the effector upon reduction of the de-activated enzyme (18).

For the intraenzymic carrier whose reduction affects the affinity for oxaloacetate, Gutman calculated from kinetic and equilibrium data on particulate preparations (15) $E_{NA} \leqslant$ -190 mV and an
n value close to one (18). In experiments of activation by reductive titrations by Ackrell et al. (16,22) showed that full activation ensues reduction of the flavin while the other redox components of the dehydrogenase have no effect on the process. Reduction
of flavin with activation of the enzyme required two electrons and,
with the estimation these authors made for the dissociation constants of oxaloacetate from the oxidized and the reduced dehydrogenase, the calculated redox potential for the flavin in the active
enzyme came to E' = -90 mV. This potential is too negative to allow
reduction by succinate, the basic event in the catalytic cycle of
the enzyme (17,23).

In collaboration with Gutman and Kroneck we tried to solve these
discrepancies by approaching directly, on the purified reconstitu-
tively active protein, the modifications in molecular parameters
and in the redox properties of the various redox active centers,
which occur in the interaction with oxaloacetate.

Relevant information on the behaviour of the redox active centers
and of the protein was obtained from optical and CD spectra (24).
Addition of oxaloacetate causes flavin oxidation to a level far
more relevant than what produced in the presence of excess fumarate:
this is clear from optical absorbancies and corresponds to the dramatic change in ellipticity at 480 nm seen in fig. 2. As will be
detailed later, this is due to the different redox potential of the
flavin in the enzyme with or without oxaloacetate added.

The dynamics of the absorbance changes following addition of
oxaloacetate are shown in fig. 3: they indicate that flavin oxidation induced by oxaloacetate consist of two reactions one accelerated if the enzyme is already oxidized by fumarate, and a slower
one which proceeds at the same rate in the presence or absence of
fumarate (24). The time course of changes in ellipticity referrable
to flavin confirms a slow linear reaction when the starting preparation was the enzyme reduced by succinate whereas preliminar oxidation by fumarate alters the time curve of the response and acce-
lerates it.

We consider than in the initial phase of oxidation accelerated
by fumarate, flavin oxidation induced by oxaloacetate may use oxidizing power potentially available in the molecule, and actually
present when fumarate is added, which facilitates it. The following
slow phase, and the overall slow process in the absence of fumarate,

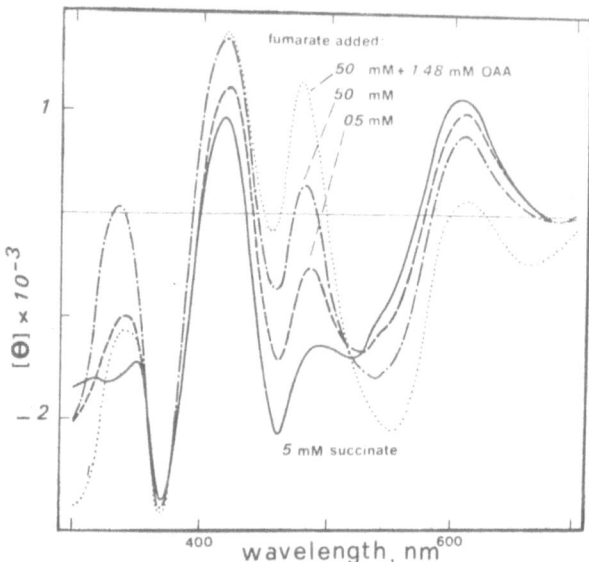

Fig. 2. Circular dichroism spectra of succinate dehydrogenase.
Modifications induced by fumarate and by oxaloacetate.
Solid line, enzyme in succinate. Sequential additions were:
dashes: 6 μl 50 mM fumarate; dashes and dots: 60 μl 0.5 M
fumarate; dots: 10 μl 0.1 M oxaloacetate. Spectra were
recorded immediately after each addition.

correspond to oxidation of the flavin made possible by a change in
redox potential consequent to modification of the protein, as detailed in a later section. The redox modifications of iron sulfur
centers which will be discussed in what follows, strongly support
the above contentions.

The decreased absorbancy at 380 nm after addition of oxaloacetate to the dehydrogenase reduced by succinate or oxidized by fumarate (fig. 4) may indicate reduction of the cubic HiPIP type
center S_3 (25). It may as well indicate disappearance of the "red"
anionic form of flavin semiquinone (26), which was suggested by
Palmer for succinate dehydrogenase (27) and recently documented
on EPR evidence by Ohnishi (28). Presence of the anionic radical
is confirmed by our spectrophotometric data shown in fig. 5 giving
the difference spectra of the enzyme in condition of maximal semiquinone formation vs the same treated with bromide which almost
totally disproportionates the radical (29). No neutral "blue" radical appears upon acidification: however no definite statement
can be made because of the decreased amount of total EPR detectable
semiquinone with decreasing pH (28,30) and of the much smaller
molar absorbancy of the neutral form (31).

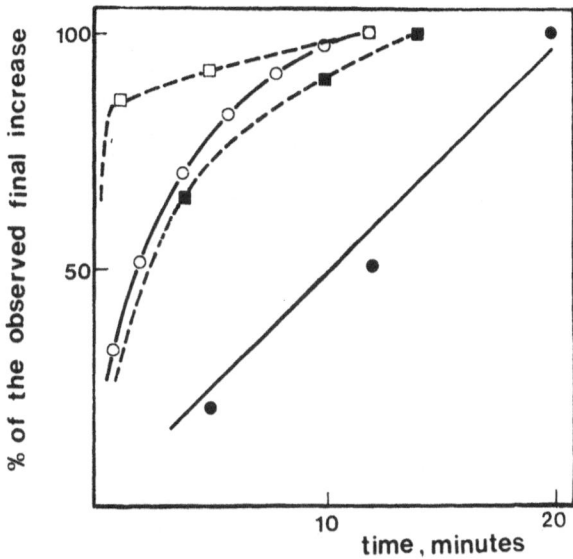

Fig. 3. Time course of oxaloacetate-induced changes in absorbance
(open symbols) and in circular dichroism (full symbols) at
470 nm. Oxaloacetate was added as in fig. 2 to the enzyme
in succinate (solid line) and to the fumarate-containing
samples (dashes).

Other evidence however certainly concerns iron sulfur centers.
The increase in ellipticity at 420 nm upon addition of fumarate
(fig. 2) corresponds to an increase of the oxidized fraction of
center S_1 (32) and conversely the reduced fraction decreases as
indicates the diminished ellipticity at 320-330 nm and at 600 nm.
Center S_2 can be excluded because of its very low potential ($E_o' =$
-400 mV) in the soluble dehydrogenase (33). Addition of oxaloacetate
induced reduction of center S_1. The full effect of the modulator
however cannot probably be appreciated from the spectra shown since
they were recorded immediately after addition of the effector and
the deactivation reaction is slow.

The chromophoric group in flavins, the isoalloxazine ring, is
optically inactive (34) and thus in our case any circular dichroism
must result from environmental perturbations induced by the optically active ribityl side chain or by the protein. Also the circular
dichroism depends on the protein via the aminoacids which participate in the clusters. Thus the ellipticity changes evidenced indicate major modification in the immediate vicinity of the flavin
and of the iron-sulfur centers and interactions occurring between
protein and flavin in the processes studied.

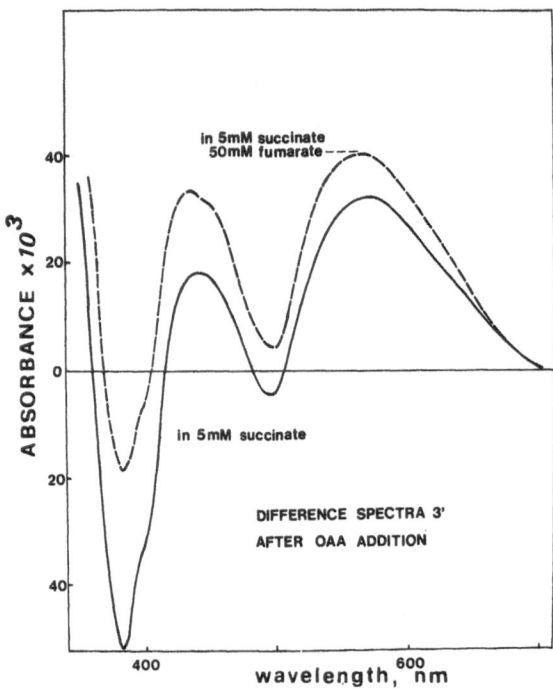

Fig. 4. Effect of oxaloacetate on the absorbance of succinate
       dehydrogenase.
       Difference spectra of succinate dehydrogenase at the DEAE-
       Sephadex purification stage (3.12 mg.ml^{-1}, 10.28 nmol his-
       FAD.mg^{-1}).

A parallel and complementary approach was to study the redox
behaviour of the redox active centers in the dehydrogenase. The
flavin absorbancy at 450 nm was titrated with the succinate-fumarate couple. After correction for the contribution of iron sulfur
centers the mid potential was $E_O^! = -3 \pm 15$ mV with n = 2 (35). The
behaviour of the radical and of binuclear iron-sulfur centers was
established by EPR redox titration and is shown in fig. 6. In active
enzyme the free radical is observed at fairly high redox potentials
reaching a maximum at an applied potential of approx. 60 mV. There
it represents about 50% of total flavin. Less radical is formed
when ferricyanide was added as the oxidant but in the same range
of potentials and with the same shape of the curve. Ferricyanide
is an electron acceptor from the flavin: very likely when it was
used, measurements were made under non-equilibrium conditions and
this explains why the radical was quantitatively less. The bell
shape response of the free radical to the redox potential is compatible with its role as an intermediate between oxidized and fully reduced flavin. This behaviour most probably is further pertur-
bed by intraenzymic interaction with other redox carriers such as
the HiPIP type center S_3 and the S_1 center (33,36). The signal of

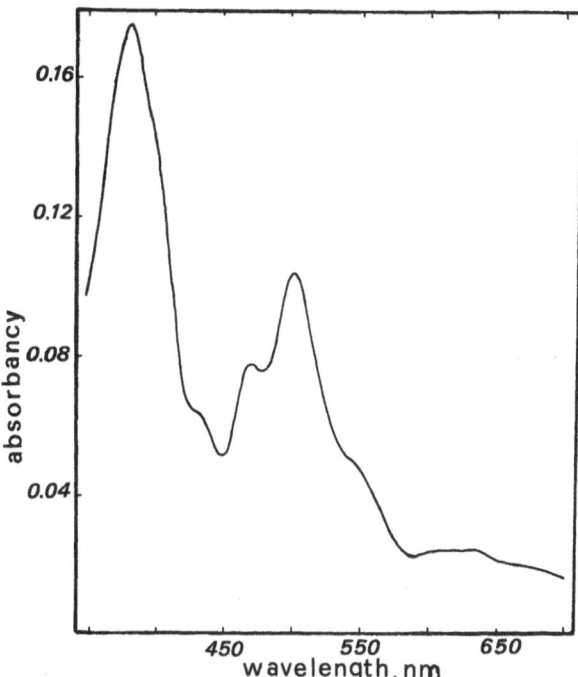

Fig. 5. Semiquinone formation in succinate dehydrogenase. Difference spectrum: Sample: Succinate dehydrogenase (8.12 nmoles
His-FAD/ml) in 50 mM Tris-acetate, 5 mM succinate, 50 mM
fumarate, pH 7.5. Reference: same, without fumarate, 833
mM KBr added.

reduced center S_1 decreases in parallel with formation of the radical from the fully reduced flavin and, as will be discussed later,
this is an indication of the redox interactions occurring in the
enzyme.

Once oxaloacetate is present the dependence of the flavin radical on redox potential is markedly changed. The potential where
it attains maximal concentration is shifted by about -200 mV and
the intensity of signal diminishes. Also in this case changes in
the amount of semiquinone correspond to the redox modifications
of iron sulfur centers. Formation of the radical from the fully
oxidized flavin (0: -110 mV range) is accompanied by a decreased
2Fe2S signal which may be attributed to oxidation of center S_1
($E_o' = 0$ mV (33)). Reduction of the semiquinone in the range from
-140 to -280 mV parallels reduction of binuclear iron sulfur which
probably corresponds to center S_2 ($E_o' = -400$ mV in soluble succinate dehydrogenase (33)).

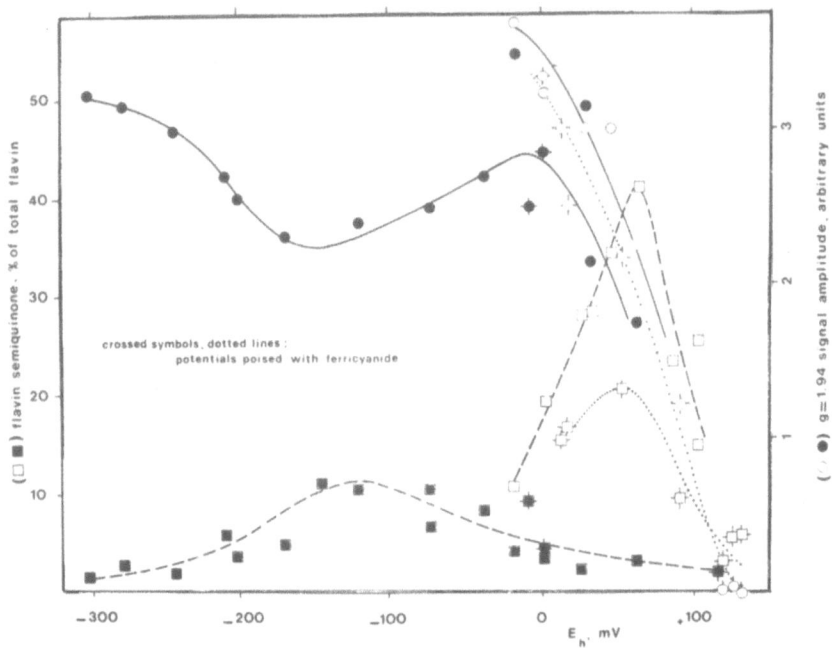

Fig. 6. Effect of the applied potential on the intensities of the
EPR signals of succinate dehydrogenase.
EPR spectra of succinate dehydrogenase were recorded at
liquid nitrogen temperature on a Brucker spectrometer.
The enzyme was at the gel eluate stage in 5 mM succinate,
50 mM Tris-acetate pH 7.5, with (full symbols) and without
(open symbols) 1.66 mM oxaloacetate added.
Potentials were either poised with fumarate, and calculated from the succinate/fumarate ratio, or with dithionite
and ferricyanide, being in this case measured with a combined platinum/silver chloride electrode. Quantitation of
the g = 2.00 signal was done by double integration and
comparison with standards of flavodoxin and glucose oxidase
radicals. EPR measurements were done by Peter Kroneck at
the Facultät Biologie, University of Konstanz, GFR.

The semiquinone forms in the reactions:

(1) $FAD + e^- + H^+$ $\rightleftharpoons$ $FADH^\bullet$

           $FADH_2$ $\rightleftharpoons$ $FADH^\bullet + e^- + H^+$

which taken jointly give the equilibrium:

           $FAD + FADH_2$ $\rightleftharpoons$ $2FADH^\bullet$

This would indicate that the maximal concentration of radical occurs at the midpotential of the two electrons couple, which, as
previously stated, was found to be $E_o' = -3 \pm 15$ mV. However maximal
concentration of semiquinone is experimentally determined at
$E_h = +60$ mV. In an attempt to solve this discrepancy we tried to
figure out on a computer the amounts of fully reduced and fully
oxidized flavin present in the active enzyme at various potentials,
together with the experimentally determined semiquinone, assuming
for the redox transition between FAD and $FADH_2$ the determined
value $E_o' = -3$ mV. The outcoming picture is shown in fig. 7.

     The fully oxidized flavin shows a biphasic increase with a
break around 60% FAD formed, due to accumulation of semiquinone.
The first part of the ascending slope represents fairly well the
$FADH_2 \longrightarrow$ FAD transition with minor disturbance by radical formation. The succinate-fumarate couple used in our titrations does
not allow potentials higher than +60 mV. Therefore the value of
oxidized FAD that in our calculations for E_o' was assumed to represent 100% oxidized flavin, in reality corresponds to only about
60% of the fully oxidized compound. If the correct 50% oxidation
is spotted on the ascending slope, the midpotential value considerably increases. This calculated value still does not attain
the found potential for maximal semiquinone concentration. This
might be due either to interplay between the various redox active
centers in the enzyme or to an effect of the protein; existence of

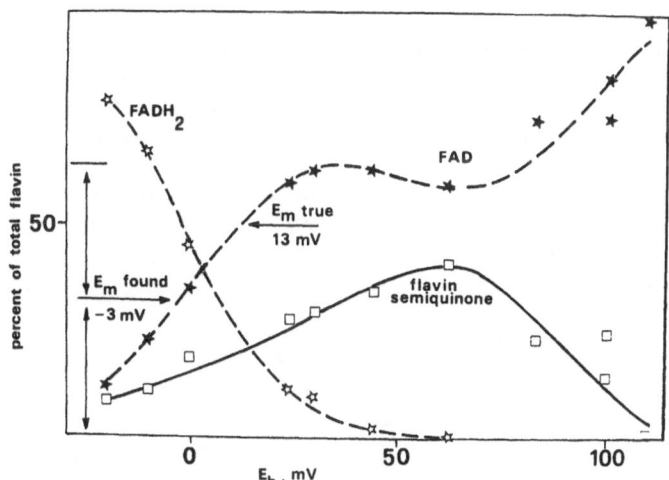

Fig. 7. Computer simulation of the levels of fully reduced and
        fully oxidized flavin accompanying the changes in concen-
        tration of flavin semiquinone determined by EPR.
        The midpoint value obtained by spectrophotometric titration
        (E_m found) (35) is compared with the value (E_m true) indi-
        cated by the calculated amount of oxidized flavin.

a mixed population of molecules can also not be excluded. The
problem is being investigated.

The behaviour shown in the figure also explains why addition
of oxaloacetate further increases flavin absorbance in the enzyme
oxidized by fumarate (24): indeed in the deactivated enzyme the
flavin recovers its midpotential of about -200 mV and this causes
full oxidation of FAD at potentials close to the physiological
ones, with concomitant increase of flavin absorbance.

The results so far discussed allow to describe the probable
physiological behaviour of the redox active groups in the activated enzyme as shown in figure 8, which also gives the measured
potentials for the transitions.In thus becomes clear from the
curves in fig. 6 that if we exclude that the iron sulfur clusters
exchange directly electrons with the succinate-fumarate couple
any transition in the bi-electronic redox equilibrium via these
substrates, modifies the concentration of FAD and FADH$_2$ and affects that of semiquinone through reactions associated with redox
changes of iron sulfur centers. This accounts for both formation
and disappearance of semiquinone. The asymmetry of the semiquinone
peak of fig. 6 suggests that two different iron sulfur centers are
involved. The semiquinone does not appear as an intermediate in
reduction of the flavin by substrate thus substantiating suggestions
by previous authors (36).

By analysing the experiments of Gutman and Silman (15,37) it
was clear (35) that responsible for reductive activation is a
redox couple with calculated potential +6 mV in the activated
enzyme and -196 mV after deactivation, and n = 1: this is likely
the flavin semiquinone in equilibrium with the fully reduced flavin. The calculated potentials closely fit with the values given
in fig. 6 by the left side inflections of the curves representing
the semiquinone measured before and after addition of oxaloacetate.

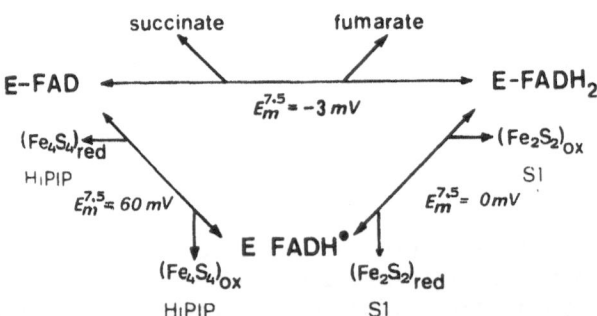

Fig. 8. Intramolecular redox reactions in succinate dehydrogenase.

The scheme proposed for reductive activation (18) can thus be modified as follows:

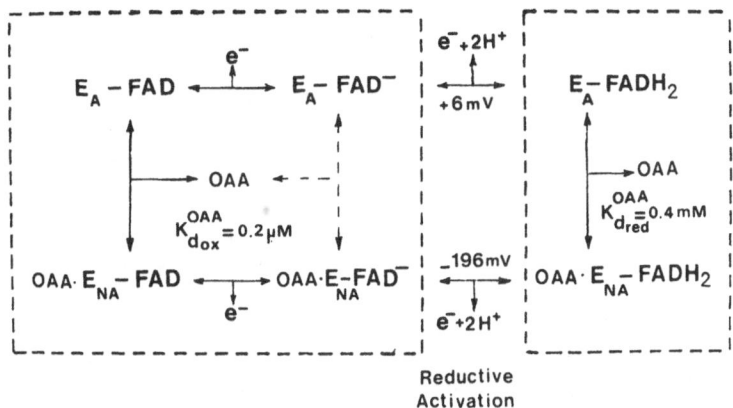

Reductive
Activation

The lowered redox potential of the flavin in the deactivated
enzyme favours its oxidation by the iron-sulfur centers, and
indeed this is the case as the results showed.

The level of activation of succinate dehydrogenase can thus be
described as a mixture of two populations: one is the active enzyme, with its histidyl flavin at a high redox potential, the
other one is the non active complex with oxaloacetate. The redox
potential of the flavin in this complex is very low, comparable
with that of free flavins (16,21).

The precise mechanism by which the protein shifts the redox
potential of the flavin in the activated dehydrogenase is still
obscure. However oxaloacetate clearly acts at this level. Our
data are in good agreement with the mentioned model of Gutman
(fig. 1) advocating distortion of the flavin in succinate dehydrogenase from its oxidized (19) or semiquinone (20) planar forms
to the bended configuration of reduced flavin (19). Theoretical
calculations of the stability of oxidized and reduced flavin
(Lindner, D.L., Branchoud, D., Doxin, B. and Lipscom, N.W., unpublished results) indicated that the energy associated with
distortion of the oxidized flavin is compatible with the measured
200 mV shift or the redox potential.

Another mechanism which accounts for the one-electron activation
and the 200 mV shift of the redox potential is based on the effect
of positive charges (either H^+ or a positively charged side chain
of the peptide structure) with region 1-2 α or N5 of the flavin
(see fig. 9).

A positive charge near the 1-2α position favours a two electron redox reaction with a high redox potential (-9 mV as measured with model compound (38)).Thus if we admit such interaction between the apo-protein and flavin in the active enzyme, we account both for the high redox potential and the two electron oxidation of succinate to fumarate. Moreover the situation mentioned will stabilize the anionic form of the flavin radical (39). The close proximity of the positive charge to the flavin can also account for the change in chirality associated with activation as measured by our CD studies (24). This state of the enzyme is the favoured conformation in the absence of oxaloacetate.

The non active enzyme is a configuration where the positive charge interacting with 1-2α position is lost. This condition is caused and stabilized by formation of a very tight complex (K_d approx. 2.10^{-7} M) (15,18,37) with oxaloacetate. Bending of the flavin, consequent to the reduction of the semiquinone, may induce a modification of the protein leading to displacement of oxaloacetate. This will remove the constraint and the enzyme will assume its active configuration, a rather slow transition characterized by a high energy of activation.

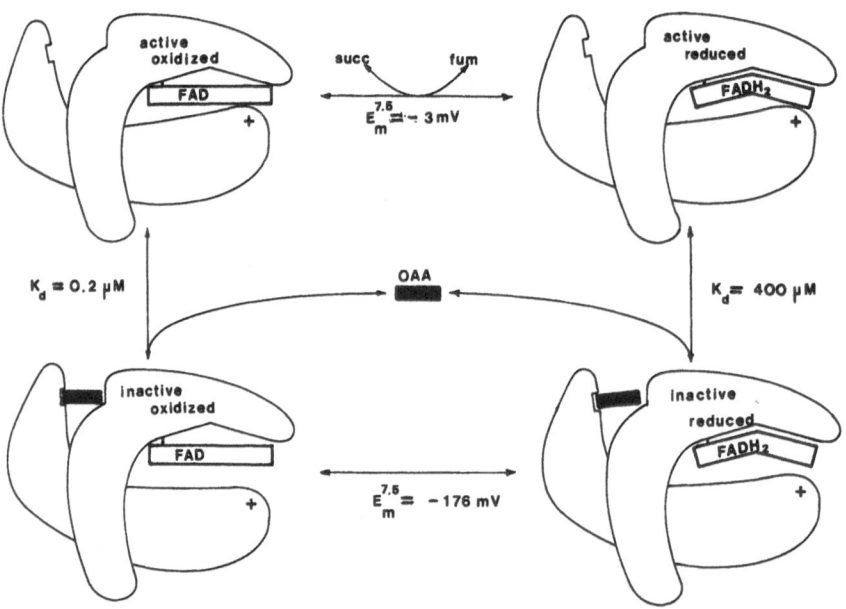

Fig. 9. Modulation of flavin redox potential in succinate dehydrogenase by a positive redidue near the 1-2α position of the isoalloxazine ring inducing the active-non active transition of the enzyme.

REFERENCES

1. K.A. Davis and Y. Hatefi, Succinate Dehydrogenase I Purification, Molecular Properties and Substructure, Biochemistry 10:2509 (1971).

2. P.G. Righetti and P. Cerletti, Molecular Parameters of the Beef Heart Succinate Dehydrogenase, FEBS Lett. 13:181 (1971).

3. W.H. Walker, T.P. Singer, S. Ghisla and P. Hemmerich, Studies on Succinate Dehydrogenase: 8α-Hystidyl-FAD as the Active Center of Succinate Dehydrogenase, Eur. J. Biochem. 26:279 (1972).

4. T. Ohnishi, Mitochondrial Iron-Sulfur Flavodehydrogenase, in "Membrane Proteins in Energy Transduction", R.A. Capaldi, ed., Dekker Inc., New York, p. 1-80 (1979).

5. M. Gutman, E.B. Kearney and T.P. Singer, Control of Succinate Dehydrogenase in Mitochondria, Biochemistry 10:4763 (1971).

6. T.P. Singer, M. Gutman and E.B. Kearney, On the Need for Regulation of Succinate Dehydrogenase, FEBS Lett. 17:11 (1971).

7. B.A.C. Ackrell, E.B. Kearney and M. Mayer, Role of Oxaloacetate in the Regulation of Mammalian Succinate Dehydrogenase, J. Biol. Chem. 249:2021 (1974).

8. M. Gutman, The Effect of Opposing Effectors on Activation Level of Succinate Dehydrogenase: Equilibrium and Kinetic Studies, Biochemistry 15:1342 (1976).

9. M. Gutman, Regulation of Mitochondrial Succinate Dehydrogenase by Substrate Type Activators, Biochemistry 16:3067 (1977).

10. L. Wojtczak, A.B. Wojtczak and L. Ernster, The Inhibition of Succinate Dehydrogenase by Oxaloacetate, Biochim. Biophys. Acta 191:10 (1969).

11. E.B. Kearney, Studies on Succinic Dehydrogenase: IV. Activation of the Beef Heart Enzyme, J. Biol. Chem. 229:363 (1957).

12. M. Gutman, E.B. Kearney and T.P. Singer, Regulation of Succinate Dehydrogenase Activity by Reduced Coenzyme Q_{10}, Biochemistry 10:2726 (1971).

13. C. Gregolin and P. Scalella, Activation of the Oxidation of Succinate by Adenosine Triphosphate in Respiratory Particles of Yeast, Biochim. Biophys. Acta 99:185 (1965).

14. P. Cerletti and A. Manzocchi, Regulation of Succinate Dehydrogenase, Acta Vitam. Enzymol. 1-4:5 (1973).

15. M. Gutman and N. Silman, The Steady State Activity of Succinate Dehydrogenase in the Presence of Opposing Effectors. Reductive Activation of Succinate Dehydrogenase in Presence of Oxaloacetate, Mol. Cell. Biochem. 7:177 (1975).

16. B.A.C. Ackrell, E.B. Kearney and D.E. Edmondson, Mechanism of Reductive Activation of Succinate Dehydrogenase, J. Biol.Chem. 250:7114 (1975).

17. E.B. Kearney, T.P. Singer and N. Zastrow, On the Requirement of Succinic Dehydrogenase for Inorganic Phosphate, Arch. Biochem. Biophys. 55:580 (1955).

18. M. Gutman, Modulation of Mitochondrial Succinate Dehydrogenase Activity; Mechanism and Function, Mol.Cell.Biochem.20:41 (1978).

19. P. Kierkegaard, P. Norrestam, P.E. Werner, I. Csoregh, M. Von-Glehn, R. Karlsson, M. Leijonmark, O. Rönquist, B. Stensland, O. Tillberg and L. Torbjornsson, X-Ray Structure Investigation of Flavin Derivatives, in: "Flavins and Flavoproteins", H. Kamin, ed., pp. 1-22, University Park Press, Baltimore (1971).

20. M. Von-Glehn, B. Stensland and P. Kierkegaard, Crystal and Molecular Structures of Two Models for Flavoprotein Inhibitor Complexes and a 5-Thiaflavin in the Radical State, in: "Flavins and Flavoproteins", K. Yagi, Y. Yamano, eds., pp. 37-44, Japan Scientific Societies Press, Tokyo (1979).

21. W.M. Clark, in: "Oxidation-Reduction Potential of Organic Systems" Williams and Wilkins, Baltimore (1960).

22. B.A.C. Ackrell, E.B. Kearney and D. Edmondson, Role of Flavin in Reductive Activation of Succinate Dehydrogenase, in: "Flavins and Flavoproteins", T.P. Singer, ed., p. 522, Elsevier, Amsterdam (1976).

23. D.V. Devartanian and C. Veeger, Studies on Succinate Dehydrogenase 1. Spectral Properties of the Purified Enzyme and Formation of Enzyme-Competitive Inhibitor Complexes, Biochim. Biophys. Acta 92:233 (1964).

24. M. Gutman, F. Bonomi, S. Pagani and P. Cerletti, The Circular Dichroism and Optical Absorbancy of the Histidyl Flavin During Active-Non Active Transition of Soluble Succinate Dehydrogenase, FEBS Lett. 104:371 (1979).

25. F. Bonomi, S. Pagani and P. Cerletti, Enzymic Restoring of the Iron-Sulfur Structure of Succinate Dehydrogenase, in: "Flavins and Flavoproteins", K. Yagi and T. Yamano, eds., Japan Scientific Societies Press, Tokyo and University Park Press, Baltimore pp. 227 (1979).

26. V. Massey and G. Palmer, On the Existence of Spectrally Distinct Classes of Flavoprotein Semiquinones. A New Method for the Quantitative Production of Flavoprotein Semiquinones, Biochemistry 5:3181 (1966).

27. G. Palmer, F. Muller and V. Massey, Electron Paramagnetic Resonance Studies on Flavoprotein Radicals, in: "Flavins and Flavoproteins", H. Kamin, ed., University Park Press, Baltimore p. 123 (1971).

28. T. Ohnishi, T.E. King, J.C. Salerno, H. Blum, J.R. Bowjer and T. Maida, Thermodynamic and Electron Paramagnetic Resonance Characterization of Flavin in Succinate Dehydrogenase, J.Biol. Chem. 256:5578 (1981).

29. F. Bonomi, S. Pagani and P. Cerletti, Regulation of Succinate Dehydrogenase Activity by Monovalent Inorganic Anions: Kinetic and Molecular Studies, in: "Flavins and Flavoproteins", V. Massey and C.H. Williams, jr., eds., Elsevier North Holland N.Y., in press.

30. F. Bonomi, S. Pagani and P. Cerletti, Studi Spettroscopici sulla Dipendenza dal pH degli Stati Redox della Flavina Covalentemente legata nella Succinato Deidrogenasi, in Presenza di Vari

Effettori, Atti del XXVII Congresso Nazionale SIB, Parma (1981).

31. V. Massey, R.G. Matthews, G.P. Foust, L.G. Howell, C.H. Williams, G. Zanetti and S. Ronchi, A New Intermediate in TPNH-linked Flavoproteins, in: "Pyridine Nucleotide-Dependent Dehydrogenase", H. Sund, ed., p. 393, Springer Verlag, Berlin (1969).

32. G. Palmer, H. Britzinger and R.W. Estabrook, Spectroscopic Studies on Spinach Ferredoxin and Adrenodoxin, Biochemistry 6:1658 (1967).

33. T. Ohnishi, J.C. Salerno, D.B. Winter, J. Lim, C.A. Yu, L. Yu and T.E. King, Thermodynamic and EPR Characteristics of two Ferredoxin-Type Iron-Sulfur Centers in the Succinate-Ubiquinone Reductase Segment of the Respiratory Chain, J. Biol. Chem. 251:2094 (1976).

34. G. Tollin, Magnetic Circular Dichroism and Circular Dichroism of Riboflavin and its Analogs, Biochemistry 7:1720 (1968).

35. M. Gutman, F. Bonomi, S. Pagani, P. Cerletti and P. Kroneck, Modulation of the Flavin Redox Potential as Mode of Regulation of Succinate Dehydrogenase, Biochim. Biophys. Acta 591:400 (1980).

36. H. Beinert, A.C. Ackrell, E.B. Kearney and T.P. Singer, Iron-Sulfur Components of Succinate Dehydrogenase Stochiometry and Kinetic behaviour in Activated Preparations, Eur. J. Biochem. 54:185 (1975).

37. M. Gutman and N. Silman, Reductive Activation of Succinate Dehydrogenase: Equilibrium and Kinetics Studies, in: "Flavins and Flavoproteins", T.P. Singer, ed., p. 537, Elsevier, Amsterdam (1976).

38. P. Hemmerich, in: "Transport by Protein", G.Bauer and H. Saund, eds., p.123, Walter de Gruyter, Berlin (1978).

39. V. Massey and P. Hemmerich, Active-Site Probes of Flavoproteins, Bioch. Soc. Trans. 8:246 (1980).

ACKNOWLEDGEMENTS

The research of our group reported in the present review was in part supported by grants of the Italian National Research Council (C.N.R.).

SULFUR METABOLISM

BIOCHEMICAL FUNCTIONS OF PERSULFIDES

John L. Wood

Department of Biochemistry
University of Tennessee Center for the Health Sciences
Memphis, TN 38163

NATURE OF PERSULFIDES

In biological systems, persulfides (RSS⁻) are reactive, unstable intermediates in the transformations of sulfur compounds. Persulfides are also referred to as polysulfides, or alkyl/aryl hydrogen disulfides. The persulfide sulfur is commonly termed, "sulfane", to designate sulfur covalently bonded only to other sulfur atoms. The scission of the sulfur-sulfur bond is enhanced by the joint action of electrophilic and nucleophilic groups[1].

Persulfides are formed in a variety of ways. Elemental sulfur and mercaptans combine in a reversible reaction.

$$RS^- + S° \rightleftharpoons RSS^- \qquad [1]$$

The reaction to the right reflects the action of mercaptans in dissolving sulfur from its suspension in water; the reverse results from the intrinsic instability of persulfides and the insolubility of free sulfur.

Hylin and Wood[2] produced persulfides by incubating flowers of sulfur with solutions of mercaptans (Table 1). If an appreciable excess of mercaptan was used, the sulfane sulfur was reduced to sulfide.

$$RSS^- + RS^- \rightleftharpoons RSSR + S^{2-} \qquad [2]$$

Table 2 shows some properties of persulfides which are useful in their detection. Cyanolysis yields thiocyanate ion which is readily detectable with ferric ion. Alkylation with iodoacetate produces a mixed disulfide[3].

$$RSS^- + ICH_2COO^- \longrightarrow RSSCH_2COO^- + I^- \qquad [3]$$

Persulfides transfer sulfur to a variety of anions.

$$RSS^- + CN^- \longrightarrow RS^- + SCN^- \qquad [4]$$

$$RSS^- + SO_3^{2-} \longrightarrow RS^- + S_2O_3^{2-} \qquad [5]$$

Reaction 4 represents cyanolysis. Where R represents an enzyme, the reactions are catalytic.

PERSULFIDES IN BIOLOGICAL SYSTEMS

Table 3 shows some persulfides that are found in biological systems. Thiocysteine has been detected by Cavallini et al.[4] as the initial product of the action of cystathionase (EC 4.4.1.1) on cystine.

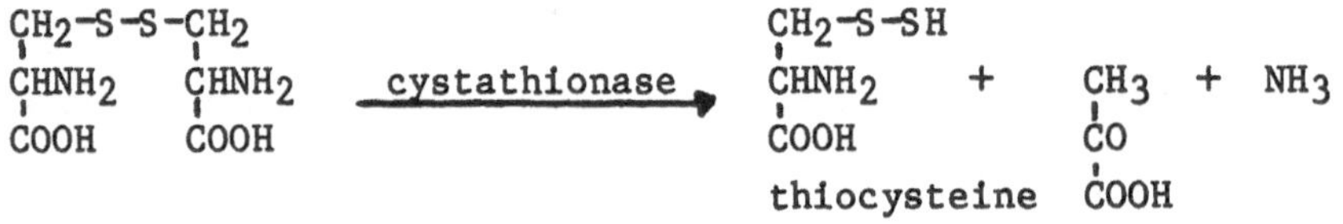

$$
\begin{array}{l}
CH_2-S-S-CH_2 \\
CHNH_2 \quad CHNH_2 \\
COOH \quad COOH
\end{array}
\xrightarrow{\text{cystathionase}}
\begin{array}{l}
CH_2-S-SH \\
CHNH_2 \quad + \quad CH_3 \quad + \quad NH_3 \\
COOH \qquad CO \qquad\qquad [6] \\
\text{thiocysteine} \quad COOH
\end{array}
$$

Table 1. Formation of Persulfides

Precursor	Sulfane Sulfur μmole/mmole
Sodium sulfide	12.8
Mercaptoethanol	9.2
Mercaptopyruvate	7.3
Cysteine	14.7

Flowers of sulfur (6 mg) and 0.1 M solutions (10 ml) of precursor in 0.1 M buffer, pH 9.1, were incubated at 37°C for 15 min. Sulfane sulfur was determined as thiocyanate after cyanolysis.

Table 2. Some Properties of Persulfides

Cyanolyse	Deposit Sulfur
Transfer Sulfur	Absorb at 330–350 nm
Alkylate	Reduce to Sulfide Ion

Table 3. Persulfides

Thiocysteine, CySSH	Thiocystine*, CySSSCy
Thiosulfate, $^-O_3SS^-$	Mercaptopyruvate*, $HSCH_2COCOOH$
Protein-SSH	Elemental Sulfur*, S_8
Thiotaurine, $NH_2CH_2CH_2SO_2SH$	

*denotes latent persulfides

Flavin[3], working with cystathionase from *Neurospora*, confirmed the structure by alkylation of the product with iodoacetate to form the mixed disulfide of cysteine and thioglycolate (Reaction 3). Cystamine similarly forms thiocysteamine when subjected to a di-amine oxidase[5].

Thiosulfate in acid solution decomposes to free sulfur and sulfite ion. Cyanolysis is catalyzed by enzymes or copper ion[6].

Alkyl thiosulfonates, exemplified by thiotaurine and a number of aryl thiosulfonates, are derivatives of thiosulfate that undergo cyanolysis and transfer the sulfane sulfur to suitable acceptors. The transfer is enhanced by catalysis by rhodanese (thiosulfate: cyanide sulfurtransferase, EC 2.8.1.1) [7,8].

Protein Persulfides

A number of proteins have been shown to form persulfides. Sörbo [9] and Schneider and Westley [10] found that colloidal sulfur, when complexed with serum albumin, was cyanolyzable. In present day terms, the complex evidently contained persulfide sulfur that formed on the protein sulfhydryl group according to Reaction 1.

The formation of persulfides on proteins by action of alkali was formulated by Tarbell and Harnish [11] as follows:

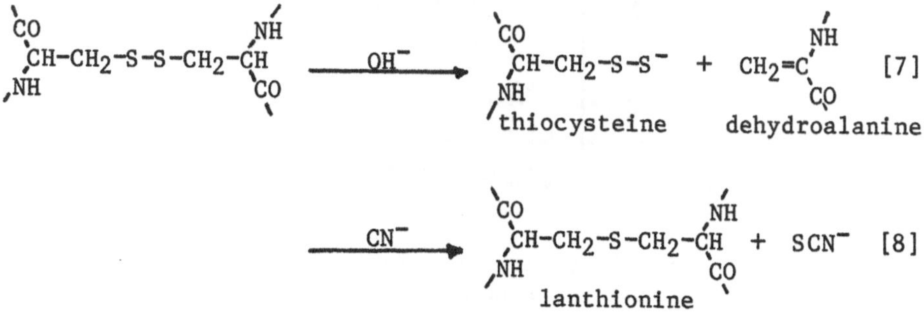

$$\underset{\text{NH}}{\overset{\text{CO}}{CH}}-CH_2-S-S-CH_2-\underset{\text{CO}}{\overset{\text{NH}}{CH}} \quad \xrightarrow{\ OH^-\ } \quad \underset{\text{NH}}{\overset{\text{CO}}{CH}}-CH_2-S-S^- \ + \ CH_2=\underset{\text{CO}}{\overset{\text{NH}}{C}} \quad [7]$$

thiocysteine   dehydroalanine

$$\xrightarrow{\ CN^-\ } \quad \underset{\text{NH}}{\overset{\text{CO}}{CH}}-CH_2-S-CH_2-\underset{\text{CO}}{\overset{\text{NH}}{CH}} \ + \ SCN^- \quad [8]$$

lanthionine

Catsimpolas and Wood[12] incubated bovine serum albumin at pH
10 for a day to produce albumin persulfide. Fig. 1 shows that
the product was rapidly cyanolyzed indicating that the persulfide had
formed thiocyanate according to Reaction 4. Some disulfide groups
had been cyanolyzed as well to form thiocyanate and lanthionine[13].

Cavallini et al.[14] obtained spectroscopic evidence for the
formation of thiocysteine persulfide groups on insulin which had been
treated with 0.025 N NaOH. The persulfide group could be cyanolyzed,
transferred to hypotaurin and alkylated with iodoacetate. Similarly, Schneider and Westley[10] used oxidized glutathione as a
model compound to show persulfide formation from the action of
0.5 N NaOH on peptide-bonded cystine.

The reverse of Reaction 2 was used by Cavallini et al.[15]
to form persulfide groups on a number of proteins that contain disulfide linkages. Bovine serum albumin, ribonuclease, casein, insulin,
ovalbumin, and chymotrypsinogen were treated with sodium sulfide for
30-40 hours. Persulfide groups resulting were detected by cyanolysis, development of an absorbance at 335 nm and transfer of sulfane
sulfur to hypotaurine to form thiotaurine. The protein persulfides
were unstable and decomposed during attempts to isolate them.

Rhodanese is an example of a protein with a relatively stable
persulfide group. Sörbo[16] predicted the persulfide nature of the
sulfane sulfur on the enzyme and Finazzi Agrò et al.[17] later demonstrated its presence by its characteristic absorption at 335 nm
and also by its inhibition of protein fluorescence. Volini and
Wang[18] had observed a persulfide property of rhodanese when they
found it slowly deposited free sulfur from solution. Ploegman et
al.[19] identified the persulfide group on rhodanese on electron density maps at 2.5 Å resolution. Evidently hydrogen bonds stabilize
this protein persulfide.

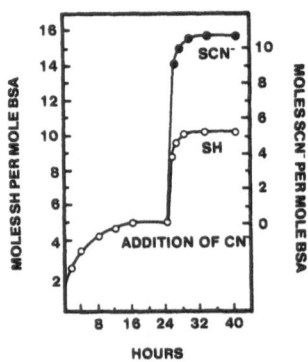

Fig. 1.   Formation of thiocyanate by action of cyanide on alkali-
          treated bovine serum albumin. BSA was incubated at pH 10
          under N_2 for 24 hours at 37°C. o--thiol groups, o--SCN⁻
          formed.

Latent Persulfides

Compounds termed "latent" ordinarily show some of the properties of a persulfide. Mercaptopyruvate is classed as a latent persulfide because it transfers its sulfur to mercaptopyruvate sulfurtransferase (EC 2.8.1.2). The enzyme evidently forms a persulfide because it readily transfers a sulfane sulfur to cyanide to form thiocyanate,[20] or to sulfite to form thiosulfate.[9] The enzyme persulfide is unstable; in the absence of an acceptor, free sulfide is unstable; in the absence of an acceptor, free sulfur precipitates.[21] Hylin et al.,[20] found that once sulfur had precipitated from the enzyme it could not be transferred to cyanide or sulfite. Investigations of Vachek and Wood[22] and recent kinetic studies by Jarabak and Westley[23] support the concept of intermediate enzyme persulfide formation.

$$HS-CH_2-\overset{O}{\overset{\|}{C}}-COOH \ + \ EnzSH \longrightarrow EnzSSH \ + \ CH_3COCOOH \quad [9]$$

$$\text{mercaptopyruvate} \qquad EnzSH \ + \ S°$$

Elemental sulfur cannot be cyanolyzed in aqueous media. If dissolved in mercaptan solutions as noted above, the cyclic sulfur polymer is opened and the persulfide readily undergoes cyanolysis (Reaction 4).

Thiocystine is stable in acid and neutral media.[24] Fletcher and Robson[25] found it was formed from proteins boiled in 5.7 N HCl for 40-100 hours. Szczepkowski and Wood[24] prepared thiocystine by the decomposition of cystine by rat liver cystathionase.

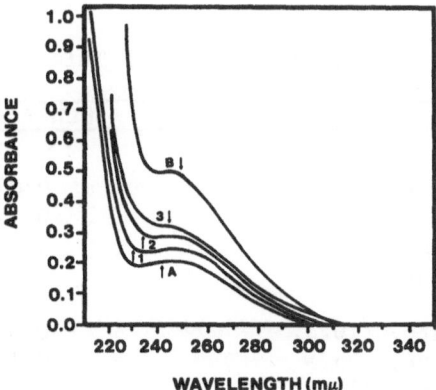

Fig. 2.   Formation of thiocystine from cystine. Cystine was treated with cystathionase at pH 7.4. Recordings were made. Curve 1, 7 min.; Curve 2, 17 min.; Curve 3, 27 min. Curve A, cystine; Curve B, thiocystine.

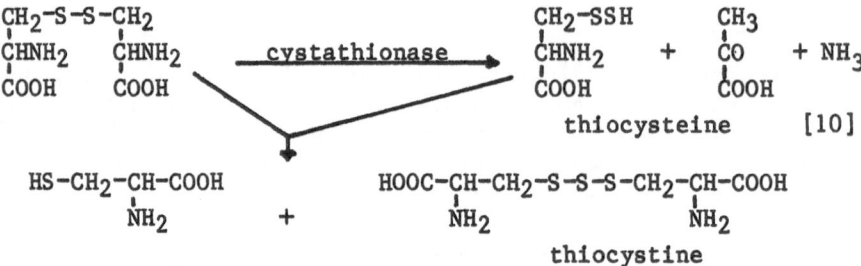

Fig. 2 shows spectral evidence for the formation of thiocystine over a period of 29 min. Thiocystine was isolated from such incubations by isotopic dilution. Sandy et al.[26] isolated thiocystine as a product of incubation of cystine with cystathionase from *Rhodopseudomonas spheroides*. They also isolated the corresponding mixed trisulfide between glutathione and cysteine. Massey et al.[27] detected the presence of the trisulfide corresponding to glutathione in commercial samples of the compound.

Thiocystine is a good substrate for production of rhodanese persulfide.[24] As seen in Table 4, its activity is several times better than thiosulfate. This was confirmed by Sandy et al.[26] with rhodanese from *R. spheroides*. Table 5 shows that the structural requirements for substrate activity by persulfides is fairly specific at pH 7.4, requiring both amino and carboxy groups.[28] At pH 8.6, the optimum for rhodanese, all the trisulfides showed some activity. This reflects the tendency of trisulfides to break down to persulfides in alkaline solution.

Table 4. Substrate Activity of Thiocystine

Substrate		SCN^- formed
	μM	$\mu mol*$
Thiocystine,	15.5	0.04
Thiocystine,	11.	0.18
Thiocystine,	16.5	0.29
Thiocystine,	16.5†	0.03
Thiosulfate,	16.5	0.03
Thiosulfate,	55.5	0.14
Thiosulfate,	111.	0.25

*Produced in 3.6 ml from 1.6 units rhodanese and substrate as shown, incubated 5 min at 25°C, pH 7.4
†Enzyme omitted

Table 5. Substrates for Rhodanese

Trisulfide	pH 7.4	pH 8.6
	μmol SCN⁻/min	
Thiocystine, $(COOH-CHNH_2-CH_2-S)_2S$	0.77	1.38
Homothiocystine, $(COOH-CHNH_2-CH_2-CH_2-S)_2S$	0.16	0.896
Diacetylthiocystamine, $(CH_3CO-NH-CH_2-CH_2-S)_2S$	0.096	2.68
Trithiodipropionic acid, $(COOH-CH_2-CH_2-S)_2S$	0.0	0.096
Trithiodiacetic acid, $(COOH-CH_2-S)_2S$	0.024	0.096
Ethyltrisulfide, $(CH_3CH_2-S)_2S$	0.0	————

For assay: substrates, 1.1 mM; rhodanese, 10 units

PERSULFIDES FROM TRISULFIDES

Data by Wood and coworkers indicate that thiocystine transfers its sulfane sulfur to rhodanese by virtue of its decomposition to thiocysteine. Szczepkowski and Wood[24] found thiocystine was fairly stable in pure solution but rapidly precipitated sulfur in the presence of sulfhydryl compounds. Fig. 3 shows the decomposition of thiocystine in the presence of mercaptosuccinate.[29] A distinct haze of precipitated sulfur developed within four minutes and rapidly thickened. The absorption fell off as particulate sulfur formed and settled. The reaction is reversible so that the rate of decomposition was reduced in the presence of disulfides. Oxidized glutathione added to the decomposition mixture was not as effective as lipoic acid.

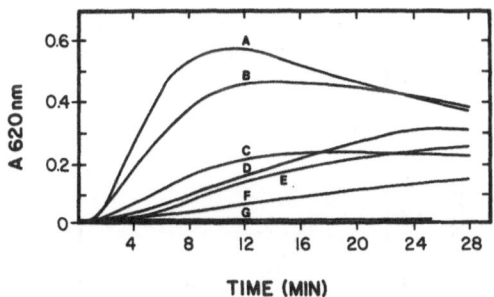

TIME (MIN)

Fig. 3. Effect of disulfide on decomposition of thiocystine by mercaptosuccinate.
A: 4.4 mM thiocystine and 0.1 mM mercaptosuccinate at pH 7.4.
B: A + 4.4 mM GSSG.
C: A + 4.4 mM lipoic acid.
D: thiocystine and mercaptosuccinate at pH 8.6.
E: D + 4.4 mM GSSG.
F: D + 4.4 mM lipoic acid.
G: 4.4 mM thiocystine at pH 7.4.

$$\text{CySSSCy} + \text{RS}^- \rightleftharpoons \text{CySS}^- + \text{RSSCy}$$

$$\text{CySS}^- \rightleftharpoons \text{CyS}^- + \text{S}° \qquad\qquad [11]$$

$$\text{CyS}^- + \text{RSSCy} \rightleftharpoons \text{CySSCy} + \text{RS}^-$$

Because of the regeneration of the sulfhydryl compound, the process is catalytic.

Further evidence for formation of thiocysteine from thiocystine is shown in Fig. 4. When thiocystine was incubated with catalytic amounts of cysteine for 15 minutes at pH 8.6, an increase in absorbance at 340 nm was observed; this slowly fell over 25 hours indicating the decomposition of thiocysteine. After 15 minutes, cyanide was added and the absorption was reduced due to transfer of the sulfane sulfur to cyanide. Conversely, when iodoacetate was added instead of cyanide the mixed disulfide between cysteine and thioglycolic acid was isolated.

As shown in Fig. 5, rhodanese acts as a catalyst to decompose thiocystine.[28] The fluorescence spectrum of rhodanese was determined by the procedure of Finazzi Agrò et al.[17] Persulfide quenching of the emission by the persulfide group (Curve A) was removed by cyanolysis (Curve B). The persulfide form was regenerated by adding an excess of thiocystine. This restored the inhibition of the fluorescence (Curve C). Additional cyanide removed the persulfide sulfur and most of its quenching effect. In further confirmation of the above, thiocystine labeled with ^{35}S in the sulfane sulfur was used

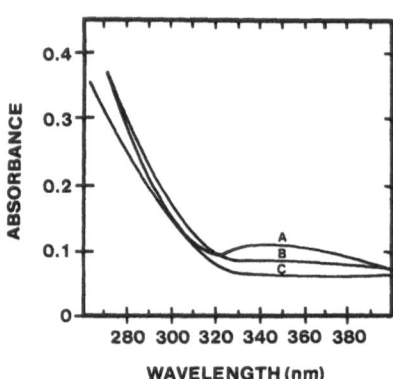

Fig. 4.   Conversion of thiocysteine to thiocystine.
          A: 0.3 mM thiocystine and 0.5 mM cysteine, pH 8.6 after
          15 min.
          B: A after 2.5 hours.
          C: 0.3 mM thiocystine.
          Curve C was also obtained when A was made 0.3 mM in cyanide.

to convert rhodanese to the persulfide form. The enzyme, purified
from excess substrate and recrysta lized, contained radioactive
sulfur. The radioactivity was entirely transferred to thiocyanate
by addition of cyanide. Conversely, when thiocystine labeled in the
two cysteine sulfurs was used, no radioactivity was transferred to
the enzyme nor was the thiocyanate produced by adding cyanide,
radioactive.

PERSULFIDES IN BIOLOGICAL SYSTEMS

The instability of the persulfide group puts limits on proposi-
tions as to its functions in biological systems. Latent persulfides
are much more stable.

The identification of thiocystine as a cystine-derived persulfide substrate, the derivation of mercaptopyruvate from cysteine and
the relative stability of rhodanese persulfide suggests that persulfides may have a central role in biochemical functions.

Although rhodanese was first identified by Lang[30] as a thiosulfate: cyanide sulfurtransferase, many investigators question the
premise that so much enzyme is present in cells solely to detoxify
cyanide. Sörbo[31] has noted, however, that the amount of an enzyme
in the cell is not particularly relevant to its functions. Also,
minor amounts of cyanide arise in vivo from ingested cyanogenic
glycosides[32] and from oxidation of thiocyanate.[33] Schievelbein et
al.[34] suggested that rhodanese is a relic from primordial times

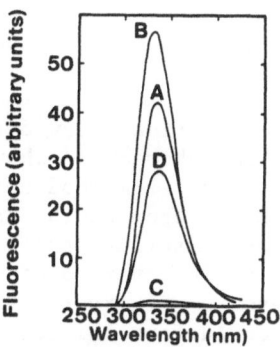

Fig. 5.   Effect of cyanide and thiocystine on rhodanese fluorescence.
          A: 5 μM rhodanese (Sulfur-containing form).
          B: 50 μM KCN added to A.
          C: 3.7 μmoles solid thiocystine added to B.
          D: 8 μmoles KCN added to C.
          Excitation at 278 nm.

when HCN was present in the atmosphere in appreciable quantities. Sörbo [35] reported that cytochrome oxidase which had been poisoned by cyanide could be reactivated by rhodanese persulfide generated from thiosulfate and Auriga and Koj[36] showed isolated mammalian mitochondria could be protected from cyanide inhibition of respiration by thiosulfate. This provided an attractive hypothesis, namely, that rhodanese is located in the mitochondrion specifically to protect the electron transport system from cyanide.

There is no known natural substrate for rhodanese that penetrates the mitochondrion as rapidly as cyanide. Excepting for thiosulfate, previously known substrates for rhodanese, i.e., arylthiosulfonates[7] and disulfide ion,[37] are compounds unnatural to mammalian systems. Recently, thiocystine, a natural substrate for rhodanese was tested;[38] it protected rats against 3-4 times the LD_{50} but the effective concentration in body fluids that could be produced by parenteral administration was limited by the relative insolubility of the compound. Mercaptopyruvate sulfurtransferase has been found in the cytosol as well as in the mitochondria.[39] Clemedson et al.[40] were able to achieve some prophylaxis with large doses of the enzyme plus mercaptopyruvate administered i.v. but they achieved little antidotal effect.

The reduction of persulfides by sulfhydryl compounds in vitro produces sulfide ion (Reaction 2). Koj and Frendo[41] and Villarejo and Westley[42] have assigned a thiosulfate reductase function to rhodanese when an acceptor such as dihydrolipoate or other disulfhydryl adjacent to the persulfide rapidly reacts to form lipoate and sulfide. However, the intermediate dihydrolipoate persulfide is active in transferring sulfide sulfur to other thiophilic acceptors and thus obviates the intracellular release of sulfide ion. It seems probable that the enzyme-dihydrolipoate persulfide complex rather than sulfide is the substrate for oxidation to sulfate in mammalian systems.[37,43]

Another function for persulfides provides a pathway for the synthesis of thiosulfate. Szczepkowski [37] showed sulfite and sulfinates will accept sulfane sulfur from rhodanese to form thiosulfate and thiosulfinates and Sörbo[35] reported the same type of thiosulfate synthesis from mercaptopyruvate sulfurtransferase persulfide. Fasth and Sörbo[44] found that the administration of mercaptopyruvate to mice resulted in an increase in urinary thiosulfate. Szczepkowski and Wood[24] showed cystine could form thiosulfate via thiocysteine and thiocystine intermediates in a coupled cystathionase and rhodanese system. This may be the principal pathway for synthesis of thiosulfate in the body.

Massey and Edmondson[45] found the inactivation of xanthine oxidase by cyanide produced thiocyanate. The inactivation could be reversed by sulfide. The sulfide entered the enzyme structure to form a persulfide that was identified by cyanolysis and spectral changes. Branzoli and Massey[46] also identified a persulfide group on aldehyde oxidase. The denaturation of ferredoxin by molecular oxygen generated persulfide sulfur in the form of trisulfide on the enzyme according to Petering et al.[47] The transformation was ascribed to the iron-bound labile sulfur moiety. If so, it must have required an intramolecular transfer.

The number of respiratory enzymes in which persulfides are involved is growing. The activation of porcine heart mitochondrial malate,[48] succinic acid dehydrogenase[49] and nitrate reductase[50] by rhodanese persulfide involved transfer of sulfane sulfur to the enzyme. Massey et al.[27] showed thiocystine and the corresponding thioglutathione stimulated the reduction of cytochrome c by glutathione. Cavallini et al.[51] found cysteamine oxygenase to be activated by thiocystine also. Wood and Cavallini[52] found the cysteamine oxygenase retained a low amount of activity in the absence of an activator. Perhaps this stemmed from the presence of a slight undetected amount of trisulfide in the cysteamine substrate.

The persulfides have been used to generate sulfide ion for reconstitution of ferredoxins. Finazzi Agrò, Cavallini et al.[53] utilized the persulfide sulfur of rhodanese (charged by thiosulfate) to produce sulfide ion by reaction with dithiothreitol. The system produced active ferredoxins when combined with apoferredoxins in the presence of ferric nitrate. Taniguchi and Kimura[54] used mercaptopyruvate sulfurtransferase persulfide and dihydrolipoate to reconstitute adrenodoxin. The conversion of rhodanese persulfide to sulfide by succinic dehydrogenase also was accompanied by incorporation of sulfide ion into the iron-sulfur protein.[55] Reduction of the persulfide sulfur was attributed to the large number of sulfhydryl groups (20) on the enzyme. Alcohol dehydrogenase with 36 sulfhydryl groups did not form as much sulfide as succinic dehydrogenase, suggesting that other factors were involved.

Wong et al.[56] isolated a sulfurtransferase from rat brain that transferred sulfur from mercaptopyruvate to tRNA. The significance of this and similar, earlier findings with bacteria have not been established.

The biochemistry of the sulfur cycle in bacteria has been reviewed by Siegel.[57] The transformations of sulfur compounds are quite complex especially in the primitive forms such as the thiobacilli, but persulfide compounds are evidently involved. For example, Suzuki[58] found an enzyme in *T. thiooxidans* which utilized the persulfide of glutathione to oxidize elemental sulfur to thiosulfate. Silver and Kelly,[59] studying *Thiobacillus A₂*, found persulfide sulfur was transferred to lipoate by rhodanese as a step in dissimilation of thiosulfate. Moriarty et al.[60] proposed that oxidation of *T. concretus* involved membrane-bound persulfide.

Studies on *Rhodopseudomonas spheroides* in Neuberger's laboratory revealed that the bacterial cells grown anaerobically contained thiocystine.[61] Sandy et al.[26] showed that the thiocystine converted an inactive form of aminolevulinate synthetase to an active form. Similar effects were obtained with polythiosulfates, potassium polythionate, and glutathione trisulfide and to a much lesser extent with thiosulfate. The destruction of thiocystine by sulfhydryl groups or by transfer by rhodanese was considered by Wider de Xifra et al.[61] to provide control of aminolevulinate synthetase activity.

The pattern that appears to be emerging is that persulfides have a considerable role in activation and control of a number of enzymes including some not yet investigated. The production of thiocysteine and thiocystine from cystine by cystathionase and the transfer of persulfide by rhodanese, and by mercaptopyruvate sulfurtransferase, may have key roles in electron transfer and in general metabolism.

REFERENCES

1. J. L. Kice, Electrophilic and nucleophilic catalysis of the scission of the sulfur-sulfur bond, Acc. Chem. Res. 1: 58 (1968).
2. J. W. Hylin and J. L. Wood, Enzymic formation of polysulfides from mercaptopyruvate, J. Biol. Chem. 234: 2141 (1959).
3. M. Flavin, Microbial transsulfuration; the mechanism of an enzymatic disulfide elimination reaction, J. Biol. Chem. 237: 768 (1962).
4. D. Cavallini, C. De Marco, B. Mondovì, and B. G. Mori, The cleavage of cystine by cystathionase and the transulfuration of hypotaurine, Enzymologia 22: 161 (1960).
5. D. Cavallini, C. De Marco, and B. Mondovì, The enzymic conversion of cystamine and thiocysteamine into thiotaurine and hypotaurine, Enzymologia 23: 101 (1961).
6. B. Sörbo, A calorimetric method for the determination of thiosulfate, Biochim. Biophys. Acta 23: 412 (1957).
7. B. H. Sörbo, On the substrate specificity of rhodanese, Acta Chem. Scand. 7: 32 (1953).
8. J. Westley and D. Heyse, Mechanism of sulfur transfer catalysis, sulfhydryl-catalyzed transfer of thiosulfonate sulfur, J. Biol. Chem. 246: 1468 (1971).
9. B. Sörbo, On the catalytic effect of blood serum on the reaction between colloidal sulfur and cyanide, Acta Chem. Scand. 9: 1656 (1955).
10. J. F. Schneider and J. Westley, Metabolic interrelations of sulfur in proteins, thiosulfate, and cystine, J. Biol. Chem. 244: 5735 (1969).

11. D. S. Tarbell and D. P. Harnish, Cleavage of the carbon-sulfur bond in divalent sulfur compounds, Chem. Rev. 49: 1 (1951).

12. N. Catsimpoolas and J. L. Wood, The reaction of cyanide with bovine serum albumin, J. Biol. Chem. 239: 4132 (1964).

13. J. M. Swan, Synthese and Lanthionin-Bildung von einigen Cystin-Derivate, Angew. Chem. 68: 215 (1956).

14. D. Cavallini, G. Federici, E. Barboni, and M. Marcucci, Formation of persulfide groups in alkaline treated insulin, FEBS Lett. 10: 125 (1970).

15. D. Cavallini, G. Federici, and E. Barboni, Interaction of proteins with sulfide, Eur. J. Biochem. 14: 169 (1970).

16. B. Sörbo, On the mechanism of rhodanese inhibition by sulfite and cyanide, Acta Chem. Scand. 16: 2455 (1962).

17. A. Finazzi Agrò, G. Federici, C. Giovagnoli, C. Cannella, and D. Cavallini, Effect of sulfur binding on rhodanese fluorescence, Eur. J. Biochem. 28: 89 (1972).

18. M. Volini and S.-F. Wang, The interdependence of substrate and protein transformations in rhodanese catalysis. III. Enzyme changes outside the catalytic cycle, J. Biol. Chem. 248: 7329 (1973).

19. J. H. Ploegman, G. Drent, K. H. Kalk and W. G. J. Hol, The structure of bovine liver rhodanese. II. The active site in sulfur-substituted and the sulfur-free enzyme, Mol. Biol. 127: 149 (1979).

20. J. W. Hylin, H. Fiedler, and J. L. Wood, Thiocyanate formation by extracts of Escherichia coli and of liver, Proc. Soc. Exp. Biol. Med. 100: 165 (1959).

21. A. Meister, P. E. Fraser and S. V. Tice, Enzymatic desulfuration of β-mercaptopyruvate to pyruvate, J. Biol. Chem. 206: 561 (1954).

22. H. Vachek and J. L. Wood, Purification and properties of mercaptopyruvate sulfurtransferase of Escherichia coli, Biochim. Biophys. Acta 258: 133 (1972).

23. R. Jarabak and J. Westley, 3-Mercaptopyruvate sulfur transferase: rapid equilibrium-ordered mechanism with cyanide as the acceptor substrate, Biochemistry 19: 900 (1980).

24. T. W. Szczepkowski and J. L. Wood, The cystathionase-rhodanese system, Biochim. Biophys. Acta 139: 469 (1967).

25. J. C. Fletcher and A. Robson, The occurrence of bis-(2-amino--2-carboxyethyl)trisulphide in hydrolysates of wool and other proteins, Biochem. J. 87: 553 (1963).

26. J. D. Sandy, R. C. Davies, and A. Neuberger, Control of 5-aminolaevulinate synthetase activity in Rhodopseudomonas spheroides: a role for trisulphides, Biochem. J., 150: 245 (1975).

27. V. Massey, C. H. Williams, Jr., and G. Palmer, The presence of S°-containing impurities in commercial samples of oxidized glutathione and their catalytic effect on the reduction of cytochrome c, Biochem. Biophys. Res. Commun. 42: 730 (1971).

28. R. Abdolrasulnia and J. L. Wood, Transfer of persulfide sulfur from thiocystine to rhodanese, Biochim. Biophys. Acta 567: 135 (1979).

29. R. Abdolrasulnia and J. L. Wood, Persulfide properties of thiocystine and related trisulfides, Bioorg. Chem., 9: 253 (1980).

30. K. Lang, Die Rhodanbildung im Tierkörper, Biochem. Z., 259: 243 (1933).

31. B. Sörbo, Thiosulfate sulfurtransferase and mercaptopyruvate sulfurtransferase, in: "Metabolic Pathways," D. M. Greenberg, ed., Academic Press, New York (1969).

32. E. E. Conn, Cyanogenic glycosides, in: "Toxicants Occurring Naturally in Foods," National Research Council, Committee on Food Protection, 2nd ed., National Academy of Sciences, Washington, D.C. (1973).

33. J. Chung and J. L. Wood, Oxidation of thiocyanate to cyanide catalyzed by hemoglobin, J. Biol. Chem. 205: 231 (1971).

34. H. Schievelbein, R. Baumeister, and R. Vogel, Comparative investigations on the activity of thiosulfate-sulfur transferase, Naturwissenschaften 56: 416 (1969).

35. B. Sörbo, Sulfite and complex-bound cyanide as sulfur acceptors for rhodanese, Acta Chem. Scand. 11: 628 (1957).

36. M. Auriga and A. Koj, Protective effect of rhodanese on the respiration of isolated mitochondria intoxicated with cyanide, Bull. Acad. Pol. Sci. 23: 305 (1975).

37. T. W. Szczepkowski, The role of rhodanese in metabolic formation of thiosulphate, Acta Biochim. Pol., 8: 251 (1961).

38. J. L. Wood, Nutritional and protective properties of thiocystine, Proc. Soc. Exp. Biol. Med. 165: 469 (1980).

39. A. Koj, J. Frendo, and L. Wojczak, Subcellular distribution and intramolecular localization of three sulfurtransferases in rat liver, FEBS Lett., 57: 42 (1975).

40. C.-J. Clemedson, T. Fredriksson, B. Hansen, H. Hultman, and B. Sörbo, On the toxicity of sodium β-mercaptopyruvate and its antidotal effect against cyanide, Acta Physiol. Scand. 42: 41 (1958).

41. A. Koj and J. Frendo, The activity of cysteine desulfhydrase and rhodanese in animal tissues, Acta Biochim. Pol. 9: 373 (1962).

42. M. Villarejo and J. Westley, Mechanism of rhodanese catalysis of thiosulfate-lipoate oxidation-reduction, J. Biol. Chem. 238: 4016 (1963).

43. A. Koj and J. Frendo, Oxidation of thiosulfate to sulfate in animal tissues, Folia Biol. Warsaw 15: 49 (1967).

44. A. Fasth and B. Sörbo, Protective effect of thiosulfate and metabolic thiosulfate precursors against toxicity of nitrogen mustard (HN_2), Biochem. Pharmacol. 22: 1337 (1973).

45. V. Massey and D. Edmondson, On the mechanism of inactivation of xanthine oxidase by cyanide, J. Biol. Chem. 245: 6595 (1970).

46. U. Branzoli and V. Massey, Evidence for an active site per-
    sulfide residue in rabbit liver aldehyde oxidase, J. Biol.
    Chem. 249: 4346 (1974).
47. D. Petering, J. A. Fee, and G. Palmer, The oxygen sensitivity
    of spinach ferredoxin and other iron-sulfur proteins, J.
    Biol. Chem. 246: 643 (1971).
48. A. Finazzi Agrò, I. Mavelli, C. Cannella, and G. Federici,
    Activation of porcine heart mitochondrial malate dehydro-
    genase by zero valence sulfur rhodanese, Biochem. Biophys.
    Res. Commun. 68: 553 (1976).
49. S. Pagani, C. Cannella, P. Cerletti, and L. Pecci, Restoration
    of reconstitutive capacity of succinate dehydrogenase by
    rhodanese, FEBS Lett. 51: 112 (1975).
50. V. Tomati, G. Giovannozzi-Sermanni, S. Dupré, and C. Cannella,
    NADH: nitrate reductase activity restoration by rhodanese,
    Phytochemistry 15: 597 (1976).
51. D. Cavallini, R. Scandurra, and C. De Marco, The role of sulfur,
    sulphide, and reducible dyes in the enzymic oxidation of
    cysteamine to hypotaurine, Biochem. J. 96: 781 (1965).
52. J. L. Wood and D. Cavallini, Enzymic oxidation of cysteamine
    to hypotaurine in the absence of a cofactor, Arch. Biochem.
    Biophys. 119: 368 (1967).
53. A. Finazzi Agrò, C. Cannella, M. T. Graziani, and D. Cavallini,
    A possible role for rhodanese: the formation of "labile"
    sulfur from thiosulfate, FEBS Lett. 16: 172 (1971).
54. T. Taniguchi and T. Kimura, Role of 3-mercaptopyruvate sulfur-
    transferase in the formation of the iron-sulfur chromophore
    of adrenal ferredoxin, Biochim. Biophys. Acta 304: 284
    (1974).
55. F. Bonomi, S. Pagani, P. Cerletti, and C. Cannella, Rhodanese-
    mediated sulfur transfer to succinate dehydrogenase, Eur.
    J. Biochem. 72: 17 (1977).
56. J.-W. Wong, M. A. Harris, and C. A. Jankowicz, Transfer ribo-
    nucleic acid sulfur-transferase isolated from rat cerebral
    hemispheres, Biochemistry 13: 2805 (1974).
57. L. M. Siegel, Biochemistry of the sulfur cycle, in: "Metabolic
    Pathways," D.M. Greenberg, ed., Academic Press, New York
    (1969).
58. I. Suzuki, Oxidation of elemental sulfur by an enzyme system
    of *Thiobacillus thiooxidans*, Biochim. Biophys. Acta 104:
    359 (1965).
59. M. Silver and D. P. Kelly, Rhodanese from *Thiobacillus A2*:
    catalysis of reactions of thiosulphate with dihydrolipoate
    and dihydrolipoamide, J. Gen. Microbiol. 97: 277 (1976).
60. D. J. W. Moriarty and D. J. D. Nicholas, Products of sulfide
    oxidation in extracts of *Thiobacillus concretivorus*, Biochim.
    Biophys. Acta 197: 143 (1970).
61. E. A. Wider de Xifra, J. D. Sandy, R. C. Davies, and A.
    Neuberger, Control of 5-aminolaevulinate synthetase activ-

ity in *Rhodopseudomonas spheroides*, Philos. Trans. R. Soc. Lond. Ser. B. 273: 79 (1976).

BIOLOGICAL UTILIZATION OF SOME SELENIUM- AND SULFUR-CONTAINING
AMINO ACIDS

Carlo De Marco[a] and Mario Di Girolamo[b]

[a]Centro di Biologia Molecolare del C.N.R.,Istituto di
Chimica Biologica
[b]Centro per lo Studio degli Acidi Nucleici del C.N.R.,
Istituto di Fisiologia Generale

Università di Roma, Rome, Italy

INTRODUCTION

It is now generally recognized that selenium is an essential
micronutrient for mammals,birds and several bacteria (1). Covalently bound selenium is naturally present in some proteins;it is
an essential constituent of at least four enzymes,the well known
mammalian glutathione peroxidase (2,3,4,8) and the bacterial formate dehydrogenase (5),glycine reductase (6) and nicotinic acid hydroxylase (7). In the first three enzymes selenium is present in
a polypeptide chain as a selenocysteine residue (6,8,9). Besides
these examples,in which the catalytic activity is strictly dependent on the presence of selenium,the unspecific occurrence of selenium in proteins is also well documented (1). For example E.coli
can utilize selenate in substitution of sulfate,and the majority
of selenium is recovered in cellular proteins as selenomethionine
(10); selenomethionine containing proteins are active enough to
support cell growth (11). Selenomethionine can substitute methionine as growth factor for a methionine requiring E.coli mutant,and
extensive replacement of protein methionine by selenomethionine
occurs (12). Thus some proteins may contain selenium in substitution of sulfur without great impairment of their biological function. A typical example is the ß-galactosidase of E.coli. The
purified enzyme obtained from bacteria grown on selenate (11) or

343

on selenomethionine (13),although having more than 70% methionine substituted by selenomethionine shows unchanged catalytic activity.

In all the above recalled instances selenium substitutes for sulfur in protein aminoacids,cysteine and methionine. It seemed interesting to study if also aminoacids containing selenium in substitution not for sulfur but for other atoms could be utilized for protein synthesis and similarly well tolerated.

It was already known that aminoacids in which a methylene group was substituted by a sulfur atom can be incorporated into proteins. An example is thialysine, a lysine analog having the γ methylene group substituted by a sulfur atom,synthesized independently by Eldjarn (14) and by Cavallini et al.(15) in 1954–55.
It was shown by various investigators that thialysine acts as a competitive inhibitor for lysine incorporation into proteins in rat bone marrow cells (16) and in rabbit reticulocytes (17). Using purified E.coli lysyl-tRNA synthetase,it was shown that thialysine is a substrate for the enzyme (18),and inhibits the lysine binding to tRNA (19). It has been reported that in the presence of thialysine anomalous proteins with a higher rate of degradation were synthesized in E.coli (20). Other examples are thiaisoleucine (21–23) and γ-thiaproline (24–26),which can be incorporated into polypeptide chains in substitution of and in competition with isoleucine and proline respectively.

In 1975 we have synthesized selenalysine, a lysine analog having the γ methylene group substituted by a selenium atom (27). Some chemical and chromatographic properties of selenalysine were investigated showing that it can be easily differentiated from lysine and thialysine by automated ion-exchange chromatography (28).

$$H_2N.CH_2.CH_2.X.CH_2.CH(NH_2).COOH$$

X = S = Thialysine
X = Se = Selenalysine

Y = S = β-Thiaproline
Y = Se = β-Selenaproline
Z = S = γ-Thiaproline
Z = Se = γ-Selenaproline

Further we have prepared other selenium—containing aminoacids, among which γ–selenaproline (29) and ß–selenaproline (30),two proline analogs with the γ or the ß methylene group substituted by selenium. It was shown that also these compounds can be differentiated from proline and from the corresponding sulfur analogs, γ– and ß–thiaproline,by paper or ion exchange chromatography.Then we have undertaken a series of studies in vitro and in vivo to test if these selenoaminoacids could be utilized for protein synthesis in substitution of their natural analogs.

In meantime it has been shown that these compounds are good substrates for some enzymes acting on their natural or sulfur-containing analogs: selenalysine,like thialysine (31,32),is α-deaminated by snake venom L-aminoacid oxidase (33,34) and ε-deaminated by pea seedlings diamineoxidase (34);it is decarboxylated by bacterial lysine decarboxylase (35); γ and ß-selenaproline,like γ and ß-thiaproline,are oxidized by D-aminoacid oxidase (36,37).

In the present rewiev the more recent results obtained studying the possible biological utilization of the above mentioned selenoaminoacids for protein synthesis will be summarized. Previous results have been already rewieved at the 3rd International Symposium on Organic Selenium and Tellurium Compounds, held in Metz in 1979 (34).

SELENALYSINE

It was first shown that in vitro selenalysine can be activated and transferred to tRNA^{lys} by aminoacyl–tRNA synthetases from E.coli,rat liver and rabbit reticulocytes,and can be incorporated into polypeptide chains by protein synthesizing systems from the same sources,in substitution of and in competition with lysine(38). It was known that thialysine acts as an inhibitor for lysine incorporation into proteins (16,17) by competing with lysine for the lysyl-tRNA synthetase (18). In comparative tests it was shown that, with respect to thialysine,selenalysine has a lower effect as inhibitor of lysine utilization by protein synthesizing systems;it is less efficient than thialysine in inhibiting the lysyl-tRNA synthesis,with either the E.coli or rat liver system,and in preventing lysine incorporation into polypeptides.

Recently the utilization of selenalysine in vivo by growing E.coli cells has been studied (39). It has been shown that selena-

lysine disappears from the culture medium and it is incorporated
into cellular proteins. The molar fraction of selenalysine in proteins of cells grown 24 hrs in the presence of different concentrations of selenalysine (up to 0.3 mM),increases with the increase
of selenalysine in the medium,reaching the value of about 1% with
respect to the total aminoacid content. In meantime the protein
lysine molar fraction decreases,while the sum selenalysine plus
lysine remains constant.

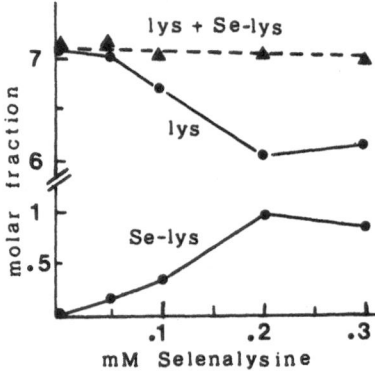

Fig.1. Lysine substitution by selenalysine in wild type E.coli
       proteins. Cell grown 24 hrs in media containing the indica-
       ted concentrations of selenalysine were washed and hydroly-
       zed. The hydrolysates were subjected to aminoacid analysis.
       The amounts of lysine (lys) and selenalysine (Se-lys) are
       expressed as molar fraction with respect to the total amino
       acid content taken as 100.

      In other words selenalysine is utilized for protein synthesis
being incorporated into polypeptides in substitution of lysine.Up
to about 14% of protein lysine is substituted by selenalysine when
this latter is present in the medium at 0.3 mM concentration. As
expected,lysine prevents selenalysine incorporation into proteins:
when lysine and selenalysine are present in the medium at equimolar concentrations,selenalysine is not incorporated at all.

      On the other hand the presence of selenalysine in the culture
medium at concentrations ranging from 0.05 to 0.3 mM inhibits E.co-

<u>li</u> growth rate and cell viability. The effect on growth rate is immediate, proportional to selenalysine concentration in the medium and it is reverted by lysine only to a limited extent.

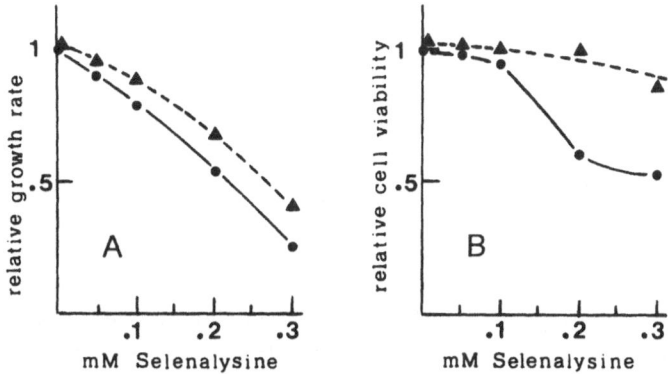

Fig.2. Effects of selenalysine on growth rate and cell viability
       in the wild type <u>E.coli</u>, and their reversion by lysine.
       A, growth rate. B, cell viability after 24 hrs growth.
       Abscissae: selenalysine concentration in the culture medium.
       Ordinates: relative values with respect to controls grown
       in the absence of selenalysine. Full lines: without added
       lysine. Broken lines: with 0.4 mM added lysine.

Growth rate inhibition can be imputed to the inhibition of protein synthesis. In fact, by determining the incorporation rate of radioactive leucine in <u>E.coli</u> cells, it has been shown that selenalysine inhibits protein synthesis and that the inhibition is immediate, being evident already 5 min. after selenalysine addition, is proportional to its concentration in the medium, and is only partially reverted by lysine. Protein synthesis inhibition can be explained by taking into account the results obtained in vitro: lysine and selenalysine show the same Km for the aminoacyl–tRNA synthetase, but the Vmax of lysine activation if five folds that of selenalysine, and also its transfer to $tRNA^{lys}$ is more efficient (38). Thus selenalysine, binding to the enzyme as well as lysine, but being utilized to a lesser extent, can inhibit competitively lysine utilization and gives rise to a decrease of protein synthesis and of growth rate.

The effect of selenalysine on cell viability can be correlated to the extent of lysine substitution by selenalysine in proteins. In fact both effects are not immediate and are almost completely reverted by lysine. About 14% of protein lysine has to
be substituted to have a marked effect on cell viability; cells
in which only 5% of lysine is substituted show normal viability.
In other words, small amounts of abnormal selenalysine-containing
proteins can be quite well tolerated by E.coli cells, larger
amounts can be lethal.

Further it has been checked if selenalysine could replace
lysine as growth factor for a lysine requiring E.coli mutant. It
has been shown that selenalysine alone is unable to support cell
growth: the presence of some lysine in the medium is necessary
to allow selenalysine utilization for growth. In this connection
it has to be recalled that also for the utilization of thialysine
by a lysine requiring E.coli mutant, the presence of some lysine
in the medium is required (18). Selenalysine gives rise to an evident increase of final cell growth with respect to lysine alone
when cells were allowed to grow for 24 hrs in a minimal medium
containing lysine at concentrations ranging from 0.005 to 0.1 mM,
and different amounts of selenalysine. The stimulatory effect is
proportional to selenalysine concentration until a maximum is reached when selenalysine and lysine are present at a ratio of about
1 to 3. On the contrary, in the presence of excess lysine, selenalysine is not utilized at all for growth.

As in the wild type selenalysine, even when it gives rise to
an increase of final cell growth, inhibits nevertheless growth rate
and affects cell viability. Growth rate inhibition can be explained by the inhibitory effect of selenalysine on protein synthesis
rate. Cell viability is much more affected in the mutant with respect to the wild type, and this can be correlated to the greater
extent of protein lysine substitution in the mutant.

An interesting result is in fact that selenalysine is utilized for protein synthesis to a greater extent by the lysine requiring mutant compared to the wild type; selenalysine can substitute
up to 30-50% of protein lysine in the mutant, while in the same
conditions the maximum substitution in the wild type does not overcome 14%. In the mutant where around 50% of lysine is substituted,
cell viability is reduced to 20%; in the wild type where no more

TABLE I. Selenalysine incorporation into proteins.
Cells were grown 24 hrs in media containing
the indicated concentration of selenalysine.
For the auxotroph 0.1 mM lysine was also
present.Whole cells hydrolysates were subjected to aminoacid analysis and the percent of
lysine substituted by selenalysine was calculated.

Selenalysine in the medium mM	% Protein lysine substituted	
	wild type	lysine auxotroph
0.05	1.8	15
0.10	4.5	33
0.20	14	54
0.30	12	51

than 14% of lysine is substituted,cell viability is reduced only
to 50%. This confirms that cell viability is strictly correlated
to the extent of protein lysine substitution by selenalysine.

The different degree of maximal lysine substitution in the
mutant compared to the wild type,points out that the utilization
of selenalysine for protein synthesis depends on the amount of
intracellular lysine;in the wild type,which synthesizes sufficient
endogenous lysine,less selenalysine is incorporated into proteins;
in the auxotroph,where intracellular lysine is present in limited
amount depending on its concentration in the medium, a greater
amount of selenalysine is utilized for protein synthesis.

It must be remarked that both in the wild type and in the
mutant,not all the selenalysine disappeared from the culture medium is found into cell proteins. Around 30–40%of selenalysine
initially present in the medium always disappears,even when it
is not utilized for growth,as for example when excess lysine is
present,thus indicating that it is at least in part metabolized
by different pathways. We have so far no data on the fate of the
selenalysine not utilized for protein synthesis;the production

of toxic compounds can not be excluded and it could be responsible
of the complete inhibition of cell growth observed when selenalysine
is present in the culture medium at 1 mM concentration.

Overall, the results obtained studying the in vivo incorporation of selenalysine in E.coli proteins and its effects on cell
growth, indicate that selenalysine can be utilized for protein synthesis in substitution of lysine, and that limited amounts of abnormal selenalysine-containing proteins can be quite well tolerated
by cells, without effect on cell viability. Thus, proteins containing
some selenium in substitution not for sulfur but for different
atoms can be active enough to support cellular functions.

In parallel to these studies on selenalysine, an investigation
on the utilization of thialysine for protein synthesis has been
undertaken. It was already reported that thialysine may be incorporated into E.coli proteins (18). It seemed interesting to study
the extent of thialysine incorporation into proteins and its effects on cell growth and viability, in comparison with selenalysine.
Preliminary results show that wild type E.coli can incorporate
thialysine into cellular proteins in substitution of lysine; a maximum of 17% can be substituted.

Like selenalysine, thialysine inhibits growth rate and protein
synthesis rate, but does not affect cell viability. The main

TABLE II. Comparison of the effects of the two lysine
          analogs on the wild type E.coli.

Analog in the medium (0.2 mM)	% inhibition of			% of protein lysine substituted
	Protein synthesis rate	Growth rate	Viability	
Selenalysine	42 –	45 –	44 +	14 +
Thialysine	20 +	20 +	0	15 +

+ : completely reverted by lysine (0.05-0.2 mM)
– : not or very partially reverted by lysine (0.4 mM)

difference between the two analogs is just that thialysine,even
if it is incorporated into proteins to the same extent as selenalysine,does not affect cell viability. Moreover,as regards protein
synthesis and growth rate,the inhibitory effects of thialysine
are much less marked and are completely reverted by lysine. These
results suggest that the substitution of a methylene group in a
protein aminoacid by a sulfur atom has less marked effect on cellular functions compared to the substitution by selenium.

γ-SELENAPROLINE

The possible utilization of γ-selenaproline for protein synthesis has been studied in vitro with protein synthesizing systems
from E.coli,rat liver and rabbit reticulocytes (40,41). It has
been demonstrated that γ-selenaproline is activated and transferred to tRNA^{pro} by either E.coli or rat liver aminoacyl-tRNA synthetases in competition with proline,and that it inhibits polypeptide
synthesizing systems. All tests were performed in comparison with
γ-thiaproline.

With regard to the activation reaction as studied by the ATP-
PPi exchange, the Km and Vmax values indicated that there are no
remarkable differences between the two proline analogs,which are
both activated by the proline-activating enzymes,but with a much
lower degree of affinity with respect to proline.

Some quantitative differences between the two compounds were
shown instead as regards their binding to tRNA and the polypeptide
synthesis. The binding of γ-thiaproline is more efficient and it
is completely reversed by proline; γ-selenaproline instead binds
to tRNA^{pro} to a lesser extent and this binding is only partially
reverted by proline. Thus, γ-selenaproline does not affect the
aminoacyl-tRNA synthetases by merely competing with proline,but
also by some unspecific still unexplained mechanisms. As regards
polypeptide synthesis it was shown that both γ-selenaproline and
γ-thiaproline may be incorporated in place of proline in the growing polypeptide chains; γ-selenaproline,once incorporated,stron-
gly impairs further chain elongation; γ-thiaproline,instead,has
a less marked inhibitory effect on the formation of a new peptide
bond. Moreover,while the effect of γ-thiaproline is completely reverted by proline, that of γ-selenaproline is only partially reverted . The results obtained in vitro showing that γ-selenaproline
inhibits protein synthesis by impairing polypeptide chain elonga-

tion,make unlikely that γ-selenaproline could be utilized in vivo for protein synthesis.

ß–SELENAPROLINE

It was first studied the effect on protein synthesis of ß–thiaproline,since no data were available on its possible utilization.It was shown that,like γ-thiaproline,also ß–thiaproline acts as a competitive inhibitor of proline in protein synthesizing sysstems (42). ß–Thiaproline is activated and transferred to tRNA^{pro} by E.coli and rat liver aminoacyl–tRNA synthetases,and inhibits proline incorporation into polypeptides in protein synthesizing systems from E.coli,rat liver and rabbit reticulocytes.Like γ-thiaproline,ß–thiaproline is incorporated into growing polypeptide chains,and once incorporated impairs further chain elongation. In the mammalian systems ß–thiaproline shows an higher inhibitory activity than γ-thiaproline in the various protein synthesis steps.

The main conclusion drawn from the results obtained with γ- and ß–thiaproline is that the presence of a sulfur atom either in ß or in γ position,besides some quantitative differences,does not substantially modify the activity of the two thiaprolines as proline competitive inhibitors in protein synthesis.

Further the behaviour of ß–selenaproline was investigated(43). It was shown that when ß–selenaproline was incubated in the usual experimental conditions with aminoacyl–tRNA synthetases from either E.coli or rat liver,no ATP–PPi exchange occurred. In order to check if the exchange did not occur for the inhability of ß–selenaproline either to bind the synthetases or to form aminoacyl–AMP once bound to the enzyme,the effect of ß–selenaproline on the proline activation was studied. It was shown that ß–selenaproline inhibits the proline–dependent ATP–PPi exchange with both E.coli and rat liver synthetases. It was also shown that ß–selenaproline inhibits specifically proline activation,since the activation of no one of the other common aminoacids was affected. Moreover,the inhibition can be completely reversed by proline,and is of fully competitive type. The specificity of the inhibitory effect of ß–selenaproline shows that it binds to prolyl–tRNA synthetase. On the other hand,its incapacity to give rise to ATP–PPi exchange shows that ß–selenaprolyl–AMP can not be formed. In this respect ß–selenaproline remarkably differs from both γ- and ß–thiaproline and from γ-selenaproline,which can form the activated complex with AMP and can be transferred to tRNA[pro]. Thus the incapacity of ß-selenaproline to bind AMP has to be ascribed to the presence of the selenium atom in ß position. The different behaviour of ß-thiaproline and ß-selenaproline as regards the activation reaction is one of greatest differences detected so far, in a biochemical reaction, between a sulfur- and a selenium-containing aminoacid analog.

The results obtained studying the interference of the different proline analogs on protein synthesis machinery show that the substitution of the methylene group in the γ position either with sulfur or selenium causes molecular changes which affect very little the bond formation with AMP and tRNA but impairs peptide bond formation, whereas the substitution of the ß methylene group yelds different effects according to the substituting atom. The substitution with a sulfur atom does not affect the binding to AMP, but reduces the linkage with tRNA and impairs peptide bond formation; the substitution with a selenium atom determines structural modifications which do not even allow the binding to AMP.

CONCLUSIONS

Some general suggestions can be tentatively drawn from the results above summarized. In an aliphatic aminoacid, the molecular modifications due to the presence of a selenium atom in substitution for a methylene group poorly affects the recognition by the activating enzymes, by the tRNA's and by the other components of the protein synthesis machinery. Thus, the selenium analog, as in the case of selenalysine, can be incorporated into growing polypeptide chains in vitro and still functioning proteins can be formed in vivo. In other words, aliphatic selenium-containing analogs can not differ so markedly from the corresponding natural aminoacids to prevent their biological utilization. When the substitution occurs in an heterocyclic aminoacid, more marked differences become evident, which can be correlated to the cyclic structure and to the necessity of well defined steric conditions for the various reactions of protein synthesis. The analog either can be incorporated into the growing polypeptide chain but impairs further chain elongation, as in the case of γ-selenaproline, either can not be incorporated at all, as in the case of ß-selenaproline.

However, the utilization in vivo of selenalysine, as the already reported utilization of selenomethionine, suggests that, when formed,abnormal selenium–containing proteins can be to some extent well tolerated by the cell,and can be sufficiently active to support cellular functions.

As regards the corresponding sulfur–containing analogs the results obtained suggest that the molecular modifications due to the presence of sulfur do not so greatly affect their possible biological utilization,and that they can be more easily utilized for protein synthesis.

REFERENCES

1. T.C. Stadtman,"Some selenium–dependent biochemical processes" Advan.Enzymol. 48:1 (1979)
2. J.T.Rotruk,A.I.Pope,H.E.Gamber,A.B.Swanson,D.G.Hafeman and W. G.Hockstra,"Selenium:Biochemical role as a component of glutathione peroxidase", Science 179:588 (1973)
3. W. Nakamura,S.Hosoda and K.Hayashy,"Purification and properties of rat liver glutathione peroxidase",Biochim.Biophys. Acta 358:251 (1974)
4. L. Flohé,W.A.Gunzler and H.H.Schock,"Glutathione peroxidase: a selenoenzyme", FEBS Lett.32:132 (1973)
5. H.G.Enoch and R.L.Lester,"The purification and properties of formate dehydrogenase and nitrate reductase from Escherichia coli", J.Biol.Chem. 250:6693 (1975)
6. J.E.Cone,R.Martin Del Rio and T.C.Stadtman,"Clostridial glycine reductase complex.Purification and characterization of the selenoprotein component", J.Biol.Chem. 252:5337 (1977)
7. D. Imhoff and J.R.Andreesen,"Nicotinic acid hydroxylase from Clostridium barkeri:selenium–dependent formation of active enzyme", FEMS Microbiol.Letts. 5:155 (1979)
8. R. Ladenstein,O.Epp,K.Bartels,A.Jones,R.Huber and A.Wendel, "Structure analysis and molecular model of the selenoenzyme glutathione peroxidase at 2.8 Å resolution", J.Mol.Biol. 134:199 (1979)
9. J.B.Jones,J.L.Dillworth and T.C.Stadtman,"Occurrence of selenocysteine in the selenium–dependent formate dehydrogenase of Methanococcus vanniellii",Archiv.Biochem.Biophys. 195: 255 (1979)
10. R.E.Huber,I.H.Segel and R.S.Criddle,"Growth of Escherichia coli on selenate", Biochim.Biophys.Acta 141:573 (1967)

11.   R.E.Huber and R.S.Criddle,"The isolation and properties of ß-
      galactosidase from Escherichia coli grown on sodium selenate",
      Biochim.Biophys.Acta 141:587 (1967)
12.   D.B.Cowie and G.N.Cohen,"Biosynthesis by Escherichia coli of
      active altered proteins containing selenium instead of sul-
      fur", Biochim.Biophys.Acta 26:252 (1957)
13.   E.H.Coch and R.C.Greene,"The utilization of selenomethionine
      by Escherichia coli", Biochim.Biophys.Acta 230:223 (1971)
14.   L. Eldjiarn,"The metabolism of cystamine and cysteamine",
      Scand.J.Clin.Lab.Invest. 6,Suppl. 13:96 (1954)
15.   D. Cavallini,C.De Marco,B.Mondovì and G.F.Azzone,"A new synthe-
      tic sulfur-containing aminoacid:S-aminoethylcysteine",
      Experientia11:61 (1955)
16.   M. Rabinowitz and K.Tuve,"Antimetabolite activity of lysine
      analogues on lysine incorporation into rat bone marrow pro
      tein in vitro", Proc.Soc.Expt.Biol.Med. 100:222 (1959)
17.   M. Rabinowitz and J.M.Fisher,"Formation of 'a ribosomal lesion
      in rabbit reticulocytes by the lysine antagonist S-(ß-ami-
      noethyl)cysteine", Biochim.Biophys.Res.Comm. 6:449 (1961)
18.   R. Stern and A.H.Mehler,"Lysil-sRNA synthetase from Escheri-
      chia coli", Biochem.Z. 342:400 (1965)
19.   F. Kalousek and I.Rychlik,"Purification and properties of ly-
      syl-sRNA synthetase from Escherichia coli", Coll.Czecho-
      slov.Chem.Comm. 30:3909 (1965)
20.   A.L.Goldberg,"Degradation of abnormal proteins in Escherichia
      coli",Proc.Natl.Acad.Sci.USA 69:422 (1972)
21.   A. Szentirmai and H.E.Umbarger,"Isoleucine and valine metabo-
      lism of Escherichia coli.XIV Effect of thiaisoleucine", J.
      Bacteriol. 95:1666 (1968)
22.   G. Treiber and M.Jaccarino,"Biochemical characterization of a
      mutant isoleucyl-transfer ribonucleic acid synthetase from
      Escherichia coli",J.Bacteriol. 107:828 (1971)
23.   V. Busiello,M.Di Girolamo and C.De Marco,"Thiaisoleucine and
      protein synthesis", Biochim.Biophys.Acta 561:206 (1979)
24.   I.J.Bekhor,Z.Mosheni and L.A.Bavetta,"Inhibition of proline
      C14 incorporation in rat liver ribosomes by thiazolidine
      4-carboxylic acid in a cell-free system",Proc.Soc.Expt.
      Biol.Med. 119:765 (1965)
25.   T.S.Papas and A.H.Mehler,"Analysis of the aminoacid binding
      to the proline transfer ribonucleic acid synthetases of
      Escherichia coli", J.Biol.Chem. 245:1588 (1970)
26.   L. Fowden,D.Lewis and H.Tristram,"Toxic aminoacids:their action

as antimetabolites", Advan.Enzymol. 29:89 (1967)

27. C. De Marco,A.Rinaldi,S.Dernini and D.Cavallini,"The synthe-
     sis of 2-aminoethyl,2-amino,3-carboxyethyl selenide(selena-
     lysine) a new analog of lysine",Gazz.Chim.It. 105:1113
     (1975)

28. C. De Marco,A.Rinaldi,S.Dernini,P.Cossu and D.Cavallini,"Chro-
     matographic separation of lysine,thialysine and selenalysine",
     J. Cromatogr. 114:291 (1975)

29. C. De Marco,R.Coccia,A.Rinaldi and D.Cavallini,"Synthesis and
     chromatographic properties of selenazolidine-4-carboxylic
     acid (selenaproline)", It.J.Biochem. 26:51 (1977)

30. C. De Marco,C.Cini,R.Coccia and C.Blarzino,"Synthesis and
     chromatographic properties of selenazolidine-2-carboxylic
     acid(ß-selenaproline)", It.J.Biochem. 28:104 (1979)

31. I. Willhardt,"Sulfur-containing aminoacid analogues as sub-
     strates of diamineoxidase and L-aminoacid oxidase", 5th
     FEBS Meeting Prague,Abstr.53 (1968)

32. C. Cini,C.Foppoli and C. De Marco,"Oxidative deamination of
     thialysine by snake venom L-aminoacid oxidase",It.J.Bio-
     chem. 27:305 (1978)

33. C. Cini and C.De Marco,"Oxidative deamination of ε-N-acetyl-
     thialysine and ε-N-acetylselenalysine by snake venom L-ami-
     noacid oxidase, It.J.Biochem. 28:221 (1979)

34. C. De Marco,"Studies on selenalysine and selenaproline" Proc.
     3rd Int.Symp.Organic Selenium Tellurium Compds.Metz(1979)
     (Cagniant,D.and Kirsch,G. Eds.,Univ.Metz)

35. C. Blarzino and C.De Marco,"Selenalysine as substrate of Lysine
     decarboxylase", It.J.Biochem. 26:444 (1977)

36. R. Coccia,C.Blarzino,C.Foppoli and C.Cini,"Oxidation of D-thi-
     azolidine-4-carboxylic acid by hog kidney D-aminoacid oxi-
     dase", It.J.Biochem. 28:252 (1979)

37. C. Foppoli,C.Cini,C.Blarzino and C.De Marco,"Oxidation of ß-
     DL-thiaproline,a possible natural substrate of D-aminoacid
     oxidase", It.J.Biochem. 30:355 (1981)

38. C. De Marco,V.Busiello,M.Di Girolamo and D.Cavallini;"Selena-
     lysine and protein synthesis", Biochim.Biophys.Acta 454:
     298 (1976)

39. C. Cini,V.Busiello,M.Di Girolamo and C.De Marco,"In vivo in-
     corporation of selenalysine in E.coli proteins and its ef-
     fects on cell growth", Biochim.Biophys.Acta 678:165 (1981)

40. C. De Marco,V.Busiello,M.Di Girolamo and D.Cavallini,"Selena-
     proline and protein synthesis", Biochim.Biophys.Acta 478:

156 (1977)

41. A. Antonucci,C.Foppoli,C.De Marco and D.Cavallini,"Inhibition of protein synthesis in rabbit reticulocytes by selenaproline", Bull.Mol.Biol.Med. 2:80 (1977)

42. V. Busiello,M. Di Girolamo,C.Cini and C.De Marco,"Action of thiazolidine-2-carboxylic acid,a proline analog,on protein synthesizing systems", Biochim.Biophys.Acta 564:311 (1979)

43. V. Busiello,M.Di Girolamo,C.Cini and C.De Marco,"ß-Selenaproline as competitive inhibitor of proline activation", Biochim.Biophys.Acta 606:347 (1980)

THE OXIDATION OF SULFUR-CONTAINING

AMINO ACIDS BY L-AMINO ACID OXIDASES

D. Cavallini, G. Ricci, G. Federici, M. Costa, B. Pensa,
R.M. Matarese and M. Achilli

Institutes of Biological Chemistry of the Universities
of Rome and Chieti, and Centro di Biologia Molecolare
del C.N.R., Rome, Italy

INTRODUCTION

L-amino acid oxidase (LAO) from various sources is known to act
on most of natural and unnatural amino acids of the L configuration
(1). Among the sulfur containing amino acids, apart methionine which
is known since long time as one of the best substrates for this enzyme,
cystine (2), homocystine (3) and thialysine (4) have also been found
more recently to be oxidized by LAO. In the course of a study on the
oxidation of cystine and lanthionine by LAO (5) we noticed the appearance of colored products not described by previous workers which
stimulated our interest to continue this investigation and to extend
it to not yet assayed sulfur containing amino acids. In the present
note we review some of the results submitted to publication elsewhere and we include also new unpublished data.

EXPERIMENTAL

The general conditions of the enzymatic assay were as follows.
Oxidation was determined manometrically in the air at 38°C. The
enzymatic mixture contained 10 μmoles of substrate, 1 mg catalase,
0.3 U of Sigma type V Crotalus adamanteus LAO, 1 ml of 1 M potassium phosphate buffer pH 7.6 and water up to 3 ml. In some assays
the snake venom LAO was substituted with turkey liver extract, known
to contain a LAO with preferential activity towards lysine (6) or
with rat kidney extract known as a mammalian source of LAO (7).
Substrates were of commercial origin except L-lanthionine and L-
homodjenkolic acid which were prepared according respectively to
ref. 8 and 9.

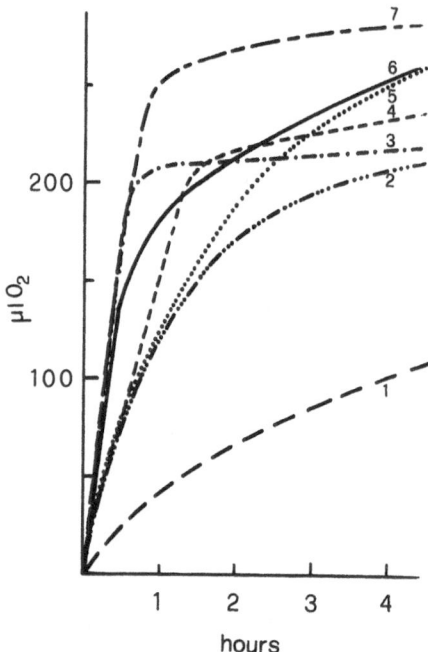

Fig. 1. Oxidation by snake venom LAO of: 1, lanthionine; 2, cystathionine; 3, djenkolic acid; 4, homocystine; 5, thialysine;
6, cystine; 7, homodjenkolic acid. All the substrates (in
the L form), 10 μmoles each. pH 7.6. Other details in the
text.

RESULTS

     Fig. 1 illustrates the oxidation of a group of sulfur contai-
ning amino acids by snake venom LAO. All the substrates are oxidized,
although at different rates. The O_2 uptake after the first hour of
incubation is higher than one mole per mole of substrate for most
of them. The ammonia released during this period, however, is only
slightly higher than that calculated for one mole per mole of substrate. This indicates a general further oxidation of the monodeamination product. Many incubation mixtures exhibit a yellow color

which, in the case of cystine and lanthionine, changed into red after long incubation. Some of the chromophores produced after 3 hours of incubation show well definite peaks in the area of 300 nm as illustrated in fig. 2. Other products show different spectral curves as reported in fig. 3.

Oxidation of the same substrates by turkey liver LAO has been studied only in part so far. A crude extract of turkey liver exhibits a preferential activity towards thialysine, which is the sulfur analog of lysine (fig. 4). A very low activity was registered by incubating the same group of amino acids with rat kidney extract, although this material is reported as one of the best sources of mammalian LAO (1). Owing to the easy supply of snake venom LAO, most of the work has been continued with this enzyme.

The appearance of the absorbancy in the area of 300 nm and the amount of ammonia released are compatible with a mono-deamination process followed by the cyclization into a ketimine ring for most of the products (fig. 2). The overoxidation is indicative of further changes of the enzymatic products. The ketimine presumed to rise by the oxidation of some of the amino acids assayed are reported in the last column of fig. 2. For the sake of simplicity we will name them by adding the name ketimine to the name of the parent amino acid. The compounds illustrated in fig. 3 are less likely to produce a ketimine ring. The instability of this series of compounds, as revealed by the overoxidation and by the change of color, and the possibility that some of them could be the substrate for novel metabolic pathways of sulfur containing products, stimulated our interest to prepare them in pure form and to study their properties. We focused initially our attention in the preparation of the ketimines of thialysine, of lanthionine and of cystine.

The easiest way to prepare thialysine and lanthionine ketimines is to add β-bromopyruvate to the respective amino-thiol, namely cysteamine and cysteine. Avi-Dor et al. (10, 11) discovered that when amino-thiols are added with halo-pyruvates an absorbtion in the range of 300 nm is produced. The reaction products have neither been isolated nor studied in detail, but the reaction was applied to the analytical determination of cysteine and glutathione. Hermann has been the first to obtain a solid product by reacting bromopyruvate with cysteamine (12). The product however was characterized as the dimer of the ketimine. Soper and Manning, more recently (13), prepared the same ketimine and were able to avoid polymerization by reducing the compound with $NaBH_4$. The final product was therefore not the ketimine but the thiamorpholine analog. Herrington et al. (14) and Cini et al. (15) prepared the monomer in dilute solutions as done by Avi-Dor and studied some of its properties; the preparation of the solid compound was not attempted.

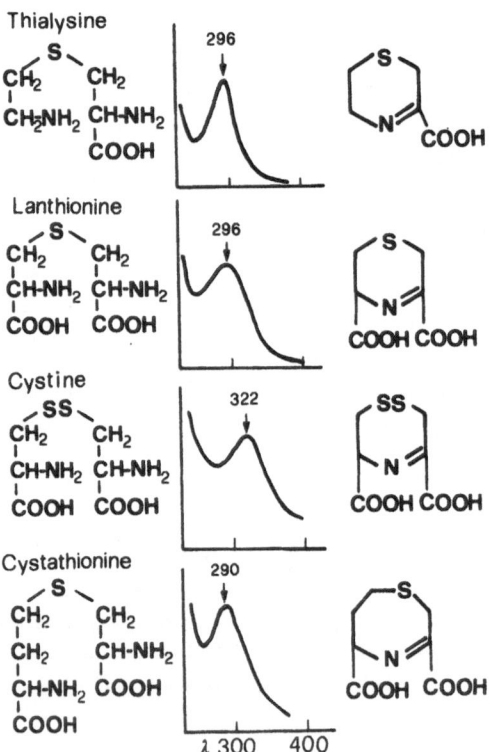

Fig. 2. First column, substrates incubated with LAO. Second column,
spectral curves after 3 hours of incubation after suitable
dilution. Third column, the presumed ketimine product.
Experimental conditions described in the text.

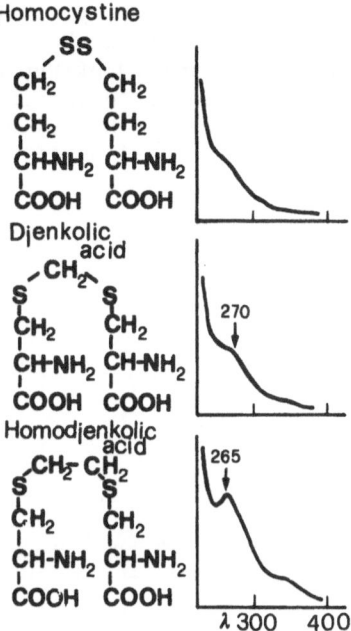

Fig. 3. First column, substrates incubated with LAO. Second column, spectral curves after 3 hours of incubation. Experimental conditions described in the text.

Preparation of thialysine ketimine

When 1 g of β-bromopyruvic acid (6 mmoles) in 2 ml water is added with a solution of 0.68 g of cysteamine hydrochloride (6 mmoles) in 2.5 ml water and with 1.5 ml 4 N NaOH, at room temp., the solution becomes hot, the pH is around 1.5 and in a short time a yellow crystalline product begins to precipitate. Crystallization is accompanied by the development of large amount of CO_2. The precipitate is filtered, washed with cold water and dried. It has been a surprise to find the alkylation of cysteamine, (and that of cysteine, as reported later) by bromopyruvic acid to occur in the acidic conditions of this procedure. Alkylation of thiols is known in fact to require alkaline conditions (16). A possible explanation could be the intermediate formation of a thiazolidine ring, known to be formed also in acidic medium (17), followed by the displacement of the halogen by internal shift of the sulfur atom from carbon 2 to carbon 3 of the pyruvic moiety of the adduct. Bromo- and fluoropyruvate are frequently used as enzyme inhibitors (18 and references therein), and this peculiar behaviour should be reminded for the correct interpretation of the inhibition.

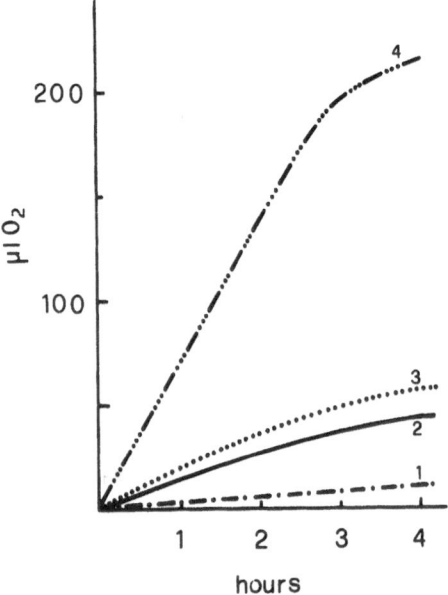

Fig. 4. Oxidation by turkey liver extract of: 1, lanthionine; 2, cystathionine; 3, cystine; 4, thialysine. All the substrates (in the L form) 10 μmoles each. Conditions identical to those described for the snake venom LAO. Any incubation mixture contained 40 mg protein of centrifuged and dialysed liver homogenate.

The spectral curve of the compound prepared as described above is reported in fig. 5 (curve 2). It exhibits a maximum at 306 nm, a minimum at 274 nm and a shoulder at 245 nm. It is different from that obtained by the product of oxidation of thialysine by LAO and from that obtained by the interaction of bromopyruvate with cysteamine in dilute solutions (14, 15), presumed to be the monomeric form. The spectrum is identical with that reported by Hermann for the dimer of the ketimine (12). By using the spectral curves as a criterion of identity, we then tried to prepare the ketimine monomer in the solid form. After a number of unsuccessful attempts to prepare the compound in water solutions under mild conditions, we finally found that it could be easily prepared by reacting bromopy-

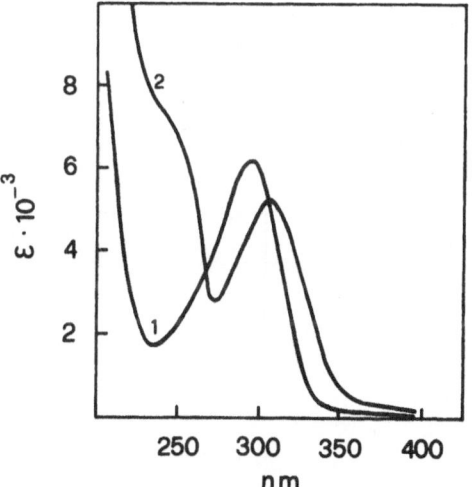

Fig. 5. Spectrophotometric curve of thialysine ketimine synthesized by reacting bromopyruvic acid with cysteamine hydrochloride in water (curve 2) and in conc. acetic acid (curve 1). Curve 1 is identical with that obtained by the oxidation of thialysine with LAO and with that obtained by adding bromopyruvate and cysteamine in dilute solutions (ketimine monomer). Curve 2 is identical with that reported by Hermann for the ketimine decarboxylated dimer (12). Solutions in phosphate buffer pH 7.6.

ruvic acid and cysteamine hydrochloride in concentrated acetic acid,
by the following simple procedure. In a test tube it is added 200
mg of bromopyruvic acid in 2 ml glacial acetic acid and 136 mg cysteamine hydrochloride in portions with shaking. After about 30 min.
at room temp. a white precipitate is formed. When dissolved in phosphate buffer pH 7.6 the product gives the spectral curve typical of
the monomer as reported in fig. 5 (curve 1). If the same preparation
is done in 4 ml in the place of 2 ml of acetic acid, crystallization
requires one day or more, but the compound appears in the form of
crystalline rosettes attached to the wall of the test tube. An interesting observation is that the supernatant acetic acid solution
becomes red upon standing for some days. The red product is formed
after many days when the preparation is done in dimethyl sulfoxide
or in absolute ethanol. In this case however no precipitate of the
monomer is formed.

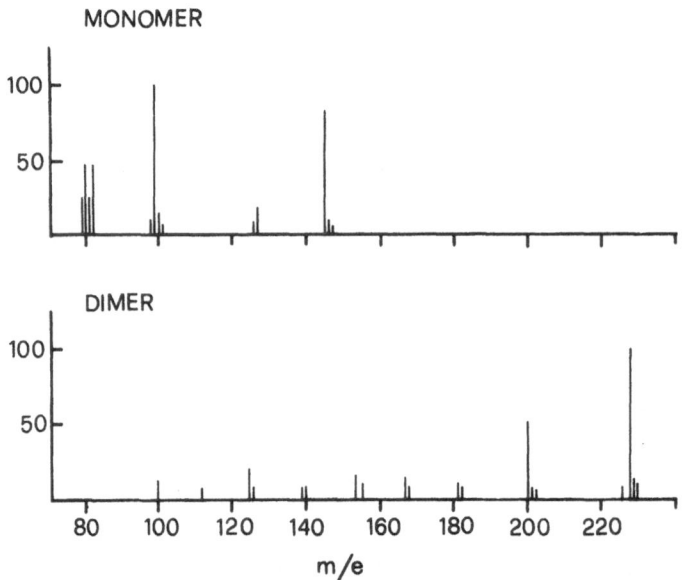

Fig. 6. Chemical ionization mass spectra of thialysine ketimine mo-
        nomer and dimer determined with a Hewlett-Packard 5980 A
        spectrometer operating at 70 eV.

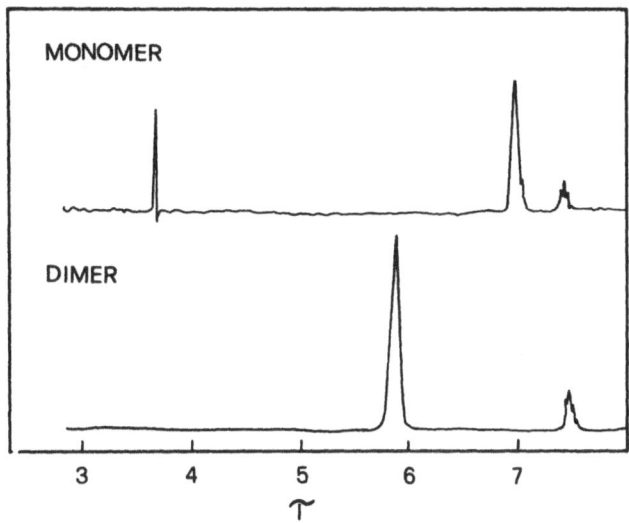

Fig. 7. NMR spectra of thialysine ketimine monomer and dimer in
   $(CD_3)_2SO$.

Thialysine ketimine in the solid form is stable for many days
when kept cold in the vacuum and dry. 100 mg dissolved in 2 ml water
slowly produce, at room temp., a yellow precipitate with the curve
of the dimer described by Hermann, reported in fig. 5 curve 1. If
the same solution is boiled as soon as prepared it is converted
into a red compound. The red compound is not produced by boiling
the yellow precipitate. Accordingly we have three different compounds to discuss: the monomer, the decarboxylated dimer and the
red product.

The main properties of the monomer are described below. Prepared as described the monomer crystallizes in the form of the
hydrobromide. The uncorr. melting point is 113-114°C. The maximum
of the UV curve is at 296 nm and the minimum at 236 nm (fig. 5 curve
1). The ketimine reacts with ninhydrin on boiling, giving a violet
color. In the amino acid analyzer the monomer runs fast in a position close to cysteic acid. It does not react with dinitrophenylhydrazine. It gives a brown color with nitroprusside in the presence
of cyanide and a positive test with the Folin-Marenzi reagent in
the absence of sulfite. Dissolved in water (the pH is around 2)
exhibits a spontaneous decarboxylation,

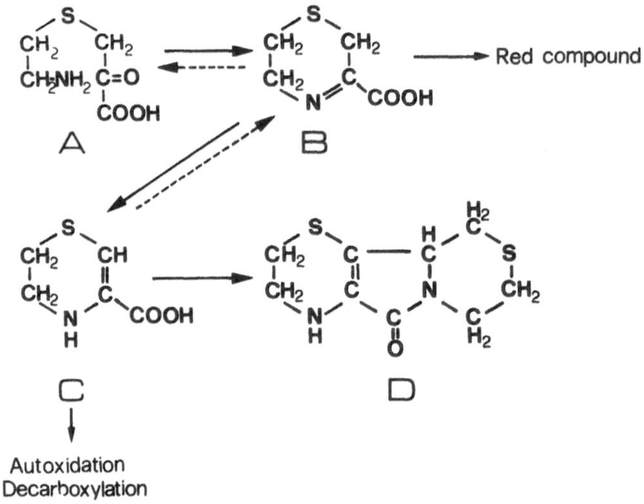

Fig. 8. A, monodeamination product of thialysine. B, ketimine derivative of A. C, product of reaction of cysteamine and bromopyruvate in conc. acetic acid. D, dimer produced by interaction of cysteamine and bromopyruvate in water.

in the Warburg apparatus, accounting for about 0.5 mole CO_2 per mole of ketimine in about 7 hours at 38°C. Near neutrality (phosphate
buffer pH 7.6) the decarboxylation is accompanied by an autoxidation accounting for about 0.5 mole O_2 per mole ketimine at the end
of 7 hours. The I.R. spectrum of the monomer in KCl disk shows absorbancies at 3420 cm^{-1} for -NH-; at 1670 cm^{-1} for -C=C- and at
1630 cm^{-1} for -COOH. The mass spectrum reveals a molecular ions at
145 (M^+ calculated for the monomer = 145) and other fragmentation
ions including m/e ions 80-82 and 79-81 indicating the hydrobromide form of the compound (fig. 6 A). The NMR spectrum in $(CD_3)_2SO$
indicates the occurrence of an olephinic proton (3.8 τ) and protons in the field of methylenic protons. The ratio is 1/4 (fig.7).
As a conclusion of the analytical data the formula reported in fig.
8 C for the monomer appears more likely than that with the carbon
nitrogen double bound (fig. 8 B).

The data obtained for the decarboxylated dimeric product obtained in water confirm the structure proposed by Hermann (12) which is rewritten in fig. 8 D. The mass spectrum gives in fact a molecular ion at 228 (M^+ calculated for the dimer = 228) and other fragmentation ions with the exclusion of ions of hydrobromide origin (fig. 6 B). The I.R. spectrum shows absorbancies at 3380 cm^{-1} for −NH−; at 1730 cm^{-1} for −CO−N=, at 1640 cm^{-1} for −C=C−. The NMR spectrum in $(CD_3)_2SO$ shows an unresolved band at 5.9 τ and the absence of the olephinic band (fig. 7).

The study of the red product is still on its beginning. We can only say at present that the dimer is not an intermediate. The red product, in fact, is formed only when the monomer is not yet converted into the dimer. An interesting feature, which is shared also by the red products formed from lanthionine and cystine ketimines, is its unusual binding to the Dowex 50 resin. When a solution of the red compound is percolated through a column of Dowex 50, H^+ form, 200−400 mesh, a red ring appears on the top of the column. The red ring is not removed by washing the resin with HCl up to 4 N, i.e., the solution generally used to regenerate the resin. The resin beads appear at the microscope covered with a red coat. The red compound is removed only by converting the resin in the Na^+ form by washing with a conc. NaOH solution. In this case the compound is eluted in a yellow form which turns back to red on acidification. This behaviour suggests a binding on the organic matrix of the resin not completely electrostatic.

Preparation of lanthionine ketimine

Cysteine hydrochloride (6 mmoles) in 5 ml water is added to β −bromopyruvic acid (6 mmoles) in 2 ml water. A precipitate is formed which is collected, washed with cold water and dried. The UV spectrum of this precipitate (fig. 9 dotted line) differs from that obtained by reacting cysteamine with bromopyruvate in water, whereas it is similar to that of thialysine ketimine monomer obtained in acetic acid. In fact in this case the maximum is centered at 296 nm and the minimum at 238 nm equal to that obtained by Avi-Dor in dilute solutions (10, 11) and attributed to the monomeric form. The solution of this compound in water slowly turns into red by staying at room temperature. By boiling a suspension of 300 mg of the ketimine in 2 ml water, the compound goes in solution, the solution then becomes deeply red and on cooling a red precipitate is formed which is washed with water and dried. The spectrum of the red product, dissolved in 0.1 M pH 7.6 phosphate buffer is reported in fig. 9 (broken line). Hermann et al. reported briefly (19, 20) that the product of interaction of cysteine and bromopyruvate dimerizes if prepared in ethanol. We have found that when 30 mg of the monomer prepared in water are dissolved in 2 ml absolute ethanol after 4-5 days deposit a brown precipitate with the

spectrum reported in fig. 9 (full line). This compound however was
not further characterized.

The UV spectrum of the solid product obtained by reacting bromopyruvic acid with cysteine is typical of the monomeric forms of these
ketimines, therefore, in contrast with thialysine ketimine, lanthionine ketimine does not dimerize in water solution in the course of

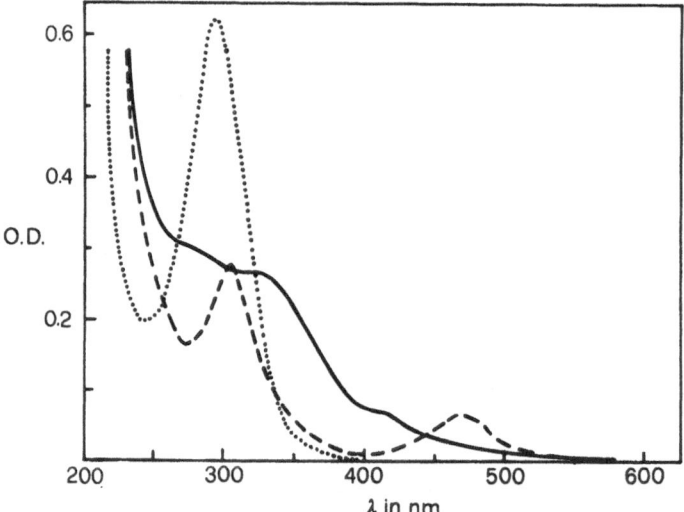

Fig. 9. Spectra of lanthionine ketimine and derivatives. The mono-
        mer as soon as crystallized .... The compound obtained
        after five days in ethanol ————. The red form obtained
        by boiling the monomer in water ———————. Approximately 19
        μg per ml of each compound in 0.1 M, pH 7.6 phosphate buf-
        fer.

its preparation. It is possible that the presence of the second
carboxyl group in this compound avoids the spontaneous dimerization
even in the concentrate conditions of its preparation. Another peculiar feature, which could be related to the higher stability of
the monomeric form, is the easy formation of the red product either
spontaneously and by boiling. Also in this case the red product
is trapped on the Dowex 50 resin as described for the one produced
from thialysine ketimine. The ketimine monomer has been characte-

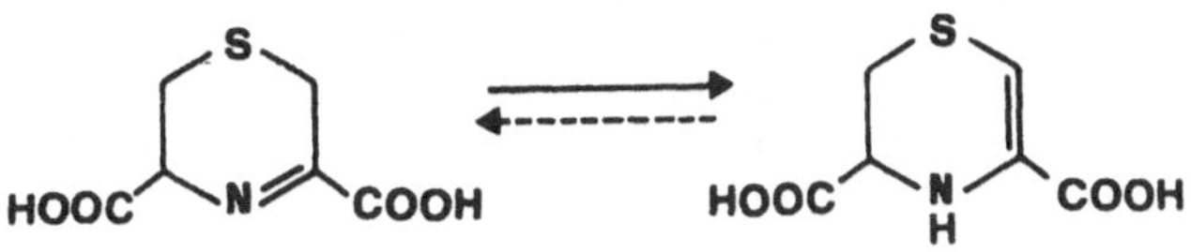

Fig. 10. Lanthionine ketimine.

rized by the same procedures described for thialysine ketimine (5). These are consistent with the carbon-carbon double bond formula reported in fig. 10.

Preparation of cystine ketimine

For the preparation of this ketimine we could make use of any of the following reactions:

1) $R-SOS-R + 2\ R'-SH \longrightarrow 2\ R-SS-R' + 2\ H_2O$

2) $R-SO_2S-R + R'-SH \longrightarrow R-SS-R' + R-SO_2H$

3) $R-SSO_3H + R'-SH \rightleftarrows R-SS-R' + H_2SO_3$

where R indicates the alanine portion and R' that of pyruvic acid. Although reactions 1 and 3 could give the best yields we have preferred reaction 2 because cystine disulfoxide is much more stable than the monosulfoxide and because reaction 3 is strongly reversible. Reaction 2 has been used by Kobayashi (21) to prepare cysteine-mercaptopyruvate mixed disulfide. We followed the same procedure and found the compound, eluted from the Dowex 50 column by 1 N HCl to give, in phosphate buffer pH 7.6, a spectrum similar to that reported in fig. 2 for the product of oxidation of cystine by LAO and consistent with the cyclized ketimine form. Attempts to get a solid product from the HCl solution have given so far unsatisfactory results and have been temporarily abandoned. The properties of this ketimine have been therefore studied in solution. Determination of its concentration has been performed as reported by Ubuka et al. (2). The results of this study have been submitted to publication elsewhere (22) and could be summarized as follows. As prepared in 1 N HCl the compound exhibits a UV spectrum with a low absorption at 350 nm. Added to phosphate buffer 1 M pH 7.6, the curve converts suddenly into one having a peak at 335 nm which then shifts slowly to 322 nm, with increased extinction. The 322 nm solution appears yellow and slowly converts into a red solution on standing at room temp. The HCl solution reacts with o-aminobenzaldehyde when brought

to pH 4.5 or to 7.7 indicating the formation of a cyclic compound
with carbon nitrogen double bond. At pH 7.6 the compound autoxidi-
zes, is decarboxylated and slowly looses the ability to react with
o-aminobenzaldehyde. The HCl solution gives a precipitate with
dinitrophenylhydrazine which, dissolved in 1 N NaOH, exhibits an
unusual spectral curve. As a result of these still preliminary observations we could conclude that cystine ketimine in 1 N HCl is
in large part in equilibrium with the open keto form and that it is
much more unstable than its homolog lanthionine ketimine. Other
properties like the spectrum indicating the monomeric form of the
compound at pH 7.6, the easy decarboxylation, oxidation and conversion into the red form, on the other hand, are similar to those
described for lanthionine ketimine. Accordingly the presumed formula of the monomer should be similar to that reported for lanthionine ketimine, (fig. 10) with a disulfide sulfur in the place of
the thioether sulfur.

DISCUSSION

      As reported above, the oxidation of a number of sulfur contai-
ning amino acids of the L configuration by LAO produces a series of
cyclic compounds easily detectable by the absorbtion in the area of
300 nm. The reaction is similar to that described for a number of
non-sulfur diamino and guanidino amino acids when they are deamina-
ted by oxidation or transamination (23 and ref. therein). It has
been suggested that cyclization could proceed through the intermediate imino acid rather than after its hydrolysis to the keto derivative (24). We do not have at present any data which could support
this hypothesis. The cyclic compounds produced by the oxidative
deamination of some of the sulfur containing diamino acids represent a group of homologous derivatives of the reduced thiazine or
dithiazine ring. The exception is cystathionine which yields a different ring. We have suggested to name these compounds by simply
adding the name ketimine to the parent amino acid even when the evidence indicates that the compound is mainly in the eneimino form.
When the carbon chain of the amino acid is over a certain length
(homocystine, djenkolic acid etc.) cyclization seems unfeasable, at
least on the basis of the UV absorption. It is likely that a similar
series of ketimines could be produced by non oxidative deamination
(transamination) thus giving a more general significance to this
group of substances. One of the preminent properties of these ketimines is the high instability, as documented by the overoxidation,
dimerization, change of color, etc. This property appears of biochemical interest because could produce new compounds used as substrates for new unsuspected metabolic reactions of sulfur containing
amino acids. This behaviour is in accord with the general enzymolo-
gical ambiguity of sulfur containing products which was stressed in
a previous note (25).

ACKNOWLEDGEMENTS

The authors are indebted to Prof. G. Lucente for Mass and NMR analysis and to Dr. C. Cini for the AA-analyzer determinations.

REFERENCES

1. A. Meister and D. Wellner, Flavoprotein amino acid oxidases, in "The Enzymes" 2nd ed. vol. 7 p. 609. Academic Press. New York 1963.

2. T. Ubuka, Y. Ishimoto and K. Kasahan, Determination of 3-mercaptopyruvate-cysteine disulfide, a product of oxidative deamination of L-cystine by L-amino acid oxidase, Anal. Biochem. 67: 66 (1975).

3. S.S. Chen, I.H. Walgate and J.A. Duerre, Oxidative deamination of sulfur amino acids by bacterial and snake venom L-amino acid oxidase, Arch. Biochem. Biophys. 146: 54 (1971).

4. C. Cini, C. Foppoli and C. De Marco, Oxidative deamination of thialysine by snake venom L-amino acid oxidase, Ital. J. Biochem. 27: 305 (1978).

5. Submitted for publication.

6. P. Boulanger and R. Osteux, Action de la L-aminoacide-deshydrogenase du foie de dindon sur les acides aminés basiques, Biochim. Biophys. Acta 21: 551 (1956).

7. M. Nakano and T.S. Danowski, Crystalline mammalian L-amino acid oxidase from rat kidney mitochondria, J. Biol. Chem. 241: 2075 (1966).

8. J. P. Greenstein and M. Winitz, "Chemistry of the amino acids", J. Wiley, New York (1961).

9. H. Zahn and K. Trautmann, Zur Kenntnis einiger aliphatischen Bisthioaether des Cystein, Ann. d. Chemie 591: 232 (1955).

10. Y. Avi-Dor and J. Mager, A spectrophotometric method for determination of cysteine and related compounds, J. Biol. Chem. 222: 249 (1956).

11. Y. Avi-Dor and R. Lipkin, A spectrophotometric method for the determination of reduced glutathione, J. Biol. Chem. 233: 69 (1958).

12. P. Hermann, Zur Reaktion von Halogenbrenztraubensaure mit Thiolaminen, Chem. Ber. 94: 442 (1961).

13. T.P. Soper and M. Manning, Beta-elimination of beta-halosubstrates by D-amino acid transaminase associated with inactivation

of the enzyme. Trapping of a key intermediate in the reaction
Biochemistry 17: 3377 (1978).

14. K.A. Herrington, K. Pointer, A. Meister and O.M. Friedmen,
Studies of latent derivatives of aminoethane thiols potentially
selective cytoprotectants, Cancer Res. 27: 130 (1967).

15. C. Cini, C. Foppoli and C. De Marco, On the product of the reaction between cysteamine and 3-bromopyruvate, Ital. J. Biochem.
27: 233 (1978).

16. M. Friedman, "The chemistry and biochemistry of the sulfhydryl
group in amino acids, etc." Pergamon Press. Oxford (1973).

17. M.P. Schubert, Compounds of thiol acids with aldehydes, J. Biol.
Chem. 114: 341 (1936).

18. P.M. Alliel, C. Mulet and F. Lederer, Bromopyruvate as an affinity label for baker's yeast flavocytochrome b_2, Eur. J.
Biochem. 105 : 343 (1980).

19. P. Hermann and I. Willhardt, Besonderheiten der Reaktion Schwefel-
haltiger Lysin und Cadaverin-Analoga mit pflanzlicher Diamino-
xydase, in: Abh. Dtsch. Akad. Wiss. Berlin (4 Int. Symp.Bio-
chem. Physiol. Alkaloide, Halle 1969) (1971).

20. P. Hermann. Chemistry and biochemistry of thia-analogues of amino
acids, in:"IUPAC, Organic sulfur chemistry," R. K. Freidlina,
and A. E. Skorova, eds., Pergamon Press, Oxford (1981).

21. K. Kobayashi, Studies on the new metabolites of cystine, Physiol.
Chem. & Phys. 2: 455 (1970).

22. G. Ricci, G. Federici, M. Achilli, R.M. Matarese and D. Cavallini, Physiol. Chem. and Phys. in press.

23. A.J.L. Cooper and A. Meister, Cyclic forms of the alfa-ketoacid
analogs of arginine, citrulline, homoarginine and homocitrul-
line, J. Biol. Chem. 253: 5407 (1978).

24. E. Hafner and D. Wellner, Reactivity of the imino acids formed
from the amino acid oxidase reaction, Biochemistry 18: 411
(1979).

25. D. Cavallini, G. Federici, S. Dupré, C. Cannella and R. Scandurra, Ambiguities in the enzymology of sulfur-containing amino
acids, in:"Natural sulfur compounds," D. Cavallini, G. E.
Gaull, and V. Zappia, eds., Plenum Press, New York (1980).

CONTRIBUTORS

Numbers in parentheses indicate the pages on which the authors'
contributions begin.

M. Achilli (359)   Institute of Biological Chemistry, 1st Uni-
                          versity of Rome, Rome, Italy

E. Antonini (67)   Institute of Medical Chemistry, 1st Univer-
                          sity of Rome, Rome, Italy

F. Ascoli (67)   Laboratory of Molecular Biology, University
                          of Camerino, Camerino, Italy

G. F. Azzone (187)   Institute of General Pathology, University
                          of Padua, Padua, Italy

A. Ballio (223)   Institute of Biological Chemistry, 1st Uni-
                          versity of Rome, Rome, Italy

D. Barra (273)   Institute of Biological Chemistry, University
                          of Camerino, Camerino, Italy

H. Beinert (123)   Institute of Enzyme Research, University of
                          Wisconsin, Madison, Wisconsin, U.S.A.

W.E. Blumberg (7)   Bell Laboratories, Murray Hill N.J. U.S.A.

C. Bonaventura (75)   Marine Biomedical Center, Duke University,
                          Marine Laboratory, Beaufort, N.C., U.S.A.

J. Bonaventura (75)   Marine Biomedical Center, Duke University,
                          Marine Laboratory, Beaufort, N.C., U.S.A.

F. Bonomi (309)   Dept. General Biochemistry, University of
                          Milan, Milan, Italy

F. Bossa (273)   Institute of Biological Chemistry, 1st Uni-
                          versity of Rome, Rome, Italy

M. Brouwer (75)   Marine Biomedical Center, Duke University
                          Marine Laboratory, Beaufort, N.C., U.S.A.

M. Brunori (111)   Department of Biochemistry, 2nd University
                          of Rome, Rome, Italy

L. Calabrese (155)   Institute of Organic Chemistry, Laboratory
                          of Biological Chemistry, University of Mes-
                          sina, Messina, Italy

F. G. Carey (49)   Woods Hole Oceanographic Institution, Woods
                          Hole, Md., U.S.A.

D. Cavallini (359)   Institute of Biological Chemistry, 1st Uni-
                          versity of Rome, Rome, Italy

P. Cerletti (309)   Department of General Biochemistry, Univer-
                          sity of Milan, Milan, Italy

B. Chance (95)   Johnson Research Foundation, University of
                          Pennsylvania, Philadelphia, PA., U.S.A.

E. Chiancone (67)   C.N.R. Center for Molecular Biology, Institu-
                          te of Biological Chemistry, University of
                          Rome, Rome, Italy

Y. Ching (95)   Bell Laboratories, Murray Hill, N.J. U.S.A.

A. Colosimo (111)   Institute of Medical Chemistry, 1st Univer-
                          sity of Rome, Rome, Italy

M. Costa (359)   Institute of Biological Chemistry, 1st Uni-
                          versity of Rome, Rome, Italy

C. Crifò (195)   Institute of Medicine and Surgery, University
                          of L'Aquila, L'Aquila, Italy

C. De Marco (343)   Institute of Biological Chemistry, 1st Uni-
                          versity of Rome, Rome, Italy

M. Di Girolamo (343)   C.N.R. Center for Nuclear Acid Research,
                          Rome, Italy

S. Doonan (273)   Department of Biochemistry, University Col-
                          lege, Cork, Ireland

P. Fasella (243)   Institute of Biological Chemistry, University
                          of Rome, Rome, Italy

G. Federici (359)   Institute of Biological Chemistry, University
                          of Chieti, Chieti, Italy.

M. E. Fielden (155)   Institute of Cancer Research, Sutton, Surrey, U.K.

A. Finazzi Agrò (141) Institute of Biological Chemistry, 1st University of Rome, Rome, Italy

A. Giartosio (283)   Institute of Biological Chemistry, University of Cagliari, Cagliari, Italy

Q. H. Gibson (49)   Department of Biochemistry, Molecular and Cell Biology, Cornell University, Ithaca, N. Y. U.S.A.

J. M. Gonzalez-Ros   Department of Biochemistry, Medical College
   (209)   of Virginia, Virginia Commonwealth University, Richmond VA.,U.S.A.

A. L. Lehninger (171) Department of Physiological Chemistry. Johns Hopkins University-School of Medicine, Baltimore, MD, U.S.A.

M. Llanillo (209)   Department of Biochemistry, Medical College of Virginia, Virginia Commonwealth University Richmond, VA., U.S.A.

B. G. Malmström (87)   Department of Biochemistry & Biophysics and Chalmers Institute of Technology, University of Göteborg, Göteborg, Sweden

M. Martinez-Carrion   Department of Biochemistry, Medical College
   (209)   of Virginia, Virginia Commonwealth University Richmond, VA., U.S.A.

V. Massey (295)   Department of Biological Chemistry, University of Michigan, MI, U.S.A.

R. M. Matarese (359)   Institute of Biological Chemistry, 1st University of Rome, Rome, Italy

B. Mondovì (141)   Institute of Applied Biochemistry, 1st University of Rome, Rome, Italy

R. M. Oliver (231)   Department of Chemistry, The University of Texas at Austin, Austin, TX, U.S.A.

S. Pagani (309)   Department of General Biochemistry, University of Milan, Milan, Italy

A. Paraschos (209)   Department of Biochemistry, Medical College of Virginia, Virginia Commonwealth University, Richmond, VA, U.S.A.

B. Pensa (359) Institute of Biological Chemistry, 1st University of Rome, Rome, Italy

M. F. Perutz (31) Medical Research Council, Laboratory of Molecular Biology, Cambridge, U.K.

D. Pietrobon (187) Institute of General Pathology, University of Padua, Padua, Italy

F. Podo (195) Istituto Superiore di Sanità, Rome, Italy

L. Powers (95) Bell Laboratories, Murray Hill, N.J., U.S.A.

L. J. Reed (231) Department of Chemistry, The University of Texas at Austin, Austin, TX., U.S.A.

A. Rigo (155) Laboratory of Biophysics, Institute of General Pathology, University of Padua, Padua, Italy

G. Ricci (359) Institute of Biological Chemistry, 1st University of Rome, Rome, Italy

F. Riva (283) Institute of Biological Chemistry, University of Camerino, Camerino, Italy

G. Rotilio (155) Department of Biology, 2nd University of Rome, Rome, Italy

C. Salerno (243) C.N.R. Center of Molecular Biology, Rome, Italy

N. Siliprandi (1) Institute of Biological Chemistry, University of Padua, Padua, Italy

E. E. Snell (257) Departments of Microbiology and Chemistry The University of Texas at Austin, Austin, TX., U.S.A.

R. Strom (195) Institute of Biological Chemistry, 1st University of Rome, Rome, Italy

C. Turano (283) Institute of Biological Chemistry, Lst University of Rome, Rome, Italy

D. Walz (187) Biozentrum, University of Basel, Basel, Switzerland

M. T. Wilson   (111)   Department of Chemistry, University of
                            Essex, Colchester, U.K.

J. L. Wood (327)   Department of Biochemistry, University of
                            Tennessee, Center of Health Sciences,-
                            Memphis, TN., U.S.A.

J. Wyman (23)   C.N.R. Center of Molecular Biology and 1st
                            University of Rome, Rome, Italy

G. Zaccai (195)   Institut Laue-Langevin, Grenoble, France.

SYMPOSIUM PARTICIPANTS